Mathematics in Industry

Volume 37

Mathematics in Industry focuses on the research and educational aspects of mathematics used in industry and other business enterprises. Books for *Mathematics in Industry* are in the following categories: research monographs, problem-oriented multi-author collections, textbooks with a problem-oriented approach, conference proceedings. Relevance to the actual practical use of mathematics in industry is the distinguishing feature of the books in the *Mathematics in Industry* series.

More information about this series at https://link.springer.com/bookseries/4650

Jong Chul Ye

Geometry of Deep Learning

A Signal Processing Perspective

Springer

Jong Chul Ye
Korea Advanced Institute of Science
and Technology (KAIST),
Daejeon, Republic of Korea

ISSN 1612-3956 ISSN 2198-3283 (electronic)
Mathematics in Industry
ISBN 978-981-16-6048-1 ISBN 978-981-16-6046-7 (eBook)
https://doi.org/10.1007/978-981-16-6046-7

Mathematics Subject Classification: 68T01, 68T07

This Springer imprint is published by the registered company Springer Nature Singapore Pte Ltd.
The registered company address is: 152 Beach Road, #21-01/04 Gateway East, Singapore 189721, Singapore

To Andy, Ella, and Joo

Preface

It was a very different, unprecedented, and weird start of the semester, and I did not know what to do. This semester, I was supposed to offer a new senior-level undergraduate class on *Advanced Intelligence* to jointly teach students at the Department of Bio/Brain Engineering and the Department of Mathematical Sciences. I had initially planned a standard method for teaching machine learning, the contents of which are practical, experience-based lectures with a lot of interaction with the students through many mini-projects and term projects. Unfortunately, the global pandemic of COVID-19 has completely changed the world and such interactive classes are no longer an option most of the time.

So, I thought about the best way to give online lectures to my students. I wanted my class to be different from other popular online machine learning courses but still provide up-to-date information about modern deep learning. However, not many options were available. Most existing textbooks are already outdated or very implementation oriented without touching the basics. One option would be to prepare presentation slides by adding all the up-to-date knowledge that I wanted to teach. However, for undergraduate-level courses, the presentation files are usually not enough for students to follow the class, and we need a textbook that students can read independently to understand the class. For this reason, I decided to write a reading material first and then create presentation files based on it, so that the students can learn independently before and after the online lectures. This was the start of my semester-long book project on *Geometry of Deep Learning*.

In fact, it has been my firm belief that a deep neural network is not a magic black box, but rather a source of endless inspiration for new mathematical discoveries. Also, I believed in the famous quote by Isaac Newton, "Standing on the shoulders of giants," and looking for a mathematical interpretation of deep learning. For me as a medical imaging researcher, this topic was critical not only from a theoretical point of view but also for clinical decision-making, because we do not want to create false features that can be recognized as diseases.

In 2017, on a street in Lisbon, I had *Eureka!* moment in understanding hidden framelet structure in encoder-decoder neural networks. The resulting interpretation of the deep convolutional framelets, published in the *SIAM Journal of Imaging*

Science, has had a significant impact on the applied math community and has been one of the most downloaded papers since its publication. However, the role of the rectified linear unit (ReLU) was not clear in this work, and one of the reviewers in a medical imaging journal consistently asked me to explain the role of the ReLU in deep neural networks. At first, this looked like a question that went beyond the scope of the medical application paper, but I am grateful to the reviewer, as during the agony of preparing the answers to the question, I realized that the ReLU determines the input space partitioning, which is automatically adapted to the input space manifold. In fact, this finding led to a 2019 ICML paper, in which we revealed the combinatorial representation of framelets, which clearly shows the crucial connection with the classic compressed sensing (CS) approaches.

Looking back, I was pretty brave to start this book project, as these are just two pieces of my geometric understanding of deep learning. However, as I was preparing the reading material for each subject of deep learning, I found that there are indeed many exciting geometric insights that have not been fully discussed.

For example, when I wrote the chapter on backpropagation, I recognized the importance of the denominator layout convention in the matrix calculus, which led to the beautiful geometry of the backpropagation. Before writing this book, the normalization and attention mechanisms looked very heuristic to me, with no evidence of a systematic understanding that is even more confusing due to their similarities. For example, AdaIN, Transformer, and BERT were like dark recipes that researchers have developed with their own secret sauces. However, an in-depth study for the preparation of the reading material has revealed a very nice mathematical structure behind their intuition, which shows a close connection between them and their relationship to optimal transport theory.

Writing a chapter on the geometry of deep neural networks was another joy that broadened my insight. During my lecture, one of my students pointed out that some partitions can lead to a low-rank mapping. In retrospect, this was already in the equation, but it was not until my students challenged me that I recognized the beautiful geometry of the partition, which fits perfectly with fascinating empirical observations of the deep neural network.

The last chapter, on generative models and unsupervised learning, is something of which I am very proud. In contrast to the conventional explanation of the generative adversarial network (GAN), variational auto-encoder (VAE), and normalizing flows with probabilistic tools, my main focus was to derive them with geometric tools. In fact, this effort was quite rewarding, and this chapter clearly unified various forms of generative model as statistical distance minimization and optimal transport problems.

In fact, the focus of this book is to give students a geometric insight that can help them understand deep learning in a unified framework, and I believe that this is one of the first deep learning books written from such a perspective. As this book is based on the materials that I have prepared for my senior-level undergraduate class, I believe that this book can be used for one-semester-long senior-level undergraduate and graduate-level classes. In addition, my class was a code-shared course for both bioengineering and math students, so that much of the content of the work is interdisciplinary, which tries to appeal to students in both disciplines.

I am very grateful to my TAs and students of the 2020 spring class of BiS400C and MAS480. I would especially like to thank my great team of TAs: Sangjoon Park, Yujin Oh, Chanyong Jung, Byeongsu Sim, Hyungjin Chung, and Gyutaek Oh. Sangjoon, in particular, has done a tremendous job as Head TA and provided organized feedback on the typographical errors and mistakes of this book. I would also like to thank my wonderful team at the Bio Imaging, Signal Processing and Learning laboratory (BISPL) at KAIST, who have produced ground-breaking research works that have inspired me.

Many thanks to my awesome son and future scientist, Andy Sangwoo, and my sweet daughter and future writer, Ella Jiwoo, for their love and support. You are my endless source of energy and inspiration, and I am so proud of you. Last, but not the least, I would like to thank my beloved wife, Seungjoo (Joo), for her endless love and constant support ever since we met. I owe you everything and you made me a good man. With my warmest thanks,

Daejeon, Korea
February, 2021

Jong Chul Ye

Contents

Part I
Basic Tools for Machine Learning

"I heard reiteration of the following claim: Complex theories do not work; simple algorithms do. I would like to demonstrate that in the area of science a good old principle is valid: Nothing is more practical than a good theory."

–Vladimir N Vapnik

Chapter 1
Mathematical Preliminaries

In this chapter, we briefly review the basic mathematical concepts that are required to understand the materials of this book.

1.1 Metric Space

A metric space $(\mathcal{X}, d)$ is a set $\mathcal{X}$ together with a metric d on the set. Here, a *metric* is a function that defines a concept of distance between any two members of the set, which is formally defined as follows.

Definition 1.1 (Metric) A metric on a set $\mathcal{X}$ is a function called the distance $d : \mathcal{X} \times \mathcal{X} \mapsto \mathbb{R}_+$, where $\mathbb{R}_+$ is the set of non-negative real numbers. For all $x, y, z \in \mathcal{X}$, this function is required to satisfy the following conditions:

1. $d(x, y) \geq 0$ (non-negativity).
2. $d(x, y) = 0$ if and only if $x = y$.
3. $d(x, y) = d(y, x)$ (symmetry).
4. $d(x, z) \leq d(x, y) + d(y, z)$ (triangle inequality).

A metric on a space induces topological properties like open and closed sets, which lead to the study of more abstract topological spaces. Specifically, about any point x in a metric space $\mathcal{X}$, we define the open ball of radius $r > 0$ about x as the set

$$B_r(x) = \{y \in \mathcal{X} : d(x, y) < r\}. \tag{1.1}$$

Using this, we have the formal definition of openness and closedness of a set.

Definition 1.2 (Open Set, Closed Set) A subset $U \in \mathcal{X}$ is called open if for every $x \in U$ there exists an $r > 0$ such that $B_r(x)$ is contained in U. The complement of an open set is called closed.

J. C. Ye, *Geometry of Deep Learning*, Mathematics in Industry 37,
https://doi.org/10.1007/978-981-16-6046-7_1

A sequence (x_n) in a metric space $\mathcal{X}$ is said to converge to the limit $x \in \mathcal{X}$ if and only if for every $\varepsilon > 0$, there exists a natural number N such that $d(x_n, x) < \varepsilon$ for all $n > N$. A subset $\mathcal{S}$ of the metric space $\mathcal{X}$ is closed if and only if every sequence in $\mathcal{S}$ that converges to a limit in $\mathcal{X}$ has its limit in $\mathcal{S}$. In addition, a sequence of elements (x_n) is a Cauchy sequence if and only if for every $\varepsilon > 0$, there is some $N \geq 1$ such that

$$d(x_n, x_m) < \varepsilon, \quad \forall \quad m, n \geq N.$$

We are now ready to define the important concepts in metric spaces.

Definition 1.3 (Completeness) A metric space $\mathcal{X}$ is said to be complete if every Cauchy sequence converges to a limit; or if $d(x_n, x_m) \rightarrow 0$ as both n and m independently go to infinity, then there is some $y \in \mathcal{X}$ with $d(x_n, y) \rightarrow 0$.

Definition 1.4 (Lipschitz Continuity) Given two metric spaces $(\mathcal{X}, d_{\mathcal{X}})$ and $(\mathcal{Y}, d_{\mathcal{Y}})$, where $d_{\mathcal{X}}$ denotes the metric on the set $\mathcal{X}$ and $d_{\mathcal{Y}}$ is the metric on set $\mathcal{Y}$, a function $f : \mathcal{X} \mapsto \mathcal{Y}$ is called Lipschitz continuous if there exists a real constant $K \geq 0$ such that, for all $x_1, x_1 \in \mathcal{X}$,

$$d_{\mathcal{Y}}(f(x_1), f(x_2)) \leq K d_{\mathcal{X}}(x_1, x_2). \tag{1.2}$$

Here, the constant K is often called the Lipschitz constant, and a function f with the Lipschitz constant K is called K-Lipschitz function.

1.2 Vector Space

A vector space $\mathcal{V}$ is a set that is closed under finite vector addition and scalar multiplication. In machine learning applications, the scalars are usually members of real or complex values, in which case $\mathcal{V}$ is called a vector space over real numbers, or complex numbers.

For example, the Euclidean n-space $\mathbb{R}^n$ is called a real vector space, and $\mathbb{C}^n$ is called a complex vector space. In the n-dimensional Euclidean space $\mathbb{R}^n$, every element is represented by a list of n real numbers, addition is component-wise, and scalar multiplication is multiplication on each term separately. More specifically, we define a column n-real-valued vector $\boldsymbol{x}$ to be an array of n real numbers, denoted by

$$\boldsymbol{x} = \begin{bmatrix} x_1 \\ x_2 \\ \vdots \\ x_n \end{bmatrix} = \begin{bmatrix} x_1 \ x_2 \ \cdots \ x_n \end{bmatrix}^\top \in \mathbb{R}^n,$$

where the superscript $^\top$ denotes the adjoint. Note that for a real vector, the adjoint is just a transpose. Then, the sum of the two vectors $\boldsymbol{x}$ and $\boldsymbol{y}$, denoted by $\boldsymbol{x}+\boldsymbol{y}$, is defined by

$$\boldsymbol{x}+\boldsymbol{y}=\begin{bmatrix} x_1+y_1 & x_2+y_2 & \cdots & x_n+y_n \end{bmatrix}^\top .$$

Similarly, the scalar multiplication with a scalar $\alpha \in \mathbb{R}$ is defined by

$$\alpha\boldsymbol{x}=\begin{bmatrix} \alpha x_1 & \alpha x_2 & \cdots & \alpha x_n \end{bmatrix}^\top .$$

In addition, we formally define the inner product and the norm in a vector space as follows.

Definition 1.5 (Inner Product) Let $\mathcal{V}$ be a vector space over $\mathbb{R}$. A function $\langle\cdot,\cdot\rangle_{\mathcal{V}}:\mathcal{V}\times\mathcal{V}\mapsto\mathbb{R}$ is an inner product on $\mathcal{V}$ if:

1. Linear: $\langle\alpha_1\boldsymbol{f}_1+\alpha_2\boldsymbol{f}_2,\boldsymbol{g}\rangle_{\mathcal{V}}=\alpha_1\langle\boldsymbol{f}_1,\boldsymbol{g}\rangle_{\mathcal{V}}+\alpha_2\langle\boldsymbol{f}_2,\boldsymbol{g}\rangle_{\mathcal{V}}$ for all $\alpha_1,\alpha_2\in\mathbb{R}$ and $\boldsymbol{f}_1,\boldsymbol{f}_2,\boldsymbol{g}\in\mathcal{V}$.
2. Symmetric: $\langle\boldsymbol{f},\boldsymbol{g}\rangle_{\mathcal{V}}=\langle\boldsymbol{g},\boldsymbol{f}\rangle_{\mathcal{V}}$.
3. $\langle\boldsymbol{f},\boldsymbol{f}\rangle_{\mathcal{V}}\geq 0$ and $\langle\boldsymbol{f},\boldsymbol{f}\rangle_{\mathcal{V}}=0$ if and only if $\boldsymbol{f}=\boldsymbol{0}$.

If the underlying vector space $\mathcal{V}$ is obvious, we usually represent the inner product without the subscript $\mathcal{V}$, i.e. $\langle\boldsymbol{f},\boldsymbol{g}\rangle$. For example, the inner product of the two vectors $\boldsymbol{f},\boldsymbol{g}\in\mathbb{R}^n$ is defined as

$$\langle\boldsymbol{f},\boldsymbol{g}\rangle=\sum_{i=1}^{n} f_i g_i=\boldsymbol{f}^\top\boldsymbol{g}.$$

Two nonzero vectors $\boldsymbol{x}, \boldsymbol{y}$ are called *orthogonal* when

$$\langle\boldsymbol{x},\boldsymbol{y}\rangle=0,$$

which we denote as $\boldsymbol{x}\perp\boldsymbol{y}$. A vector $\boldsymbol{x}$ is orthogonal to a subset $\mathcal{S}\subset\mathcal{V}$, denoted by $\boldsymbol{x}\perp\mathcal{S}$, if it is orthogonal to every element of $\mathcal{S}$. The orthogonal complement of $\mathcal{S}$, denoted by $\mathcal{S}^\perp$, consists of all vectors in $\mathcal{V}$ that are orthogonal to every vector in $\mathcal{S}$, i.e.

$$\mathcal{S}^\perp=\{\boldsymbol{x}\in\mathcal{V}:\langle\boldsymbol{v},\boldsymbol{x}\rangle=0,\ \forall\boldsymbol{v}\in\mathcal{S}\}.$$

Definition 1.6 (Norm) A norm $\|\cdot\|$ is a real-valued function defined on the vector space that has the following properties:

1. $\|\boldsymbol{x}\|\geq 0$, and $\|\boldsymbol{x}\|=0$ if and only if $\boldsymbol{x}=\boldsymbol{0}$.
2. $\|\alpha\boldsymbol{x}\|=|\alpha|\|\boldsymbol{x}\|$ for any scalar α.
3. Triangular inequality: $\|\boldsymbol{x}+\boldsymbol{y}\|\leq\|\boldsymbol{x}\|+\|\boldsymbol{y}\|$ for any vectors $\boldsymbol{x}$ and $\boldsymbol{y}$.

From the inner product, we can obtain the so-called induced norm:

$$\|\boldsymbol{x}\| = \sqrt{\langle \boldsymbol{x}, \boldsymbol{x} \rangle}.$$

Similarly, the definition of the metric in Sect. 1.1 informs us that a norm in a vector space $\mathcal{V}$ induces a metric, i.e.

$$d(\boldsymbol{x}, \boldsymbol{y}) = \|\boldsymbol{x} - \boldsymbol{y}\|, \quad \boldsymbol{x}, \boldsymbol{y} \in \mathcal{V}. \tag{1.3}$$

The norm and inner product in a vector space have special relations. For example, for any two vectors $\boldsymbol{x}, \boldsymbol{y} \in \mathcal{V}$, the following Cauchy–Schwarz inequality always holds:

$$|\langle \boldsymbol{x}, \boldsymbol{y} \rangle| \leq \|\boldsymbol{x}\| \|\boldsymbol{y}\|. \tag{1.4}$$

1.3 Banach and Hilbert Space

An *inner product space* is defined as a vector space that is equipped with an inner product. A *normed space* is a vector space on which a norm is defined. An inner product space is always a normed space since we can define a norm as $\|\boldsymbol{f}\| = \sqrt{\langle \boldsymbol{f}, \boldsymbol{f} \rangle}$, which is often called the induced norm. Among the various forms of the normed space, one of the most useful normed spaces is the *Banach space*.

Definition 1.7 The Banach space is a complete normed space.

Here, the "completeness" is especially important from the optimization perspective, since most optimization algorithms are implemented in an iterative manner so that the final solution of the iterative method should belong to the underlying space $\mathcal{H}$. Recall that the convergence property is a property of a metric space. Therefore, the Banach space can be regarded as a vector space equipped with desirable properties of a metric space. Similarly, we can define the *Hilbert space*.

Definition 1.8 The Hilbert space is a complete inner product space.

We can easily see that the Hilbert space is also a Banach space thanks to the induced norm. The inclusion relationship between vector spaces, normed spaces, inner product spaces, Banach spaces and Hilbert spaces is illustrated in Fig. 1.1.

As shown in Fig. 1.1, the Hilbert space has many nice mathematical structures such as inner product, norm, completeness, etc., so it is widely used in the machine learning literature. The following are well-known examples of Hilbert spaces:

- $l^2(\mathbb{Z})$: a function space composed of square summable discrete-time signals, i.e.

$$l^2(\mathbb{Z}) = \left\{ \boldsymbol{x} = \{x_l\}_{l=-\infty}^{\infty} \mid \sum_{l=-\infty}^{\infty} |x_l|^2 < \infty \right\}.$$

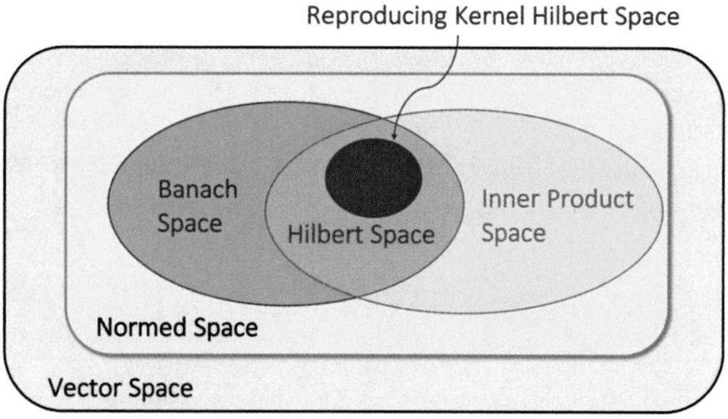

Fig. 1.1 RKHS, Hilbert space, Banach space, and vector space

Here, the inner product is defined as

$$\langle \boldsymbol{x}, \boldsymbol{y} \rangle_{\mathcal{H}} = \sum_{l=-\infty}^{\infty} x_l y_l, \quad \forall \boldsymbol{x}, \boldsymbol{y} \in \mathcal{H}. \tag{1.5}$$

- $L^2(\mathbb{R})$: a function space composed of square integrable continuous-time signals, i.e.

$$L^2(\mathbb{R}) = \left\{ x(t) \mid \int_{-\infty}^{\infty} |x(t)|^2 dt < \infty \right\}.$$

Here, the inner product is defined as

$$\langle x, y \rangle_{\mathcal{H}} = \int x(t)y(t)dt. \tag{1.6}$$

Among the various forms of the Hilbert space, the reproducing kernel Hilbert space (RKHS) is of particular interest in the classical machine learning literature, which will be explained later in this book. Here, the readers are reminded that the RKHS is only a subset of the Hilbert space as shown in Fig. 1.1, i.e. the Hilbert space is more general than the RKHS.

1.3.1 Basis and Frames

The set of vectors $\{\boldsymbol{x}_1, \cdots, \boldsymbol{x}_k\}$ is said to be *linearly independent* if a *linear combination* denoted by

$$\alpha_1 \boldsymbol{x}_1 + \alpha_2 \boldsymbol{x}_2 + \cdots + \alpha_k \boldsymbol{x}_k = \boldsymbol{0}$$

implies that

$$\alpha_i = 0, \quad i = 1, \cdots, k.$$

The set of all vectors *reachable* by taking linear combinations of vectors in a set $\mathcal{S}$ is called the *span* of $\mathcal{S}$. For example, if $\mathcal{S} = \{\boldsymbol{x}_i\}_{i=1}^k$, then we have

$$\text{span}(\mathcal{S}) = \left\{ \sum_{i=1}^k \alpha_i \boldsymbol{x}_i, \forall \alpha_i \in \mathbb{R} \right\}.$$

A set $\mathcal{B} = \{\boldsymbol{b}_i\}_{i=1}^m$ of elements (vectors) in a vector space $\mathcal{V}$ is called a basis, if every element of $\mathcal{V}$ may be written in a unique way as a linear combination of elements of $\mathcal{B}$, that is, for all $\boldsymbol{f} \in \mathcal{V}$, there exists unique coefficients $\{c_i\}$ such that

$$\boldsymbol{f} = \sum_{i=1}^m c_i \boldsymbol{b}_i. \tag{1.7}$$

A set $\mathcal{B}$ is a basis of $\mathcal{V}$ if and only if every element of $\mathcal{B}$ is linearly independent and $\text{span}(\mathcal{B}) = \mathcal{V}$. The coefficients of this linear combination are referred to as expansion coefficients, or coordinates on $\mathcal{B}$ of the vector. The elements of a basis are called basis vectors. In general, for m-dimensional spaces, the number of basis vectors is m. For example, when $\mathcal{V} = \mathbb{R}^2$, the following two sets are some examples of a basis:

$$\left\{ \begin{bmatrix} 1 \\ 0 \end{bmatrix}, \begin{bmatrix} 0 \\ 1 \end{bmatrix} \right\}, \quad \left\{ \begin{bmatrix} 1 \\ 1 \end{bmatrix}, \begin{bmatrix} 1 \\ -1 \end{bmatrix} \right\}. \tag{1.8}$$

For function spaces, the number of basis vectors can be infinite. For example, for the space V_T composed of periodic functions with the period of T, the following complex sinusoidals constitute its basis:

$$B = \{\varphi_n(t)\}_{n=-\infty}^{\infty}, \quad \varphi_n(t) = e^{i\frac{2\pi nt}{T}}, \tag{1.9}$$

so that any function $x(t) \in V_T$ can be represented by

$$x(t) = \sum_{n=-\infty}^{\infty} a_n \varphi_n(t), \tag{1.10}$$

where the expansion coefficient is given by

$$a_n = \frac{1}{T} \int_T x(t) \varphi_n^*(t) dt. \tag{1.11}$$

In fact, this basis expansion is often called the *Fourier series*.

Unlike the basis, which leads to the unique expansion, the frame is composed of redundant basis vectors, which allows multiple representations. For example, consider the following frame in $\mathbb{R}^2$:

$$\{\boldsymbol{v}_1, \boldsymbol{v}_2, \boldsymbol{v}_3\} = \left\{ \begin{bmatrix} 1 \\ 0 \end{bmatrix}, \begin{bmatrix} 0 \\ 1 \end{bmatrix}, \begin{bmatrix} 1 \\ 1 \end{bmatrix} \right\}. \tag{1.12}$$

Then, we can easily see that the frame allows multiple representations of, for example, $\boldsymbol{x} = [2, 3]^\top$ as shown in the following:

$$\boldsymbol{x} = 2\boldsymbol{v}_1 + 3\boldsymbol{v}_2 = \boldsymbol{v}_2 + 2\boldsymbol{v}_3. \tag{1.13}$$

Frames can also be extended to deal with function spaces, in which case the number of frame elements is infinite.

Formally, a set of functions

$$\boldsymbol{\Phi} = [\boldsymbol{\phi}_k]_{k \in \Gamma} = \left[\cdots \ \boldsymbol{\phi}_{k-1} \ \boldsymbol{\phi}_k \ \cdots \right]$$

in a Hilbert space H is called a *frame* if it satisfies the following inequality [1]:

$$\alpha \|\boldsymbol{f}\|^2 \leq \sum_{k \in \Gamma} |\langle \boldsymbol{f}, \boldsymbol{\phi}_k \rangle|^2 \leq \beta \|\boldsymbol{f}\|^2, \quad \forall \boldsymbol{f} \in H, \tag{1.14}$$

where $\alpha, \beta > 0$ are called the frame bounds. If $\alpha = \beta$, then the frame is said to be tight. In fact, the basis is a special case of tight frames.

1.4 Probability Space

We now start with a formal definition of a probability space and related terms from the measure theory [2].

Definition 1.9 (Probability Space) A probability space is a triple $(\Omega, \mathcal{F}, \mu)$ consisting of the sample space Ω, an event space $\mathcal{F}$ composed of a subset of Ω (which is often called σ-algebra), and the *probability measure (or distribution)* $\mu : \mathcal{F} \mapsto [0, 1]$, a function such that:

- μ must satisfy the countable additivity property that for all countable collections $\{E_i\}$ of pairwise disjoint sets:

$$\mu(\cup_i E_i) = \cup_i \mu(E_i);$$

- the measure of the entire sample space is equal to one: $\mu(\Omega) = 1$.

In fact, the probability measure is a special case of the general "measure" in measure theory [2]. Specifically, the general term "measure" is defined similarly to the probability measure defined above except that only positivity and the countable additivity property are required. Another important special case of a measure is the *counting measure* $\nu(A)$, which is the measure that assigns its value as the number of elements in the set A.

To understand the concept of a probability space, we give two examples: one for the discrete case, the other for the continuous one.

Example (Discrete Probability Space)
If the experiment consists of just one flip of a fair coin, then the outcome is either heads or tails: {H, T}. Hence, the sample space is $\Omega = \{\text{H}, \text{T}\}$. The σ-algebra or the event space contains $2^2 = 4$ events, namely: {H} ("heads"), {T} ("tails"), $\emptyset$ ("neither heads nor tails"), and {H, T} ("either heads or tails"); in other words, $\mathcal{F} = \{\emptyset, \{\text{H}\}, \{\text{T}\}, \{\text{H}, \text{T}\}\}$. There is a 50% chance of tossing heads and 50% for tails, so the probability measure in this example is $P(\emptyset = 0)$, $P(\{\text{H}\}) = 0.5$, $P(\{\text{T}\}) = 0.5$, $P(\{\text{H}, \text{T}\}) = 1$.

Example (Continuous Probability Space)
A number between 0 and 1 is chosen at random, uniformly. Here $\Omega = [0, 1]$. In this case, the event space $\mathcal{F}$ can be generated by: (i) the open intervals (a, b) on $[0, 1]$; (ii) the closed intervals $[a, b]$; (iii) the closed half-lines $[0, a]$, and their union, intersection, complement, and so on. Finally, the measure μ is the Lebesgue measure, defined as the sum of the lengths of the intervals contained in $\mathcal{F}$, i.e. $\mu([0.2, 0.5]) = 0.3$, $\mu([0, 0.2) \cup [0.5, 0.8]) = 0.5$, $\mu(\{0.5\}) = 0$.

We now define the *Radon–Nikodym derivative,* which is a mathematical tool to derive the probability density function (pdf) for the continuous domain, or probability mass function (pmf) for the discrete domain in a rigorous setting. This is also important in deriving the statistical distances, in particular, the divergences. For this, we need to understand the concept of an *absolutely continuous measure.*

Definition 1.10 (Absolutely Continuous Measure) If μ and ν are two measures on any event set $\mathcal{F}$ of Ω, we say that ν is absolutely continuous with respect to μ, or $\nu \ll \mu$, if for every measurable set A, $\mu(A) = 0$ implies $\nu(A) = 0$.

Theorem 1.1 (Radon–Nikodym Theorem) *Let λ and ν be two measures on any event set $\mathcal{F}$ of Ω. If $\lambda \ll \nu$, then there exists a non-negative function g on Ω such that*

$$\lambda(A) = \int_A d\lambda = \int_A g d\nu, \quad A \in \mathcal{F}. \tag{1.15}$$

The function g is called the Radon–Nikodym derivative or density of λ w.r.t. ν and is denoted by $d\lambda/d\nu$. One of the popular Radon–Nikodym derivatives in probability theory is the probability density function (pdf) or probability mass function (pmf) as discussed below.

For a probability space $(\Omega, \mathcal{F}, \mu)$, a *random variable* is defined as a function $X : \Omega \mapsto M$ from a set of possible outcomes Ω to a measurable space M. For the random variable X, we can now define the mean for its functions:

$$\mathbb{E}_\mu[g(X)] = \int_X g(x) d\mu(x). \tag{1.16}$$

1.5 Some Matrix Algebra

In the following, we introduce some matrix algebra that is useful in understanding the materials in this book.

A *matrix* is a rectangular array of numbers, denoted by an upper case letter, say $\boldsymbol{A}$. A matrix with m rows and n columns is called an $m \times n$ matrix given by

$$\boldsymbol{A} = \begin{bmatrix} a_{11} & a_{12} & \cdots & a_{1n} \\ a_{21} & a_{22} & \cdots & a_{2n} \\ \vdots & \vdots & \ddots & \vdots \\ a_{m1} & a_{m2} & \cdots & a_{mn} \end{bmatrix}.$$

The k-th column of matrix $\boldsymbol{A}$ is often denoted by $\boldsymbol{a}_k$. The maximal number of linearly independent columns of $\boldsymbol{A}$ is called the *rank* of the matrix $\boldsymbol{A}$. It is easy to show that

$$\text{Rank}(\boldsymbol{A}) = \dim \text{span}\left([\boldsymbol{a}_1, \cdots, \boldsymbol{a}_n]\right).$$

The trace of a square matrix $\boldsymbol{A} \in \mathbb{R}^{n \times n}$, denoted $\text{Tr}(\boldsymbol{A})$ is defined to be the sum of elements on the main diagonal (from the upper left to the lower right) of $\boldsymbol{A}$:

$$\text{Tr}(\boldsymbol{A}) = \sum_{i=1}^{n} a_{ii}.$$

Definition 1.11 (Range Space) The range space of a matrix $\boldsymbol{A} \in \mathbb{R}^{m \times n}$, denoted by $\mathcal{R}(\boldsymbol{A})$, is defined by $\mathcal{R}(\boldsymbol{A}) := \{\boldsymbol{A}\boldsymbol{x} \mid \forall \boldsymbol{x} \in \mathbb{R}^n\}$.

Definition 1.12 (Null Space) The null space of a matrix $\boldsymbol{A} \in \mathbb{R}^{m \times n}$, denoted by $\mathcal{N}(\boldsymbol{A})$, is defined by $\mathcal{N}(\boldsymbol{A}) := \{\boldsymbol{x} \in \mathbb{R}^n \mid \boldsymbol{A}\boldsymbol{x} = \boldsymbol{0}\}$.

A subset of a vector space is called a *subspace* if it is closed under both addition and scalar multiplication. We can easily see that the range and null spaces are subspaces. Moreover, we can show the following fundamental property:

$$\mathcal{R}(\boldsymbol{A})^{\perp} = \mathcal{N}(\boldsymbol{A}^{\top}), \quad \mathcal{N}(\boldsymbol{A})^{\perp} = \mathcal{R}(\boldsymbol{A}^{\top}). \tag{1.17}$$

If a vector space $\mathcal{V}$ is Hilbert space, then it is known that for a subspace $\mathcal{S} \in \mathcal{V}$ and the vector $\boldsymbol{y} \in \mathcal{V}$, the point in $\mathcal{S}$ that is closest to $\boldsymbol{y}$ exists and is unique, and given by

$$\hat{\boldsymbol{y}} = \mathcal{P}_{\mathcal{S}}\boldsymbol{y}$$

where $\mathcal{P}_{\mathcal{S}}$ is the *projector* associated with the subspace $\mathcal{S}$. In particular, if the subspace $\mathcal{S}$ has a basis $\boldsymbol{B}$, then the projector for $\mathcal{S}$ is given by

$$\mathcal{P}_{\mathcal{S}} = \boldsymbol{B}(\boldsymbol{B}^{\top}\boldsymbol{B})^{-1}\boldsymbol{B}^{\top}.$$

The eigen-decomposition of a square matrix is defined as follows.

Definition 1.13 (Eigen-Decomposition) A (nonzero) vector $\boldsymbol{v} \in \mathbb{C}^n$ is an eigenvector of a square matrix $\boldsymbol{A} \in \mathbb{C}^{n\times n}$ if it satisfies the linear equation

$$\boldsymbol{A}\boldsymbol{v} = \lambda\boldsymbol{v}, \tag{1.18}$$

where λ is a scalar, termed the eigenvalue corresponding to $\boldsymbol{v}$.

We now define the singular value decomposition (SVD) of $\boldsymbol{A}$.

Theorem 1.2 (SVD Theorem) *If $\boldsymbol{A} \in \mathbb{C}^{m\times n}$ is a rank r matrix, then there exist matrices $\boldsymbol{U} \in \mathbb{C}^{m\times r}$ and $\boldsymbol{V} \in \mathbb{C}^{n\times r}$ such that $\boldsymbol{U}^{\top}\boldsymbol{U} = \boldsymbol{V}^{\top}\boldsymbol{V} = \boldsymbol{I}_r$ and $\boldsymbol{A} = \boldsymbol{U}\boldsymbol{\Sigma}\boldsymbol{V}^{\top}$, where $\boldsymbol{I}_r$ is the $r \times r$ identity matrix and $\boldsymbol{\Sigma}$ is an $r \times r$ diagonal matrix whose diagonal entries, called* singular values, *satisfy*

$$\sigma_1 \geq \sigma_2 \geq \cdots \geq \sigma_r > 0.$$

The decomposition can be written as

$$\boldsymbol{A} = \begin{bmatrix}\boldsymbol{u}_1 & \cdots & \boldsymbol{u}_r\end{bmatrix} \begin{bmatrix} \sigma_1 & 0 & \cdots & 0 \\ 0 & \sigma_2 & \ddots & \vdots \\ \vdots & \ddots & \ddots & 0 \\ 0 & \cdots & 0 & \sigma_r \end{bmatrix} \begin{bmatrix}\boldsymbol{v}_1 & \cdots & \boldsymbol{v}_r\end{bmatrix}^{\top} = \sum_{k=1}^{r} \sigma_k \boldsymbol{u}_k \boldsymbol{v}_k^{\top},$$

where $\boldsymbol{u}_k$ and $\boldsymbol{v}_k$ are called left singular vectors and right singular vectors, respectively.

Using the SVD, we can easily show the following:

$$\mathcal{P}_{\mathcal{R}(\boldsymbol{A})} = \boldsymbol{U}\boldsymbol{U}^\top, \quad \mathcal{P}_{\mathcal{R}(\boldsymbol{A}^\top)} = \boldsymbol{V}\boldsymbol{V}^\top. \tag{1.19}$$

Using the SVD, we can define the matrix norm. Among the various forms of matrix norms for a matrix $\boldsymbol{X} \in \mathbb{R}^{n \times n}$, the spectral norm $\|\boldsymbol{X}\|_2$ and the nuclear norm $\|\boldsymbol{X}\|_*$ are quite often used, which are defined by

$$\|\boldsymbol{X}\|_2 = \sigma_{\max}(\boldsymbol{X}) = (\lambda_{\max}(\boldsymbol{X}^\top \boldsymbol{X}))^{1/2}, \tag{1.20}$$

$$\|\boldsymbol{X}\|_* = \sum_i \sigma_i(\boldsymbol{X}) = \sum_i (\lambda_i(\boldsymbol{X}^\top \boldsymbol{X}))^{1/2}, \tag{1.21}$$

where $\sigma_{\max}(\cdot)$ and $\lambda_{\max}(\cdot)$ denote the largest singular value and eigenvalue, respectively.

The following matrix inversion lemma [3] is quite useful.

Lemma 1.1 (Matrix Inversion Lemma)

$$(\boldsymbol{I} + \boldsymbol{U}\boldsymbol{C}\boldsymbol{V})^{-1} = \boldsymbol{I} - \boldsymbol{U}\left(\boldsymbol{C}^{-1} + \boldsymbol{V}\boldsymbol{U}\right)^{-1}\boldsymbol{V}, \tag{1.22}$$

$$(\boldsymbol{A} + \boldsymbol{U}\boldsymbol{C}\boldsymbol{V})^{-1} = \boldsymbol{A}^{-1} - \boldsymbol{A}^{-1}\boldsymbol{U}\left(\boldsymbol{C}^{-1} + \boldsymbol{V}\boldsymbol{A}^{-1}\boldsymbol{U}\right)^{-1}\boldsymbol{V}\boldsymbol{A}^{-1}. \tag{1.23}$$

1.5.1 Kronecker Product

In mathematics, the Kronecker product, sometimes denoted by ⊗, is an operation on two matrices of arbitrary size resulting in a block matrix. The formal definition is given as follows.

Definition 1.14 (Kronecker Product) If $\boldsymbol{A}$ is an $m \times n$ matrix and $\boldsymbol{B}$ is a $p \times q$ matrix, then the Kronecker product $\boldsymbol{A} \otimes \boldsymbol{B}$ is the $pm \times qn$ block matrix:

$$\boldsymbol{A} \otimes \boldsymbol{B} = \begin{bmatrix} a_{11}\boldsymbol{B} & \cdots & a_{1n}\boldsymbol{B} \\ \vdots & \ddots & \vdots \\ a_{m1}\boldsymbol{B} & \cdots & a_{mn}\boldsymbol{B} \end{bmatrix}. \tag{1.24}$$

The Kronecker product has many important properties, which can be exploited to simplify many matrix-related operations. Some of the basic properties are provided in the following lemma. The proofs of the lemmas are straightforward, which can easily be found from a standard linear algebra textbook [4].

Lemma 1.2

$$A \otimes (B + C) = A \otimes B + A \otimes C. \tag{1.25}$$

$$(B + C) \otimes A = B \otimes A + C \otimes A. \tag{1.26}$$

$$A \otimes B \neq B \otimes A. \tag{1.27}$$

$$(A \otimes B) \otimes C = A \otimes (B \otimes C). \tag{1.28}$$

$$(A \otimes B)^\top = A^\top \otimes B^\top. \tag{1.29}$$

$$(A \otimes B)^{-1} = A^{-1} \otimes B^{-1}. \tag{1.30}$$

Lemma 1.3 *If A, B, C and D are matrices of such a size that one can form the matrix products AC and BD, then*

$$(A \otimes B)(C \otimes D) = AC \otimes BD. \tag{1.31}$$

One of the important usages of the Kronecker product comes from the vectorization operation of a matrix. For this we first define the following two operations.

Definition 1.15 If $A = \begin{bmatrix} a_1 & \cdots & a_n \end{bmatrix} \in \mathbb{R}^{m \times n}$, then

$$\mathrm{VEC}(A) = \begin{bmatrix} a_1 \\ \vdots \\ a_n \end{bmatrix} \in \mathbb{R}^{mn}, \tag{1.32}$$

$$\mathrm{UNVEC}(\mathrm{VEC}(A)) = \mathrm{UNVEC}\left(\begin{bmatrix} a_1 \\ \vdots \\ a_n \end{bmatrix} \right) = A. \tag{1.33}$$

From these definitions, we can obtain the following two lemmas which will be extensively used here.

Lemma 1.4 ([4]) *For the matrices A, B, C with appropriate sizes, we have*

$$\mathrm{VEC}(CAB) = (B^\top \otimes C)\mathrm{VEC}(A), \tag{1.34}$$

where $\mathrm{VEC}(\cdot)$ *is the column-wise vectorization operation.*

Lemma 1.5 *For the vectors $x \in \mathbb{R}^m$, $y \in \mathbb{R}^n$, we have*

$$\mathrm{VEC}(xy^\top) = (y \otimes I_m)x, \tag{1.35}$$

where I_m denotes the $m \times m$ identity matrix.

Proof By plugging $\boldsymbol{C} = \boldsymbol{I}_m$, $\boldsymbol{A} = \boldsymbol{x}$ and $\boldsymbol{B} = \boldsymbol{y}^\top$ into (1.34), we conclude the proof. □

1.5.2 Matrix and Vector Calculus

In computing a derivative of a scalar, vector, or matrix with respect to a scalar, vector, or matrix, we should be consistent with the notation. In fact, there are two different conventions: numerator layout and denominator layout. For example, for a given scalar y and a column vector $\boldsymbol{x} = [x_1, \cdots, x_n]^\top \in \mathbb{R}^n$, the numerator layout has the following convention:

$$\frac{\partial y}{\partial \boldsymbol{x}} = \begin{bmatrix} \frac{\partial y}{\partial x_1} & \cdots & \frac{\partial y}{\partial x_n} \end{bmatrix}, \quad \frac{\partial \boldsymbol{x}}{\partial y} = \begin{bmatrix} \frac{\partial x_1}{\partial y} \\ \vdots \\ \frac{\partial x_n}{\partial y} \end{bmatrix},$$

implying that the number of the row follows that of the numerator. On the other hand, the denominator layout notation provides

$$\frac{\partial y}{\partial \boldsymbol{x}} = \begin{bmatrix} \frac{\partial y}{\partial x_1} \\ \vdots \\ \frac{\partial y}{\partial x_n} \end{bmatrix}, \quad \frac{\partial \boldsymbol{x}}{\partial y} = \begin{bmatrix} \frac{\partial x_1}{\partial y} & \cdots & \frac{\partial x_n}{\partial y} \end{bmatrix},$$

where the number of resulting rows follows that of the denominator. Either layout convention is okay, but we should be consistent in using the convention.

Here, we will follow the denominator layout convention. The main motivation for using the denominator layout is from the derivative with respect to the matrix. More specifically, for a given scalar c and a matrix $\boldsymbol{W} \in \mathbb{R}^{m\times n}$, according to the denominator layout, we have

$$\frac{\partial c}{\partial \boldsymbol{W}} = \begin{bmatrix} \frac{\partial c}{\partial w_{11}} & \cdots & \frac{\partial c}{\partial w_{1n}} \\ \vdots & \ddots & \vdots \\ \frac{\partial c}{\partial w_{m1}} & \cdots & \frac{\partial c}{\partial w_{mn}} \end{bmatrix} \in \mathbb{R}^{m\times n}. \tag{1.36}$$

Furthermore, this notation leads to the following familiar result:

$$\frac{\partial \boldsymbol{a}^\top \boldsymbol{x}}{\partial \boldsymbol{x}} = \frac{\partial \boldsymbol{x}^\top \boldsymbol{a}}{\partial \boldsymbol{x}} = \boldsymbol{a}. \tag{1.37}$$

Accordingly, for a given scalar c and a matrix $\boldsymbol{W} \in \mathbb{R}^{m\times n}$, we can show that

$$\frac{\partial c}{\partial \boldsymbol{W}} := \textsc{UnVec}\left(\frac{\partial c}{\partial \textsc{Vec}(\boldsymbol{W})}\right) \in \mathbb{R}^{m\times n}, \tag{1.38}$$

in order to be consistent with (1.36). Under the denominator layout notation, for given vectors $\boldsymbol{x} \in \mathbb{R}^m$ and $\boldsymbol{y} \in \mathbb{R}^n$, the derivative of a vector with respect to a vector is given by

$$\frac{\partial \boldsymbol{y}}{\partial \boldsymbol{x}} = \begin{bmatrix} \frac{\partial y_1}{\partial x_1} & \cdots & \frac{\partial y_n}{\partial x_1} \\ \vdots & \ddots & \vdots \\ \frac{\partial y_1}{\partial x_m} & \cdots & \frac{\partial y_n}{\partial x_m} \end{bmatrix} \in \mathbb{R}^{m\times n}. \tag{1.39}$$

Then, the chain rule can be specified as follows:

$$\frac{\partial c(\boldsymbol{g}(\boldsymbol{u}))}{\partial \boldsymbol{x}} = \frac{\partial \boldsymbol{u}}{\partial \boldsymbol{x}}\frac{\partial \boldsymbol{g}(\boldsymbol{u})}{\partial \boldsymbol{u}}\frac{\partial c(\boldsymbol{g})}{\partial \boldsymbol{g}}. \tag{1.40}$$

Eq. (1.37) also leads to

$$\frac{\partial \boldsymbol{A}\boldsymbol{x}}{\partial \boldsymbol{x}} = \boldsymbol{A}^\top. \tag{1.41}$$

Finally, the following result is useful.

Lemma 1.6 *Let $\boldsymbol{A} \in \mathbb{R}^{m\times n}$ and $\boldsymbol{x} \in \mathbb{R}^n$. Then, we have*

$$\frac{\partial \boldsymbol{A}\boldsymbol{x}}{\partial \textsc{Vec}(\boldsymbol{A})} = \boldsymbol{x} \otimes \boldsymbol{I}_m. \tag{1.42}$$

Proof Using Lemma 1.4, we have $\boldsymbol{A}\boldsymbol{x} = \textsc{Vec}(\boldsymbol{A}\boldsymbol{x}) = (\boldsymbol{x}^\top \otimes \boldsymbol{I}_m)\textsc{Vec}(\boldsymbol{A})$. Thus,

$$\begin{aligned} \frac{\partial \boldsymbol{A}\boldsymbol{x}}{\partial \textsc{Vec}(\boldsymbol{A})} &= \frac{\partial (\boldsymbol{x}^\top \otimes \boldsymbol{I}_m)\textsc{Vec}(\boldsymbol{A})}{\partial \textsc{Vec}(\boldsymbol{A})} \\ &= (\boldsymbol{x}^\top \otimes \boldsymbol{I}_m)^\top \\ &= \boldsymbol{x} \otimes \boldsymbol{I}_m, \end{aligned} \tag{1.43}$$

where we use (1.37) and (1.29) for the second and the third equalities, respectively. Q.E.D. □

Lemma 1.7 ([5]) *Let $\boldsymbol{x}, \boldsymbol{a}$ and $\boldsymbol{B}$ denote vectors and a matrix with appropriate sizes, respectively. Then, we have*

$$\frac{\partial \boldsymbol{x}^\top \boldsymbol{a}}{\partial \boldsymbol{x}} = \frac{\partial \boldsymbol{a}^\top \boldsymbol{x}}{\partial \boldsymbol{x}} = \boldsymbol{a}, \tag{1.44}$$

$$\frac{\partial \boldsymbol{x}^\top \boldsymbol{B}\boldsymbol{x}}{\partial \boldsymbol{x}} = (\boldsymbol{B} + \boldsymbol{B}^\top)\boldsymbol{x}. \tag{1.45}$$

For a given scalar function $\ell : \boldsymbol{x} \in \mathbb{R}^n \mapsto \mathbb{R}$, the derivative is often called the gradient, which can be represented by the denominator layout:

$$\nabla \ell := \frac{\partial \ell}{\partial \boldsymbol{x}} \in \mathbb{R}^n.$$

1.6 Elements of Convex Optimization

1.6.1 Some Definitions

Let $\mathcal{X}$, $\mathcal{Y}$ and $\mathcal{Z}$ be non-empty sets. The identity operator on $\mathcal{H}$ is denoted by $\mathcal{I}$, i.e. $\mathcal{I}\boldsymbol{x} = \boldsymbol{x}, \forall \boldsymbol{x} \in \mathcal{H}$. Let $\mathcal{D} \subset \mathcal{H}$ be a non-empty set. The set of the *fixed points* of an operator $\mathcal{T} : D \mapsto D$ is denoted by

$$\mathrm{Fix}\mathcal{T} = \{\boldsymbol{x} \in \mathcal{D} \mid \mathcal{T}\boldsymbol{x} = \boldsymbol{x}\}.$$

Let $\mathcal{X}$ and $\mathcal{Y}$ be real normed vector space. As a special case of an operator, we define a set of *linear operators*:

$$\mathcal{B}(\mathcal{X}, \mathcal{Y}) = \{\mathcal{T} : \mathcal{X} \mapsto \mathcal{Y} \mid \mathcal{T} \text{ is linear and continuous}\}$$

and we write $\mathcal{B}(\mathcal{X}) = \mathcal{B}(\mathcal{X}, \mathcal{X})$. Let $f : \mathcal{X} \mapsto [-\infty, \infty]$ be a function. The *domain* of f is

$$\mathrm{dom} f = \{\boldsymbol{x} \in \mathcal{X} | f(\boldsymbol{x}) < \infty\},$$

the *graph* of f is

$$\mathrm{gra} f = \{(\boldsymbol{x}, y) \in \mathcal{X} \times \mathbb{R} | f(\boldsymbol{x}) = y\},$$

and the *epigraph* of f is

$$\mathrm{epi} f = \{(\boldsymbol{x}, y) : \boldsymbol{x} \in \mathcal{X}, y \in \mathbb{R}, y \geq f(\boldsymbol{x})\}.$$

The *indicator function* $\iota_C : \mathcal{X} \mapsto [-\infty, \infty]$ of $C \subset \mathcal{X}$ is defined as

$$\iota_C(\boldsymbol{x}) = \begin{cases} 0, & \text{if } \boldsymbol{x} \in C, \\ \infty, & \text{otherwise}. \end{cases} \tag{1.46}$$

We often use another definition of the indicator function:

$$\chi_C(\boldsymbol{x}) = \begin{cases} 1, & \text{if } \boldsymbol{x} \in C, \\ 0, & \text{otherwise}. \end{cases} \tag{1.47}$$

The *support function* of a set C is defined as

$$S_C(\boldsymbol{x}) = \sup\{\langle \boldsymbol{x}, \boldsymbol{y} \rangle | \boldsymbol{y} \in C\}.$$

An *affine function* is denoted by

$$\boldsymbol{x} \mapsto \mathcal{T}\boldsymbol{x} + \boldsymbol{b}, \quad \boldsymbol{x} \in \mathcal{X}, \boldsymbol{y} \in \mathcal{Y}, \mathcal{T} \in \mathcal{B}(\mathcal{X}, \mathcal{Y}).$$

A function f is called *lower semicontinuous* at $\boldsymbol{x}_0$ if for every $\varepsilon > 0$ there exists a neighbourhood $\mathcal{U}$ of $\boldsymbol{x}_0$ such that $f(\boldsymbol{x}) \geq f(\boldsymbol{x}_0) - \varepsilon$ for all $\boldsymbol{x} \in \mathcal{U}$. This is expressed as

$$\liminf_{\boldsymbol{x} \to \boldsymbol{x}_0} f(\boldsymbol{x}) \geq f(\boldsymbol{x}_0).$$

A function is lower semicontinuous if and only if all of its lower level sets $\{\boldsymbol{x} \in \mathcal{X} : f(\boldsymbol{x}) \leq \alpha\}$ are closed. Alternatively, f is lower semicontinuous if and only if the epigraph of f is closed. A function is *proper* if $-\infty \notin f(\mathcal{X})$ and $\mathrm{dom} f \neq \emptyset$ (Fig. 1.2).

An operator $\mathcal{A} : \mathcal{H} \mapsto \mathcal{H}$ is positive semidefinite if and only if

$$\langle \boldsymbol{x}, \mathcal{A}\boldsymbol{x} \rangle \geq 0, \quad \forall \boldsymbol{x} \in \mathcal{H}.$$

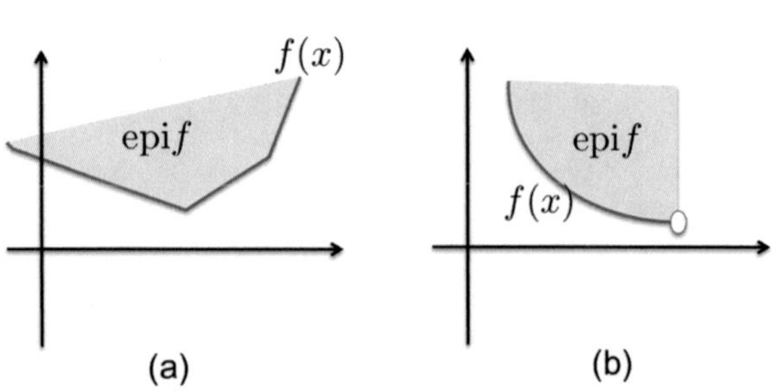

Fig. 1.2 Epigraphs for (**a**) a lower semicontinuous function, and (**b**) a function which is not lower semicontinuous

An operator $\mathcal{A} : \mathcal{H} \mapsto \mathcal{H}$ is positive definite if and only if

$$\langle \boldsymbol{x}, \mathcal{A}\boldsymbol{x} \rangle > 0, \quad \forall \boldsymbol{x} \in \mathcal{H}.$$

For simplicity, we denote $\mathcal{A} \succeq 0$ (resp. $\mathcal{A} \succ 0$) for positive semidefinite (resp. positive definite) operators. If $\mathcal{A} : \mathbb{C}^n \mapsto \mathbb{C}^n$, then $\mathbb{S}^n_{++}$ and $\mathbb{S}^n_{+}$ denote the set of $n \times n$ positive definite and semipositive definite matrices, respectively. Here, the eigenvalues of positive semidefinite (resp. positive definite) are all real and non-negative (resp. positive).

1.6.2 Convex Sets, Convex Functions

A function $f(\boldsymbol{x})$ is a *convex function* if $\mathrm{dom} f$ is a convex set and

$$f(\theta \boldsymbol{x}_1 + (1-\theta)\boldsymbol{x}_2) \le \theta f(\boldsymbol{x}_1) + (1-\theta) f(\boldsymbol{x}_1)$$

for all $\boldsymbol{x}_1, \boldsymbol{x}_2 \in \mathrm{dom} f$, $0 \le \theta \le 1$. A *convex set* is a set that contains every line segment between any two points in the set (see Fig. 1.3). Specifically, a set C is convex if $\boldsymbol{x}_1, \boldsymbol{x}_2 \in C$, then $\theta \boldsymbol{x}_1 + (1-\theta)\boldsymbol{x}_2 \in C$ for all $0 \le \theta \le 1$. The relation between a convex function and a convex set can also be stated using its epigraph. Specifically, a function $f(\boldsymbol{x})$ is convex if and only if its epigraph $\mathrm{epi} f$ is a convex set.

Convexity is preserved under various operations. For example, if $\{f_i\}_{i \in I}$ is a family of convex functions, then, $\sup_{i \in I} f_i$ is convex. In addition, a set of convex functions is closed under addition and multiplication by strictly positive real numbers. Moreover, the limit point of a convergent sequence of convex functions is also convex. Important examples of convex functions are summarized in Table 1.1.

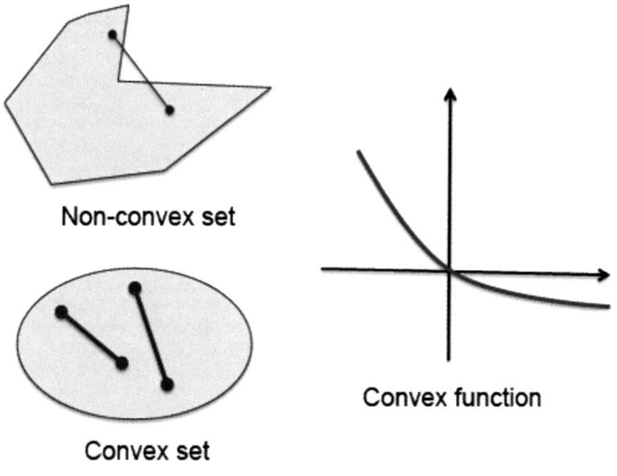

Fig. 1.3 A convex set and a convex function

Table 1.1 Examples of convex functions

Name	$f(x)$
Exponential	$e^{ax}, \quad \forall a \in \mathbb{R}$
Quadratic over linear	$x^2/y, \quad (x, y) \in \mathbb{R} \times \mathbb{R}_{++}$
Huber function	$\begin{cases} \lvert x\rvert^2/2\mu, & \text{if } \lvert x\rvert < \mu \\ \lvert x\rvert - \mu/2, & \text{if } \lvert x\rvert \geq \mu \end{cases}$
Relative entropy	$y \log y - y \log x, \quad (x, y) \in \mathbb{R}_{++} \times \mathbb{R}_{++}$
Indicator function	$\iota_C(\boldsymbol{x}), \quad C:$ convex set
Support function	$S_C(\boldsymbol{x}) = \sup\{\langle \boldsymbol{x}, \boldsymbol{y}\rangle \mid \boldsymbol{y} \in C\}$
Distance to a set	$d(\boldsymbol{x}, \mathcal{S}) = \inf_{\boldsymbol{y}\in\mathcal{S}} \lVert\boldsymbol{x} - \boldsymbol{y}\rVert$
Affine function	$\boldsymbol{T}\boldsymbol{x} + \boldsymbol{b}, \quad \boldsymbol{x} \in \mathbb{R}^n.$
Quadratic function	$\boldsymbol{x}^\top \boldsymbol{Q}\boldsymbol{x}/2, \quad \boldsymbol{x} \in \mathbb{R}^n, \boldsymbol{Q} \in \mathbb{S}_+$
p-norms	$\lVert\boldsymbol{x}\rVert_p = \left(\sum_i \lvert x_i\rvert^p\right)^{1/p}, p \geq 1$
l_∞-norm	$\lVert\boldsymbol{x}\rVert_\infty = \max_i \lvert x_i\rvert$
Max function	$\max\{x_1, \cdots, x_n\}$
Log-sum-exponential	$\log\left(\sum_{i=1}^n e^{x_i}\right), x = (x_1, \cdots, x^{(k)}) \in \mathbb{R}^n.$
Gaussian data fidelity	$\lVert\boldsymbol{y} - \boldsymbol{A}\boldsymbol{x}\rVert^2, \quad \boldsymbol{x} \in \mathcal{H}$
Poisson data fidelity	$\langle \mathbf{1}, \boldsymbol{A}\boldsymbol{x}\rangle - \langle \boldsymbol{y}, \log(\boldsymbol{A}\boldsymbol{x})\rangle, \quad \boldsymbol{x} \in \mathbb{R}^n, \mathbf{1} = (1, \cdots, 1) \in \mathbb{R}^n$
Spectral norm	$\lVert\boldsymbol{X}\rVert_2 = \sigma_{\max}(\boldsymbol{X}) = (\lambda_{\max}(\boldsymbol{X}^\top\boldsymbol{X}))^{1/2}, \boldsymbol{X} \in \mathbb{R}^{n\times n}$
Nuclear norm	$\lVert\boldsymbol{X}\rVert_* = \sum_i \sigma_i(\boldsymbol{X}) = \sum_i (\lambda_i(\boldsymbol{X}^\top\boldsymbol{X}))^{1/2}, \boldsymbol{X} \in \mathbb{R}^{n\times n}$

Table 1.2 Examples of concave functions

Name	$f(x)$
Powers	$x^p, \quad 0 \leq p \leq 1, \quad x \in \mathbb{R}_{++}$
Geometric mean	$\left(\prod_{i=1}^n x_i\right)^{\frac{1}{n}}$
Logarithm	$\log x, \quad x \in \mathbb{R}_{++}$
Log determinant	$\log\det(\boldsymbol{X}), \quad \boldsymbol{X} \in \mathbb{S}_{++}$

A function f is *concave* if $-f$ is convex. It is easy to show that an affine function $f(\boldsymbol{x}) = \boldsymbol{A}\boldsymbol{x} + \boldsymbol{b}$ is both convex and concave. Examples of concave functions that are often used in this textbook can be found in Table 1.2.

1.6.3 Subdifferentials

The *directional derivative* of f at $\boldsymbol{x} \in \mathrm{dom} f$ in the direction of $\boldsymbol{y} \in \mathcal{H}$ is defined by

$$f'(\boldsymbol{x}; \boldsymbol{y}) = \lim_{\alpha\downarrow 0} \frac{f(\boldsymbol{x} + \alpha\boldsymbol{y}) - f(\boldsymbol{x})}{\alpha} \tag{1.48}$$

if the limit exists. If the limit exists for all $\boldsymbol{y} \in \mathcal{H}$, then one says that f is *Gãteaux differentiable* at $\boldsymbol{x}$. Suppose $f'(\boldsymbol{x}; \cdot)$ is linear and continuous on $\mathcal{H}$. Then, there exist a unique gradient vector $\nabla f(\boldsymbol{x}) \in \mathcal{H}$ such that

$$f'(\boldsymbol{x}; \boldsymbol{y}) = \langle \boldsymbol{y}, \nabla f(\boldsymbol{x}) \rangle, \quad \forall \boldsymbol{y} \in \mathcal{H}.$$

If a function is differentiable, the convexity of a function can easily be checked using the first- and second-order differentiability, as stated in the following:

Proposition 1.1 *Let* $f : \mathcal{H} \mapsto (-\infty, \infty]$ *be proper. Suppose that* $\mathrm{dom} f$ *is open and convex, and* f *is Gâteux differentiable on* $\mathrm{dom} f$. *Then, the followings are equivalent:*

1. f *is convex.*
2. *(First-order):* $f(\boldsymbol{y}) \geq f(\boldsymbol{x}) + \langle \boldsymbol{y} - \boldsymbol{x}, \nabla f(\boldsymbol{x}) \rangle, \quad \forall \boldsymbol{x}, \boldsymbol{y} \in \mathcal{H}.$
3. *(Monotonicity of gradient):* $\langle \boldsymbol{y} - \boldsymbol{x}, \nabla f(\boldsymbol{y}) - \nabla f(\boldsymbol{x}) \rangle \geq 0, \quad \forall \boldsymbol{x}, \boldsymbol{y} \in \mathcal{H}.$

If the convergence in (1.48) is uniform with respect to $\boldsymbol{y}$ on bounded sets, i.e.

$$\lim_{\boldsymbol{0} \neq \boldsymbol{y} \to \boldsymbol{0}} \frac{f(\boldsymbol{x} + \boldsymbol{y}) - f(\boldsymbol{x}) - \langle \boldsymbol{y}, \nabla f(\boldsymbol{x}) \rangle}{\|\boldsymbol{y}\|} = 0, \tag{1.49}$$

then f is *Fréchet differentiable* and $\nabla f(\boldsymbol{x})$ is called the *Fréchet gradient* of f at $\boldsymbol{x}$. If f is differentiable and convex, then it is clear that

$$\boldsymbol{x} \in \arg\min f \Leftrightarrow \nabla f(\boldsymbol{x}) = \boldsymbol{0}.$$

However, if f is not differentiable, we need a more general framework to characterize the minimizers. The *sub-differential* of f is a set-valued operator defined as

$$\partial f(\boldsymbol{x}) = \{\boldsymbol{u} \in \mathcal{H} : f(\boldsymbol{y}) \geq f(\boldsymbol{x}) + \langle \boldsymbol{y} - \boldsymbol{x}, \boldsymbol{u} \rangle, \quad \forall \boldsymbol{y} \in \mathcal{H}\}. \tag{1.50}$$

The elements of sub-differential $\partial f(\boldsymbol{x})$ are called *sub-gradients* of f at $\boldsymbol{x}$. Another important role of the subdifferentials comes from *Fermat's rule* that characterizes the global minimizers (Fig. 1.4):

Theorem 1.3 (Fermat's Rule) *Let* $f : \mathcal{H} \mapsto (-\infty, \infty]$ *be proper. Then,*

$$\arg\min f = \mathrm{zer} \partial f := \{\boldsymbol{x} \in \mathcal{H} \mid \boldsymbol{0} \in \partial f(\boldsymbol{x})\}. \tag{1.51}$$

1.6.4 Convex Conjugate

A *convex conjugate* or *convex dual* is very important concept for both classical and mordern convex optimization techniques. Formally, the *conjugate* function f^* :

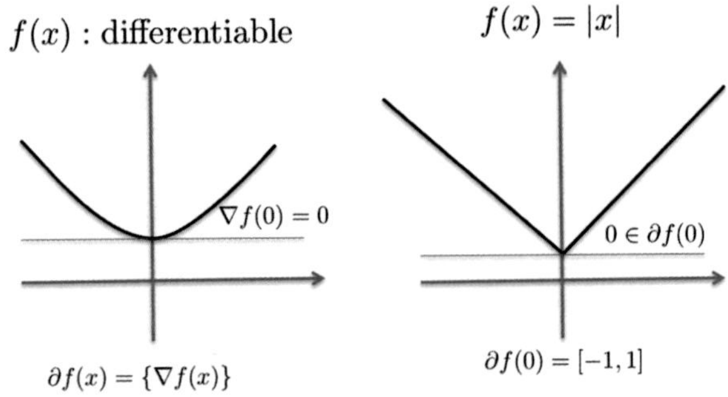

Fig. 1.4 Fermat's rule for the global minimizer

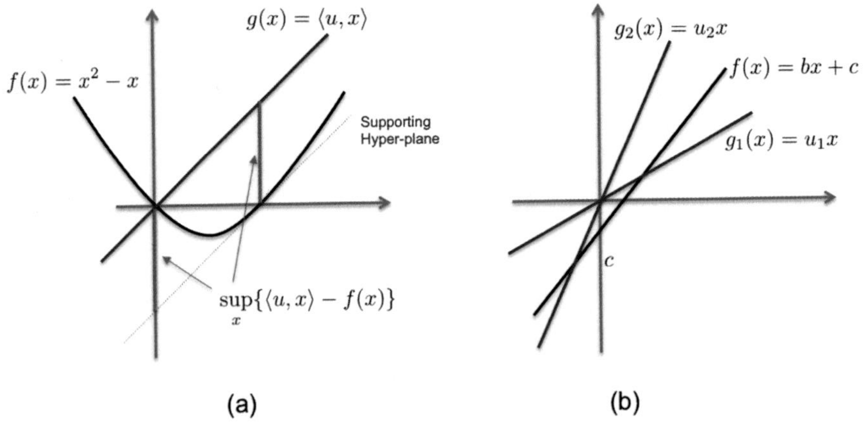

Fig. 1.5 (**a**) Geometry of convex conjugate. (**b**) Examples of finding convex conjugate for $f(x) = bx + c$

$\mathcal{H} \mapsto [-\infty, \infty]$ of a function $f : \mathcal{H} \mapsto [-\infty, \infty]$ is defined as

$$f^*(u) = \sup_{x \in \mathcal{H}} \{\langle \boldsymbol{u}, \boldsymbol{x}\rangle - f(\boldsymbol{x})\}. \tag{1.52}$$

The transform in (1.52) is often called *Legendre-Fenchel transform.*

Figure 1.5a shows a geometric interpretation of the convex conjugate when $\mathcal{H} = \mathbb{R}$. For example, when $f(x) = x^2 - x$, the convex conjugate $f^*(u)$ at $u = 1$ is the maximum difference between $g(x) = x$ and $f(x) = x^2 - x$, which occurs at $x = 1$ in this example. The difference is also equal to the magnitude of the y-intercept of the supporting hyerplane of $f(x)$ at $x = 1$. Figure 1.5b shows another intuitive example. Here, $f(x) = bx + c$. In this case, the difference between the line $g_1(x) = u_1 x$ and $f(x)$ becomes infinite at $x \to -\infty$. Similarly, the difference between the line $g_2(x) = u_2 x$ and $f(x)$ becomes infinite at $x \to \infty$. Only when $u = b$ does, the maximum distance becomes finite and is equal to $-c$. Therefore,

Table 1.3 Examples of convex conjugate pairs used often in imaging problems. Here, $D \subset \mathcal{H}$ and we use the interpretation $0 \log 0 = 0$

$f(x)$	dom f	$f^*(u)$	dom f^*
$f(ax)$	D	$f^*(u/a)$	D
$f(x+b)$	D	$f^*(u) - \langle b, u\rangle$	D
$af(x), a > 0$	D	$af^*(u/a)$	D
$bx + c$	D	$\begin{cases} -c & y = a \\ +\infty, & u \neq a \end{cases}$	$\{a\}$
$1/x$	$\mathbb{R}_{++}$	$-2\sqrt{-u}$	$-\mathbb{R}_+$
$-\log x$	$\mathbb{R}_{++}$	$-(1 + \log(-u))$	$-\mathbb{R}_{++}$
$x \log x$	$\mathbb{R}_+$	e^{u-1}	$\mathbb{R}$
$\sqrt{1+x^2}$	$\mathbb{R}$	$-\sqrt{1-u^2}$	$[-1, 1]$
e^x	$\mathbb{R}$	$u \log(u) - u$	$\mathbb{R}_+$
$\log(1+e^x)$	$\mathbb{R}$	$u \log(u) + (1-u)\log(1-u)$	$[0, 1]$
$-\log(1-e^x)$	$\mathbb{R}_{--}$	$u \log(u) + (1+u)\log(1+u)$	$\mathbb{R}_+$
$\frac{\lvert x\rvert^p}{p},\ p > 1$	$\mathbb{R}$	$\frac{\lvert u\rvert^q}{q},\ \frac{1}{p} + \frac{1}{q} = 1$	$\mathbb{R}$
$\lVert x\rVert_1$	$\mathbb{R}^n$	$\begin{cases} 0, & \lVert u\rVert_2 \leq 1 \\ \infty & \lVert u\rVert_2 > 1 \end{cases}$	$\{u \in \mathbb{R}^n : \lVert u\rVert_2 < 1\}$
$\langle \boldsymbol{a}, \boldsymbol{x}\rangle + b$	$\mathbb{R}^n$	$\begin{cases} -b, & u = b \\ \infty, & u \neq a \end{cases}$	$\{b\} \subset \mathbb{R}^n$
$\frac{1}{2}\boldsymbol{x}^\top \boldsymbol{Q}\boldsymbol{x},\quad \boldsymbol{Q} \in \mathbb{S}_{++}$	$\mathbb{R}^n$	$\frac{1}{2}\boldsymbol{u}^\top \boldsymbol{Q}^{-1}\boldsymbol{u}$	$\mathbb{R}^n$
$\iota_C(x)$	C	$S_C(u)$	$\mathcal{H}$
$\log\left(\sum_{i=1}^n e^{x_i}\right)$	$\mathbb{R}^n$	$\sum_{i=1}^n u_i \log u_i,\quad \sum_{i=1}^n u_i = 1$	$\mathbb{R}^n_+$
$-\log\det \boldsymbol{X}^{-1}$	$\mathbb{S}^n_{++}$	$\log\det(-\boldsymbol{U})^{-1} - n$	$-\mathbb{S}^n_{++}$

the convex conjugate of $f(x) = bx + c$ is

$$f^*(u) = \begin{cases} -c, & u = b, \\ \infty, & u \neq b. \end{cases}$$

Table 1.3 summarizes these findings for a variety of functions that are often used in applications.

It is clear that f^* is convex since f^* is a point-wise supremum of a convex function of y. In general, if $f : \mathcal{H} \mapsto [-\infty, \infty]$, then the following hold:

1. For $\alpha \in \mathbb{R}_{++}$, we have

$$(\alpha f)^* = \alpha f^*(\cdot/\alpha). \tag{1.53}$$

2. *Fenchel–Young inequality*:

$$f(\boldsymbol{x}) + f^*(\boldsymbol{y}) \geq \langle \boldsymbol{y}, \boldsymbol{x}\rangle, \quad \forall \boldsymbol{x}, \boldsymbol{y} \in \mathcal{H}. \tag{1.54}$$

3. Let f, g be proper functions from $\mathcal{H}$ to $(-\infty, \infty]$. Then,

$$f(\boldsymbol{x}) + g(\boldsymbol{x}) \geq -f^*(\boldsymbol{u}) - g^*(-\boldsymbol{u}), \quad \forall \boldsymbol{x}, \boldsymbol{u} \in \mathcal{H}. \tag{1.55}$$

If f is convex, proper, and lower semicontinuous, then the following properties hold:

$$f^{**} = f, \tag{1.56}$$

$$\boldsymbol{y} \in \partial f(\boldsymbol{x}) \Longleftrightarrow f(\boldsymbol{x}) + f^*(\boldsymbol{y}) = \langle \boldsymbol{x}, \boldsymbol{y} \rangle \Longleftrightarrow \boldsymbol{x} \in \partial f^*(\boldsymbol{y}). \tag{1.57}$$

1.6.5 Lagrangian Dual Formulation

Perhaps one of the most important uses of convex conjugate is to obtain the dual formulation. More specifically, for a given *primal problem* (P),

$$(P): \quad \min_{\boldsymbol{x} \in \mathcal{H}} f(\boldsymbol{x}) + g(\boldsymbol{x}), \tag{1.58}$$

we can obtain the associated *dual problem* using (1.55):

$$(D): \quad -\min_{\boldsymbol{u} \in \mathcal{H}} f^*(\boldsymbol{u}) + g^*(-\boldsymbol{u}). \tag{1.59}$$

The gap between the primal and dual problem is called the *duality gap*.

Example: Dual for Composite Function
For the given primal problem:

$$(P): \quad \min_{\boldsymbol{x} \in \mathbb{R}^n} f(\boldsymbol{x}) + g(\boldsymbol{A}\boldsymbol{x}), \tag{1.60}$$

with $\boldsymbol{A} \in \mathbb{R}^{n \times m}$, the dual problem is given by

$$(D): -\min_{\boldsymbol{u} \in \mathbb{R}^m} f^*(\boldsymbol{A}^\top \boldsymbol{u}) + g^*(-\boldsymbol{u}).$$

Proof Note that (P) is equivalent to the following constraint minimization problem:

$$\min_{\boldsymbol{x}, \boldsymbol{y}} f(\boldsymbol{x}) + g(\boldsymbol{y})$$
$$\text{subject to} \quad \boldsymbol{A}\boldsymbol{x} = \boldsymbol{y},$$

(continued)

which provides

$$\begin{aligned}\min_{x\in\mathbb{R}^n} f(\boldsymbol{x}) + g(\boldsymbol{A}\boldsymbol{x}) &\leq \min_{\boldsymbol{x},\boldsymbol{y}} f(\boldsymbol{x}) + g(\boldsymbol{y}) + \boldsymbol{u}^\top(\boldsymbol{A}\boldsymbol{x}) - \boldsymbol{u}^\top \boldsymbol{y} \\ &\leq \min_{\boldsymbol{x}}\{f(\boldsymbol{x}) + (\boldsymbol{A}^\top \boldsymbol{u})^\top \boldsymbol{x})\} + \min_{\boldsymbol{y}}\{g(\boldsymbol{y}) - \boldsymbol{u}^\top \boldsymbol{y}\} \\ &= -f^*(\boldsymbol{A}^\top \boldsymbol{u}) - g^*(-\boldsymbol{u}).\end{aligned}$$

Therefore, the dual problem is

$$-\min_{\boldsymbol{u}\in\mathbb{R}^m} f^*(\boldsymbol{A}^\top \boldsymbol{u}) + g^*(-\boldsymbol{u}).$$

This concludes the proof. □

Example: Quadratic Programming Under Affine Constraint
Consider the following optimization problem:

$$P : \min \frac{1}{2}\boldsymbol{x}^\top \boldsymbol{x} \quad \text{subject to} \quad \boldsymbol{b} = \boldsymbol{A}\boldsymbol{x}$$

with $\boldsymbol{A} \in \mathbb{R}^{n\times n}$. Now, we define $C = \{\boldsymbol{0}\}$ such that $\boldsymbol{b} - \boldsymbol{A}\boldsymbol{x} \in C$. Then, the original minimization problem becomes

$$\begin{aligned}&\min_{\boldsymbol{x},\boldsymbol{y}} \iota_C(\boldsymbol{y}) + \tfrac{1}{2}\boldsymbol{x}^\top \boldsymbol{x} \\ &\text{subject to } \ \boldsymbol{y} = \boldsymbol{b} - \boldsymbol{A}\boldsymbol{x}.\end{aligned}$$

Therefore, we have

$$\begin{aligned}\min_{\boldsymbol{x}} \iota_C(\boldsymbol{A}\boldsymbol{x} - \boldsymbol{b}) + \frac{1}{2}\boldsymbol{x}^\top \boldsymbol{x} &\leq \min_{\boldsymbol{x},\boldsymbol{y}} \iota_C(\boldsymbol{y}) + \frac{1}{2}\boldsymbol{x}^\top \boldsymbol{x} + \boldsymbol{u}^\top(\boldsymbol{A}\boldsymbol{x} - \boldsymbol{b} - \boldsymbol{y}) \\ &\leq \min_{\boldsymbol{y}} \iota_C(\boldsymbol{y}) - \boldsymbol{u}^\top \boldsymbol{y} + \min_{\boldsymbol{x}} \frac{1}{2}\boldsymbol{x}^\top \boldsymbol{x} - \boldsymbol{u}^\top \boldsymbol{A}\boldsymbol{x} + \boldsymbol{u}^\top \boldsymbol{b} \\ &\leq \min_{\boldsymbol{y}\in\{\boldsymbol{0}\}} -\boldsymbol{u}^\top \boldsymbol{y} + \min_{\boldsymbol{x}} \frac{1}{2}\boldsymbol{x}^\top \boldsymbol{x} - \boldsymbol{u}^\top \boldsymbol{A}\boldsymbol{x} + \boldsymbol{u}^\top \boldsymbol{b} \\ &= \frac{1}{2}\boldsymbol{u}^\top \boldsymbol{A}\boldsymbol{A}^\top \boldsymbol{u} + \boldsymbol{u}^\top \boldsymbol{b},\end{aligned}$$

(continued)

where the last equality comes from $\boldsymbol{x} = \boldsymbol{A}^\top \boldsymbol{u}$ at the minimizer. Hence, the dual problem becomes

$$D: \min_{\boldsymbol{u}\in\mathbb{R}^m} \frac{1}{2}\boldsymbol{u}^\top AA^\top \boldsymbol{u} + \boldsymbol{u}^\top \boldsymbol{b}.$$

Why is this dual formulation useful? Suppose that A is highly ill-posed, say that $n = 1000$ and $m = 1$. Then, the dual problem (D) is a one-dimensional problem which is computationally much less expensive than the primal problem (P) of the dimension $n = 1000$. After the dual solution $\hat{\boldsymbol{u}}$ is obtained, the primal solution is just $\hat{\boldsymbol{x}} = \boldsymbol{A}^\top \hat{\boldsymbol{u}}$.

We formally define a Lagrangian dual problem.

Definition 1.16 ([6]) Suppose that a primal problem is given by

$$\min_{\boldsymbol{x}} \quad f_0(\boldsymbol{x})$$

$$\text{subject to } f_i(\boldsymbol{x}) \leq 0,\ i = 1, \cdots, n, \tag{1.61}$$

$$h_i(\boldsymbol{x}) = 0,\ i = 1, \cdots, p. \tag{1.62}$$

Then, the associated Lagrangian dual problem is defined by

$$\max_{\boldsymbol{\alpha},\boldsymbol{\nu}} g(\boldsymbol{\alpha}, \boldsymbol{\nu}) \tag{1.63}$$

$$\text{subject to } \boldsymbol{\alpha} \geq \boldsymbol{0}, \tag{1.64}$$

where $\boldsymbol{\alpha} = [\alpha_1, \cdots, \alpha_n]$ and $\boldsymbol{\nu} = [\nu_1, \cdots, \nu_p]$ are referred to as the *dual variables* or *Lagrangian multipliers*, $\boldsymbol{\alpha} \geq \boldsymbol{0}$ implies that each element is non-negative, and the Lagrangian $g(\boldsymbol{\alpha}, \boldsymbol{\nu})$ is defined by

$$g(\boldsymbol{\alpha}, \boldsymbol{\nu}) := \inf_{\boldsymbol{x}} \left\{ f_0(\boldsymbol{x}) + \sum_{i=1}^{n} \alpha_i f_i(\boldsymbol{x}) + \sum_{j=1}^{p} \nu_j h_j(\boldsymbol{x}) \right\}. \tag{1.65}$$

One of the important findings in convex optimization theory [6] is that if the primal problem is convex, then we have the following strong duality:

$$g(\boldsymbol{\alpha}^*, \boldsymbol{\nu}^*) = f_0(\boldsymbol{x}^*), \tag{1.66}$$

where $\boldsymbol{x}^*$ and $\boldsymbol{\alpha}^*, \boldsymbol{\nu}^*$ are the optimal solutions for the primal and dual problems, respectively. Often, the dual formulation is easier to solve than the primal problem. Additionally, there is also interesting an geometric interpretation, which will be explained later.

1.7 Exercises

1. Show that an l_p norm with $0 < p < 1$ is *not* a norm.
2. Prove the equalities in (1.17).
3. Prove the matrix inversion lemma, Eq. (1.23).
4. Let $\boldsymbol{x} \in \mathbb{R}^n$, $\boldsymbol{y} \in \mathbb{R}^m$ and $\boldsymbol{A} \in \mathbb{R}^{m \times n}$. Then, show the following:

$$\begin{aligned}\hat{\boldsymbol{x}} &= \arg\min_{\boldsymbol{x}\in\mathbb{R}^n} \|\boldsymbol{y} - \boldsymbol{A}\boldsymbol{x}\|^2 + \lambda\|\boldsymbol{x}\|^2 \\ &= (\boldsymbol{A}^\top\boldsymbol{A} + \lambda \boldsymbol{I})^{-1}\boldsymbol{A}^\top\boldsymbol{y} \\ &= \boldsymbol{A}^\top(\boldsymbol{A}\boldsymbol{A}^\top + \lambda\boldsymbol{I})^{-1}\boldsymbol{y},\end{aligned}$$

   where $\boldsymbol{A}^\top$ denotes the transpose of $\boldsymbol{A}$, and $\boldsymbol{I}$ is an appropriate size identity matrix. (Hint: for the last equality, you need to use the matrix inversion lemma.)
5. Prove Lemma 1.2.
6. Prove (1.31).
7. Prove Lemma 1.4.
8. Prove Lemma 1.7.
9. Show that if $\mathcal{L}$ is an affine mapping and f is convex, then $f \circ \mathcal{L}$ is also convex, where $\circ$ refers to the composite function.
10. Find at least three examples of functions that are not semicontinuous.
11. In Table 1.1, show that the relative entropy, indicator function, support function, p-norm (with $p \geq 1$) and max functions are convex.
12. Let $f : \mathcal{H} \mapsto (-\infty, \infty]$ be proper. Suppose that $\mathrm{dom} f$ is open and convex, and f is Gâteux differentiable on $\mathrm{dom} f$. Then, show that the following are equivalent:

    a. f is convex.
    b. $f(\boldsymbol{y}) \geq f(\boldsymbol{x}) + \langle \boldsymbol{y} - \boldsymbol{x}, \nabla f(\boldsymbol{x})\rangle, \quad \forall \boldsymbol{x}, \boldsymbol{y} \in \mathcal{H}$.
    c. $\langle \boldsymbol{y} - \boldsymbol{x}, \nabla f(\boldsymbol{y}) - \nabla f(\boldsymbol{x})\rangle \geq 0, \quad \forall \boldsymbol{x}, \boldsymbol{y} \in \mathcal{H}$.
    d. Moreover, if f is twice Gâteux differentiable on $\mathrm{dom} f$,

$$\nabla^2 f(\boldsymbol{x}) \succeq 0, \quad \forall \boldsymbol{x} \in \mathrm{dom} f.$$

13. Let $f(x) = |x|$ with $x \in [-1, 1]$. Find its subdifferential $\partial f(x)$.
14. Prove Fermat's rule in Theorem 1.3.
15. Show that the following properties hold for the subdifferentials:

    a. If f is differentiable, then $\partial f(\boldsymbol{x}) = \{\nabla f(\boldsymbol{x})\}$.
    b. Let f be proper. Then, $\partial f(\boldsymbol{x})$ is closed and convex for any $\boldsymbol{x} \in \mathrm{dom} f$.
    c. Let $\lambda \in \mathbb{R}_{++}$. Then, $\partial(\lambda f) = \lambda \partial f$.
    d. Let f, g be convex, and lower semicontinuous functions, and $\mathcal{L}$ is a linear operator. Then

$$\partial(f + g \circ \mathcal{L}) = \partial f + \mathcal{L}^* \circ (\partial g) \circ \mathcal{L}. \tag{1.67}$$

16. Prove Eq. (1.53).
17. Let $f(\boldsymbol{x}) = \frac{1}{2}(x_1^2 + x_2^2) - x_1 - x_2$. Derive the convex conjugate $f^*(\boldsymbol{x})$.
18. Let f be a proper function from $\mathcal{H}$ to $(-\infty, \infty]$. Show that

$$f(\boldsymbol{x}) + f^*(\boldsymbol{y}) \geq \langle \boldsymbol{y}, \boldsymbol{x} \rangle, \quad \forall \boldsymbol{x}, \boldsymbol{y} \in \mathcal{H}.$$

19. If f is convex and lower semicontinuous, then show that

$$(\partial f)^{-1} = \partial f^*.$$

20. We often have the following form of the primal problem:

$$(P): \quad \min_{\boldsymbol{x} \in \mathbb{R}^n} f(\boldsymbol{x}) + g(\boldsymbol{A}\boldsymbol{x}), \tag{1.68}$$

where

$$g(\boldsymbol{A}\boldsymbol{x}) = \|\boldsymbol{A}\boldsymbol{x}\|_1, \quad f(\boldsymbol{x}) = \|\boldsymbol{y} - \boldsymbol{x}\|_2^2$$

with the operator $\boldsymbol{A} : \mathbb{R}^n \mapsto \mathbb{R}^m$. Show that the associated *dual problem* is given by

$$-\min_{\boldsymbol{u} \in \mathbb{R}^m} \boldsymbol{u}^\top \boldsymbol{A}\boldsymbol{A}^\top \boldsymbol{u} + \boldsymbol{y}^\top \boldsymbol{A}^\top \boldsymbol{u}$$

$$\text{subject to} \qquad \|\boldsymbol{u}\|_2 \leq 1.$$

Chapter 2
Linear and Kernel Classifiers

2.1 Introduction

Classification is one of the most basic tasks in machine learning. In computer vision, an image classifier is designed to classify input images in corresponding categories. Although this task appears trivial to humans, there are considerable challenges with regard to automated classification by computer algorithms.

For example, let us think about recognizing "dog" images. One of the first technical issues here is that a dog image is usually taken in the form of a digital format such as JPEG, PNG, etc. Aside from the compression scheme used in the digital format, the image is basically just a collection of numbers on a two-dimensional grid, which takes integer values from 0 to 255. Therefore, a computer algorithm should read the numbers to decide whether such a collection of numbers corresponds to a high-level concept of "dog". However, if the viewpoint is changed, the composition of the numbers in the array is totally changed, which poses additional challenges to the computer program. To make matters worse, in a natural setting a dog is rarely found on a white background; rather, the dog plays on the lawn or takes a nap in the living room, hides underneath furniture or chews with her eyes closed, which makes the distribution of the numbers very different depending on the situation. Additional technical challenges in computer-based recognition of a dog come from all kinds of sources such as different illumination conditions, different poses, occlusion, intra-class variation, etc., as shown in Fig. 2.1. Therefore, designing a classifier that is robust to such variations was one of the important topics in computer vision literature for several decades.

In fact, the ImageNet Large Scale Visual Recognition Challenge (ILSVRC) [7] was initiated to evaluate various computer algorithms for image classification at large scale. ImageNet is a large visual database designed for use in visual object recognition software research [8]. Over 14 million images have been hand-annotated in the project to indicate which objects are depicted, and at least one million of the images also have bounding boxes. In particular, ImageNet contains more than

J. C. Ye, *Geometry of Deep Learning*, Mathematics in Industry 37,
https://doi.org/10.1007/978-981-16-6046-7_2

Fig. 2.1 Technical challenges in recognizing a dog from digital images. Figures courtesy of Ella Jiwoo Ye

20,000 categories made up of several hundred images. Since 2010, the ImageNet project has organized an annual software competition, the ImageNet Large Scale Visual Recognition Challenge (ILSVRC), in which software programs compete for the correct classification and recognition of objects and scenes. The main motivation is to allow researchers to compare progress in classification across a wider variety of objects. Since the introduction of AlexNet in 2012 [9], which was the first deep learning approach to win the ImageNet Challenge, the state-of-the art image classification methods are all deep learning approaches, and now their performance even surpasses human observers.

Before we discuss in detail recent deep learning approaches, we revisit the classical classifier, in particular the support vector machine (SVM) [10], to discuss its mathematical principles. Although the SVM is already an old classical technique, its review is important since the mathematical understanding of the SVM allows readers to understand how the modern deep learning approaches are closely related to the classical ones.

Specifically, consider binary classification problems where data sets from two different classes are distributed as shown in Fig. 2.2a,b,c. Note that in Fig. 2.2a, the two sets are perfectly separable with linear hyperplanes. For the case of Fig. 2.2b, there exists no linear hyperplane that perfectly separates two data sets, but one could find a linear boundary where only a small set of data are incorrectly classified. However, the situation in Fig. 2.2c is much different, since there exists no linear boundary that can separate the majority of elements of the two classes. Rather, one could find a nonlinear class boundary that can separate the two sets with small errors. The theory of the SVM deals with all situations in Fig. 2.2a,b,c using a hard-margin linear classifier, soft-margin linear classifier, and kernel SVM method, respectively. In the following, we discuss each topic in detail.

Fig. 2.2 Examples of binary classification problems: (**a**) linear separable case, (**b**) approximately linear separable case, and (**c**) linear non-separable case

2.2 Hard-Margin Linear Classifier

2.2.1 Maximum Margin Classifier for Separable Cases

For the linear separable case in Fig. 2.2a, there can be an infinite number of choices of linear hyperplanes. Among them, one of the most widely used choices of the classification boundary is to maximize the margin between the two classes. This is often called the maximum margin linear classifier [10].

To derive this, we introduce some notations. Let $\{\boldsymbol{x}_i, y_i\}_{i=1}^N$ denote the set of the data $\boldsymbol{x}_i \in \mathcal{X} \subset \mathbb{R}^d$ with the binary label y_i such that $y_i \in \{1, -1\}$. We now define a hyperplane in $\mathbb{R}^d$:

$$\langle \boldsymbol{w}, \boldsymbol{x} \rangle + b = \boldsymbol{w}^\top \boldsymbol{x} + b = 0, \tag{2.1}$$

where $^\top$ denotes the transpose, $\langle \cdot, \cdot \rangle$ is the inner product, $b \in \mathbb{R}$ is a bias term. See Fig. 2.3 for more details. If the two classes are separable, then there exist sets S_1 and S_{-1} such that the data set with $y_i = 1$ and $y_1 = -1$ belongs to the sets S_1 and S_{-1},

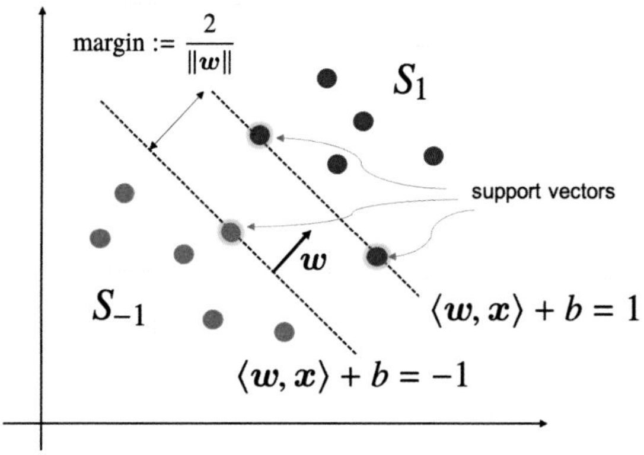

Fig. 2.3 Geometric structure of hard-margin linear support vector machine classifier

respectively:

$$S_1 = \{x \in \mathbb{R}^d \mid \langle \boldsymbol{w}, \boldsymbol{x} \rangle + b \geq 1\}, \tag{2.2}$$

$$S_{-1} = \{x \in \mathbb{R}^d \mid \langle \boldsymbol{w}, \boldsymbol{x} \rangle + b \leq -1\}. \tag{2.3}$$

Then, the margin between the two sets is defined as the minimum distance between the two linear boundaries of S_1 and S_{-1}. To calculate this, we need the following lemma:

Lemma 2.1 *The distance between two parallel hyperplanes* $\ell_1 : \langle \boldsymbol{w}, \boldsymbol{x} \rangle + c_1 = 0$ *and* $\ell_2 : \langle \boldsymbol{w}, \boldsymbol{x} \rangle + c_2 = 0$ *is given by*

$$m := \frac{|c_1 - c_2|}{\|\boldsymbol{w}\|}. \tag{2.4}$$

Proof Let m be the distance between the two parallel hyperplanes ℓ_1 and ℓ_2, then there exists two points $\boldsymbol{x} \in \ell_1$ and $\boldsymbol{x}_2 \in \ell_2$ such that $\|\boldsymbol{x}_1 - \boldsymbol{x}_2\| = m$. Then, using the Pythagoras theorem, the vector $\boldsymbol{v} := \boldsymbol{x}_1 - \boldsymbol{x}_2$ should be along the normal direction of the hyperplanes. Accordingly,

$$m = \|\boldsymbol{x}_1 - \boldsymbol{x}_2\| = \|\langle \boldsymbol{w}/\|\boldsymbol{w}\|, \boldsymbol{x}_1 \rangle - \langle \boldsymbol{w}/\|\boldsymbol{w}\|, \boldsymbol{x}_2 \rangle\|,$$

since $\boldsymbol{w}/\|\boldsymbol{w}\|$ is the unit normal vector of the hyperplanes. Therefore, we have

$$m = \frac{\|\langle \boldsymbol{w}, \boldsymbol{x}_1 \rangle - \langle \boldsymbol{w}, \boldsymbol{x}_2 \rangle\|}{\|\boldsymbol{w}\|} = \frac{|c_1 - c_2|}{\|\boldsymbol{w}\|}.$$

Q.E.D. □

Since $\langle \boldsymbol{w}, \boldsymbol{x} \rangle + b - 1 = 0$ and $\langle \boldsymbol{w}, \boldsymbol{x} \rangle + b + 1 = 0$ correspond to the linear boundaries of S_1 and S_{-1}, Lemma 2.1 informs us that the margin between the two classes is given by

$$\text{margin} := \frac{2}{\|\boldsymbol{w}\|}. \tag{2.5}$$

Therefore, for the given training data set $\{\boldsymbol{x}_i, y_i\}_{i=1}^n$ with $\boldsymbol{x}_i \in \mathcal{X} \subset \mathbb{R}^d$ and the binary label $y_i \in \{1, -1\}$, the maximum margin linear binary classifier design problem can be formulated as follows:

$$\text{(P)} \quad \min_{\boldsymbol{w}} \quad \frac{1}{2}\|\boldsymbol{w}\|^2 \tag{2.6}$$

$$\text{subject to } 1 - y_i \left(\langle \boldsymbol{w}, \boldsymbol{x}_i \rangle + b\right) \leq 0, \quad \forall i. \tag{2.7}$$

Note that the minimization of $\|\boldsymbol{w}\|^2/2$ in (2.6) is equivalent to the maximization of the margin $2/\|\boldsymbol{w}\|^2$, and by noting that $y_i = 1$ and -1 for the sets S_1 and S_{-1}, respectively, we can see that (2.7) corresponds to the desirable constraints. Another thing to note here is that although the cost minimization in (P) is with respect to $\boldsymbol{w}$, the dependency on b is hidden in this formulation. The explicit dependency on b becomes more evident in its dual formulation described in the following.

2.2.2 *Dual Formulation*

The optimization problem (P) is a constrained optimization problem under inequality constraints. A standard method for the constrained optimization problem is to use the Lagrangian dual formulation [6]. In the following, we formally define a Lagrangian dual problem.

Definition 2.1 [6] Suppose that a primal problem is given by

$$\min_{\boldsymbol{x}} \quad f_0(\boldsymbol{x})$$

$$\text{subject to } f_i(\boldsymbol{x}) \leq 0, \; i = 1, \cdots, n \tag{2.8}$$

$$h_i(\boldsymbol{x}) = 0, \; i = 1, \cdots, p. \tag{2.9}$$

Then, the associated Lagrangian dual problem is defined by

$$\max_{\boldsymbol{\alpha}, \boldsymbol{\nu}} g(\boldsymbol{\alpha}, \boldsymbol{\nu}) \tag{2.10}$$

$$\text{subject to } \boldsymbol{\alpha} \geq \boldsymbol{0}, \tag{2.11}$$

where $\boldsymbol{\alpha} = [\alpha_1, \cdots, \alpha_n]$ and $\boldsymbol{\nu} = [\nu_1, \cdots, \nu_p]$ are referred to the *dual variables* or *Lagrangian multipliers*, $\boldsymbol{\alpha} \geq \boldsymbol{0}$ implies that each element is non-negative, and the Lagrangian $g(\boldsymbol{\alpha}, \boldsymbol{\nu})$ is defined by

$$g(\boldsymbol{\alpha}, \boldsymbol{\nu}) := \inf_{\boldsymbol{x}} \left\{ f_0(\boldsymbol{x}) + \sum_{i=1}^{n} \alpha_i f_i(\boldsymbol{x}) + \sum_{j=1}^{p} \nu_j h_j(\boldsymbol{x}) \right\}. \tag{2.12}$$

One of the important findings in convex optimization theory [6] is that if the primal problem is convex, then we have the following strong duality:

$$g(\boldsymbol{\alpha}^*, \boldsymbol{\nu}^*) = f_0(\boldsymbol{x}^*), \tag{2.13}$$

where $\boldsymbol{x}^*$ and $\boldsymbol{\alpha}^*, \boldsymbol{\nu}^*$ are the optimal solutions for the primal and dual problems, respectively. Often, the dual formulation is easier to solve than the primal problem. Additionally, there is also interesting geometric interpretation.

Our binary classification problem (P) in (2.6) is a convex optimization problem with respect to $\boldsymbol{w} \in \mathbb{R}^d$, since both the objective function and the constraint sets are convex. Therefore, using Definition 2.1, the original problem can be converted to a dual problem:

$$\begin{aligned} \text{(D)} \quad & \max_{\boldsymbol{\alpha}} \; g(\boldsymbol{\alpha}) \\ & \text{subject to } \boldsymbol{\alpha} \geq \mathbf{0}, \end{aligned}$$

where $\boldsymbol{\alpha} = [\alpha_1, \cdots, \alpha_n]$ is a dual variable with respect to the primal variable $\boldsymbol{w}$ and b, and

$$g(\boldsymbol{\alpha}) = \min_{\boldsymbol{w},b} \frac{\|\boldsymbol{w}\|^2}{2} + \sum_{i=1}^{n} \alpha_i \left(1 - y_i(\langle \boldsymbol{w}, \boldsymbol{x}_i \rangle + b)\right). \tag{2.14}$$

At the minimizers of (2.14), the derivatives with respect to $\boldsymbol{w}$ and b should be zero, which leads to the following first-order necessary conditions (FONC):

$$\boldsymbol{w} = \sum_{i=1}^{n} \alpha_i y_i \boldsymbol{x}_i, \qquad \sum_{i=1}^{n} \alpha_i y_i = 0. \tag{2.15}$$

The FONCs in Eq. (2.15) have very important geometric interpretations. For example, the first equation in (2.15) clearly shows how the normal vector for the hyperplanes can be constructed using the dual variables. The second equation leads to the balancing conditions. These will be explained in more detail later.

By plugging these FONCs into (2.14), the dual problem (D) becomes

$$\begin{aligned} \max_{\boldsymbol{\alpha}} \quad & \sum_{i=1}^{n} \alpha_i - \frac{1}{2} \sum_{i=1}^{n} \sum_{j=1}^{n} \alpha_i \alpha_j y_i y_j \langle \boldsymbol{x}_i, \boldsymbol{x}_j \rangle \\ \text{subject to} \quad & \sum_{i=1}^{n} \alpha_i y_i = 0, \quad \alpha_i \geq 0, \quad \forall i. \end{aligned} \tag{2.16}$$

Let $\boldsymbol{w}^*$, b^* and $\boldsymbol{\alpha}^*$ denote the solutions for the primal and dual problems. Then, the resulting binary classifier is given by

$$y \leftarrow \text{sign}(\langle \boldsymbol{w}^*, \boldsymbol{x} \rangle + b^*) \tag{2.17}$$

for the case of the primal formulation, or

$$y \leftarrow \text{sign}\left(\sum_{i=1}^{n} \alpha_i^* y_i \langle \boldsymbol{x}_i, \boldsymbol{x} \rangle + b^*\right) \tag{2.18}$$

for the case of the dual formulation, where $\text{sign}(x)$ denotes the sign of x.

2.2.3 KKT Conditions and Support Vectors

To achieve the strong duality in (2.13), the so-called Karush–Kuhn–Tucker (KKT) conditions should be satisfied [6]. More details on the KKT conditions can be found in the standard convex optimization textbook [6], so here we briefly introduce the core condition that is directly related to geometric understanding of the maximum margin linear classifier.

More specifically, suppose that $\boldsymbol{x}^*$ and $\boldsymbol{\alpha}^*$, $\boldsymbol{\nu}^*$ denote the optimal solutions for the primal and dual problems, respectively. Then, we have

$$\begin{aligned} g(\boldsymbol{\alpha}^*, \boldsymbol{\nu}^*) &= f_0(\boldsymbol{x}^*) + \sum_{i=1}^{n} \alpha_i^* f_i(\boldsymbol{x}^*) + \sum_{j=1}^{p} \nu_j^* h_j(\boldsymbol{x}^*) \\ &= f_0(\boldsymbol{x}^*) + \sum_{i=1}^{n} \alpha_i^* f_i(\boldsymbol{x}^*), \end{aligned} \tag{2.19}$$

where the last equality comes from the constraint $h_j(\boldsymbol{x}^*) = 0$ in the primal problem. In order to make (2.19) equal to $f_0(\boldsymbol{x}^*)$, which corresponds to the strong duality (2.13), the following condition should be satisfied:

$$\alpha_i^* > 0 \Longrightarrow f_i(\boldsymbol{x}^*) = 0 \quad \text{or} \quad f_i(\boldsymbol{x}^*) < 0 \Longrightarrow \alpha_i^* = 0. \tag{2.20}$$

This is the key KKT condition.

If (2.20) is applied to our classifier design problem, we have

$$\alpha_i^* > 0 \Longrightarrow y_i(\langle \boldsymbol{w}^*, \boldsymbol{x}_i \rangle + b) = 1, \tag{2.21}$$

which implies that in constructing the normal vector direction $\boldsymbol{w}^*$ of the hyperplane using (2.15), only the training data at the class boundaries contribute:

$$\boldsymbol{w}^* = \sum_{i=1}^{n} \alpha_i^* y_i \boldsymbol{x}_i = \sum_{i \in I^+} \alpha_i^* \boldsymbol{x}_i - \sum_{i \in I^-} \alpha_i^* \boldsymbol{x}_i, \tag{2.22}$$

where I^+ and I^- are index sets such that

$$I^+ = \{i \in [1, \cdots, n] \mid \langle \boldsymbol{w}^*, \boldsymbol{x}_i \rangle + b = 1\}, \tag{2.23}$$

$$I^- = \{i \in [1, \cdots, n] \mid \langle \boldsymbol{w}^*, \boldsymbol{x}_i \rangle + b = -1\}. \tag{2.24}$$

On the other hand, for the case of the training data $\boldsymbol{x}_i$ inside the class boundaries, $y_i(\langle \boldsymbol{w}, \boldsymbol{x}_i \rangle + b) > 1$. Therefore, the corresponding Lagrangian variable α_i becomes zero. This situation is illustrated in Fig. 2.3. Here, the set of the training data $\boldsymbol{x}_i$ with $i \in I^+$ or $i \in I^-$ is often called the *support vector*, which is why the corresponding classifier is often called the *support vector machine (SVM)* [10].

Finally, the second equation in (2.15) leads to additional geometric relationship between nonzero dual variables:

$$\sum_{i \in I^+} \alpha_i^* = \sum_{i \in I^-} \alpha_i^*,$$

which states the balancing condition between dual variables. In other words, the weighting parameters for the support vectors should be balanced for each class boundary.

2.3 Soft-Margin Linear Classifiers

2.3.1 Maximum Margin Classifier with Noise

As shown in Fig. 2.2b, many practical classification problems often contain data sets that cannot be perfectly separable by a hyperplane. When the two classes are not linearly separable (e.g., due to noise), the condition for the optimal hyperplane can be relaxed by including extra terms:

$$y_i(\langle \boldsymbol{w}, \boldsymbol{x}_i \rangle + b) \geq 1 - \xi_i, \quad \xi_i \geq 0 \quad \forall i, \tag{2.25}$$

where ξ_i are often called the *slack variables*. The role of the slack variables is to allow errors in the classification. Then, the optimization goal is to find the classifier with the maximum margin with the minimum errors as shown in Fig. 2.4.

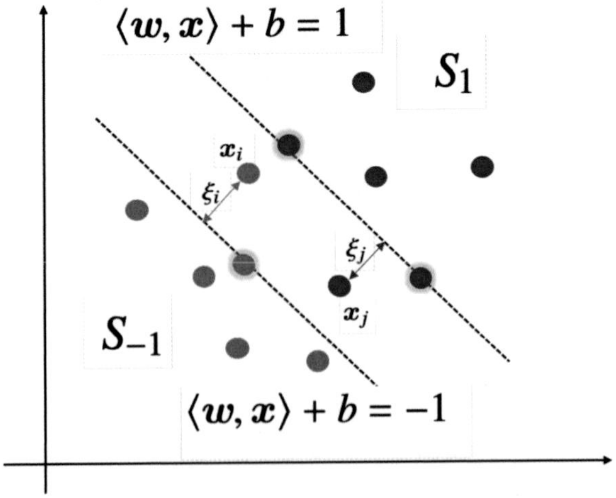

Fig. 2.4 Geometric structure of soft-margin linear support vector machine classifier

The corresponding primal problem is then given by

$$
\begin{aligned}
(\mathrm{P}') \quad \min_{\boldsymbol{w},\boldsymbol{\xi}} \quad & \frac{1}{2}\|\boldsymbol{w}\|^2 + C\sum_{i=1}^{n}\xi_i \\
\text{subject to } & 1 - y_i\left(\langle \boldsymbol{w}, \boldsymbol{x}_i\rangle + b\right) \le \xi_i, \\
& \xi_i \ge 0, \quad \forall i,
\end{aligned} \tag{2.26}
$$

where the optimization problem again has implicit dependency on the bias term b. The following theorem shows that the corresponding dual problem has a form very similar to the hard-margin classifier in (2.16) with the exception of the differences in the constraint for the dual variables.

Theorem 2.1 *The Lagrangian dual formulation of the primal problem in* (2.26) *is given by*

$$
\max_{\boldsymbol{\alpha}} \quad \sum_{i=1}^{n}\alpha_i - \frac{1}{2}\sum_{i=1}^{n}\sum_{j=1}^{n}\alpha_i\alpha_j y_i y_j \langle \boldsymbol{x}_i, \boldsymbol{x}_j\rangle \tag{2.27}
$$

$$
\textit{subject to} \quad \sum_{i=1}^{n}\alpha_i y_i = 0, \quad 0 \le \alpha_i \le C, \quad \forall i.
$$

Proof For the given primal problem in (2.26), the corresponding Lagrangian dual is given by

$$
\begin{aligned}
\max_{\boldsymbol{\alpha},\boldsymbol{\gamma}} \quad & g(\boldsymbol{\alpha}, \boldsymbol{\gamma}) \\
\text{subject to} \quad & \boldsymbol{\alpha} \ge 0, \ \boldsymbol{\gamma} \ge 0,
\end{aligned} \tag{2.28}
$$

$$
g(\boldsymbol{\alpha}, \boldsymbol{\gamma}) = \max_{\boldsymbol{w},b,\boldsymbol{\xi}} \left\{ \frac{1}{2}\|\boldsymbol{w}\|^2 + C\sum_{i=1}^{n}\xi_i \right. \tag{2.29}
$$

$$
\left. + \sum_{i=1}^{n}\alpha_i\left(1 - y_i\left(\langle \boldsymbol{w}, \boldsymbol{x}_i\rangle + b\right) - \xi_i\right) - \sum_{i=1}^{n}\gamma_i\xi_i \right\}.
$$

The first-order necessary conditions (FONCs) with respect to $\boldsymbol{w}$, b and $\boldsymbol{\xi}$ lead to the following equations:

$$
\boldsymbol{w} = \sum_{i=1}^{n}\alpha_i y_i \boldsymbol{x}_i \tag{2.30}
$$

and

$$\sum_{i=1}^{n} \alpha_i y_i = 0, \qquad \alpha_i + \gamma_i = C. \tag{2.31}$$

By plugging (2.30) and (2.31) into Eq. (2.29), we have

$$g(\boldsymbol{\alpha}, \boldsymbol{\gamma}) = \sum_{i=1}^{n} \alpha_i - \frac{1}{2}\sum_{i=1}^{n}\sum_{j=1}^{n} \alpha_i \alpha_j y_i y_j \langle \boldsymbol{x}_i, \boldsymbol{x}_j \rangle,$$

where $0 \leq \alpha_i \leq C$, since $\gamma_i = C - \alpha_i \geq 0$. This concludes the proof. □

Another way of representing the primal problem in (2.26) is using the so-called *hinge loss* [10, 11]:

$$\ell_{hinge}(y, \hat{y}) = \max\{0, 1 - y\hat{y}\}, \tag{2.32}$$

of which a pictorial description is given in Fig. 2.5. Specifically, we define the slack variable:

$$\xi_i := 1 - y_i(\langle \boldsymbol{w}, \boldsymbol{x}_i \rangle + b).$$

To make the slack variable represent the classification error for the data set $(\boldsymbol{x}_i, y_i)$ within the class boundary, ξ_i should be zero when the data is already well classified, but positive when there exists a classification error. This leads to the following definition of the slack variable:

$$\xi_i = \max\{0, 1 - y_i(\langle \boldsymbol{w}, \boldsymbol{x}_i \rangle + b)\} = \ell_{hinge}\left(y_i, \langle \boldsymbol{w}, \boldsymbol{x}_i \rangle + b\right). \tag{2.33}$$

Then, the primal problem in (2.26) can be represented by

$$\min_{\boldsymbol{w}, b} \tfrac{1}{2}\|\boldsymbol{w}\|^2 + C \textstyle\sum_{i=1}^{n} \ell_{hinge}\left(y_i, \langle \boldsymbol{w}, \boldsymbol{x}_i \rangle + b\right). \tag{2.34}$$

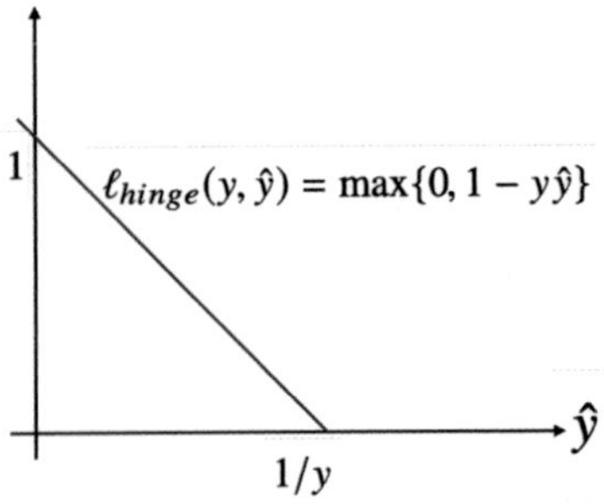

Fig. 2.5 Pictorial description of hinge loss $\ell_{hinge}(y, \hat{y}) = \max\{0, 1 - y\hat{y}\}$

Later, we will show that this representation is closely related to the so-called *representer theorem* [11].

2.4 Nonlinear Classifier Using Kernel SVM

2.4.1 Linear Classifier in the Feature Space

Now consider a classification problem in $\mathbb{R}^2$ as shown in Figs. 2.6 or 2.2c, where there exists no linear hyperplane that can separate two classes. Specifically, the data in class 1 are within an ellipse:

$$S_1 = \{\boldsymbol{x} = (x_1, x_2) \mid (x_1 + x_2)^2 + x_2^2 \leq 2\}, \tag{2.35}$$

whereas class 2 data are located outside of the ellipse. This implies that although the two classes of data cannot be separated by a single hyperplane, the nonlinear boundary in (2.35) can separate the two classes.

Interestingly, the existence of the nonlinear boundary implies that we can find the corresponding linear hyperplane in the higher-dimensional space. Specifically, suppose we have a nonlinear mapping $\boldsymbol{\varphi} : \boldsymbol{x} = [x_1, x_2]^\top \mapsto \boldsymbol{\varphi}(\boldsymbol{x})$ to the feature space in $\mathbb{R}^3$ such that

$$\boldsymbol{\varphi}(\boldsymbol{x}) = [\varphi_1, \varphi_2, \varphi_2]^\top = \left[x_1^2,\ x_2^2,\ \sqrt{2}x_1x_2\right]^\top . \tag{2.36}$$

Then, we can easily see that S_1 can be represented in the feature space by

$$S_1 = \{(\varphi_1, \varphi_2, \varphi_3) \mid \varphi_1 + 2\varphi_2 + \sqrt{2}\varphi_3 \leq 2\}. \tag{2.37}$$

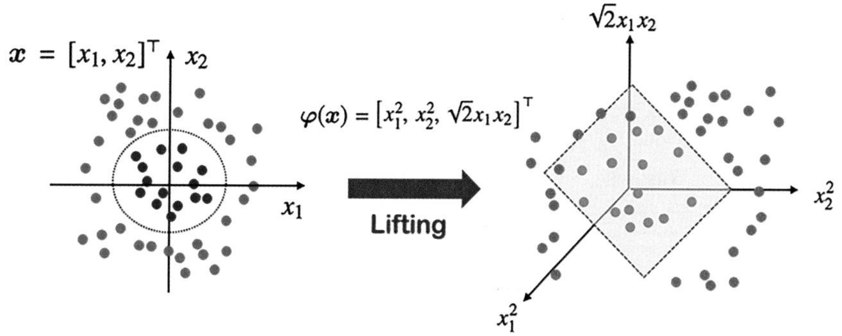

Fig. 2.6 Lifting to a high-dimensional feature space for linear classifier design

Therefore, there exists a linear classifier in $\mathbb{R}^3$ using the feature space mapping $\varphi(\boldsymbol{x})$ as shown in Fig. 2.6.

In general, to allow the existence of a linear classifier, the feature space should be in a higher-dimensional space than the ambient input space. In this sense, the feature mapping $\boldsymbol{\varphi}(\boldsymbol{x})$ works as a *lifting* operation that lifts up the dimension of the data to a higher-dimensional one. In the lifted feature space by the feature mapping $\boldsymbol{\varphi}(\boldsymbol{x})$, the binary classifier design problem in (2.27) can be defined as

$$\max_{\boldsymbol{\alpha}} \quad \sum_{i=1}^{n} \alpha_i - \frac{1}{2}\sum_{i=1}^{n}\sum_{i=j}^{n} \alpha_i \alpha_j y_i y_j \langle \boldsymbol{\varphi}(\boldsymbol{x}_i), \boldsymbol{\varphi}(\boldsymbol{x}_j)\rangle \tag{2.38}$$

$$\text{subject to} \quad \sum_{i=1}^{n} \alpha_i y_i = 0, \quad 0 \le \alpha_i \le C, \quad \forall i.$$

By extending (2.18) from the linear classifier, the associated nonlinear classifier with respect to the optimization problem (2.38) can be similarly defined by

$$y \leftarrow \text{sign}\left(\sum_{i=1}^{n} \alpha_i^* y_i \langle \boldsymbol{\varphi}(\boldsymbol{x}_i), \boldsymbol{\varphi}(\boldsymbol{x})\rangle + b\right), \tag{2.39}$$

where α_i^* and b are the solutions for the dual problem.

2.4.2 *Kernel Trick*

Although (2.38) and (2.39) are nice generalizations of (2.27) and (2.18), there exist several technical issues. One of the most critical issues is that for the existence of a linear classifier, the lifting operation may require a very-high-dimensional or even infinite-dimensional feature space. Therefore, an explicit calculation of the feature vector $\boldsymbol{\varphi}(\boldsymbol{x})$ may be computationally intensive or not possible.

The so-called *kernel trick* may overcome this technical issue by bypassing the explicit construction of the lifting operation [11]. Specifically, as shown in (2.38) and (2.39), all we need for the calculation of the linear classifier is the inner product between the two feature vectors. Specifically, if we define the kernel function $K : \mathcal{X} \times \mathcal{X} \mapsto \mathbb{R}$ as follows:

$$K(\boldsymbol{x}, \boldsymbol{x}') := \langle \boldsymbol{\varphi}(\boldsymbol{x}), \boldsymbol{\varphi}(\boldsymbol{x}')\rangle \tag{2.40}$$

then (2.38) and (2.39) can be converted to

$$\max_{\boldsymbol{\alpha}} \quad \sum_{i=1}^{n} \alpha_i - \frac{1}{2}\sum_{i=1}^{n}\sum_{j=1}^{n} \alpha_i \alpha_j y_i y_j K(\boldsymbol{x}_i, \boldsymbol{x}_j) \tag{2.41}$$

$$\text{subject to} \quad \sum_{i=1}^{n} \alpha_i y_i = 0, \quad 0 \le \alpha_i \le C, \quad \forall i$$

and the resulting classifier is

$$y \leftarrow \operatorname{sign}\left(\sum_{i=1}^{n} \alpha_i^* y_i K(\boldsymbol{x}_i, \boldsymbol{x}) + b\right). \tag{2.42}$$

For example of (2.36), the corresponding kernel is given by

$$K(\boldsymbol{x}, \boldsymbol{y}) = x_1^2 y_1^2 + x_2^2 y_2^2 + 2x_1 x_2 y_1 y_2 = (\langle \boldsymbol{x}, \boldsymbol{y} \rangle)^2,$$

which corresponds to a polynomial function with degree 2. Therefore, the common practice in SVM literature is to design the kernel directly rather than to obtain it from the underlying feature mapping. The following are representative examples of kernels that are often used in the kernel SVM.

- Polynomial kernel with degree exactly p:

$$K(\boldsymbol{x}, \boldsymbol{y}) = (\boldsymbol{x}^\top \boldsymbol{y})^p.$$

- Polynomial kernel with degree up to p:

$$K(\boldsymbol{x}, \boldsymbol{y}) = (\boldsymbol{x}^\top \boldsymbol{y} + 1)^p.$$

- Radial basis function kernel with width σ:

$$K(\boldsymbol{x}, \boldsymbol{y}) = \exp(-\|\boldsymbol{x} - \boldsymbol{y}\|^2/(2\sigma^2)).$$

- Sigmoid kernel:

$$\tanh(\eta \boldsymbol{x}^\top \boldsymbol{y} + \nu).$$

However, care should be taken since not all kernels can be used for SVM. To be a viable option, a kernel should originate from the feature space mapping $\boldsymbol{\varphi}(\boldsymbol{x})$. In fact, there exists an associated feature mapping if the kernel function satisfies the so-called Mercer's condition [11]. The kernel that satisfies Mercer's condition is often called the positive definite kernel. The details of Mercer's condition can be found from standard SVM literature [11] and will be explained later in the context of the representer theorem.

2.5 Classical Approaches for Image Classification

Although the SVM and its kernel extension are beautiful convex optimization frameworks devoid of local minimizers, there are fundamental challenges in using these methods for image classification. In particular, the ambient space $\mathcal{X}$ should not be significantly large in the SVM due to the computationally extensive optimization procedure. Accordingly, one of the essential steps of using the SVM framework is *feature engineering*, which pre-processes the input images to obtain significantly smaller dimensional vector $\boldsymbol{x} \in \mathcal{X}$ that can capture all essential information of the input images. For example, a classical pipeline for the image classification task can be summarized as follows (see Fig. 2.7):

- Process the data set to extract hand-crafted features based on some knowledge of imaging physics, geometry, and other analytic tools,
- or extract features by feeding the data into a standard set of feature extractors such as SIFT (the Scale-Invariant Feature Transform) [12], or SURF (the Speeded-Up Robust Features) [13], etc.
- Choose the kernels based on your domain expertise.
- Put the training data composed of hand-crated features and labels into a kernel SVM to learn a classifier.

Here, the main technical innovations usually comes from the feature extraction, often based on the serendipitous discoveries of lucky graduate students. Moreover, kernel selection also requires domain expertise that was previously the subject of extensive research. We will see later that one of the main innovations in the modern deep learning approach is that this hand-crafted feature engineering and kernel design are no longer required as they are automatically learned from the training data. This simplicity can be one of the main reasons for the success of deep learning, which led to the deluge of new deep tech companies.

So far we have mainly discussed the binary classification problems. Note that more general forms of the classifiers beyond the binary classifier are of importance in practice: for example, ImageNet has more than 20,000 categories. The extension of the linear classifier for such a setup is important, but will be discussed later.

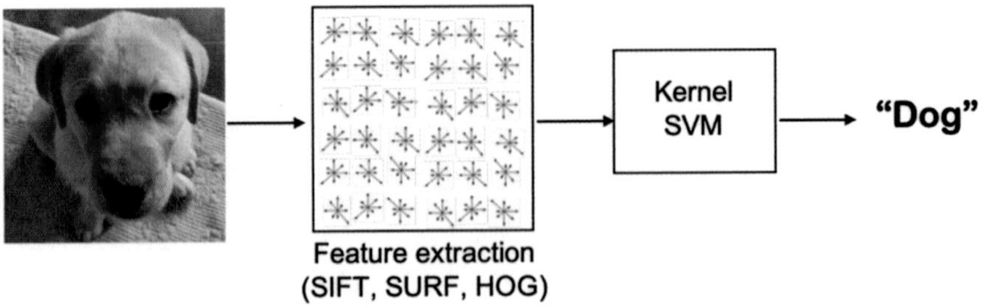

Fig. 2.7 Classical classifier design flowchart

2.6 Exercises

1. For a given polynomial kernel up to degree 2,

$$k(\boldsymbol{x}, \boldsymbol{y}) = (\boldsymbol{x}^\top \boldsymbol{y} + c)^2, \quad \boldsymbol{x}, \boldsymbol{y} \in \mathbb{R}^2,$$

what is the corresponding feature mapping $\varphi(\boldsymbol{x})$ such that $k(\boldsymbol{x}, \boldsymbol{y}) = \langle \boldsymbol{\varphi}(\boldsymbol{x}), \boldsymbol{\varphi}(\boldsymbol{y}) \rangle$?

2. Show that the feature space dimension for the radial basis function is infinite.
3. Suppose we are given the following positively labeled data points:

$$\boldsymbol{x}_1 = [2, 1]^\top, \boldsymbol{x}_2 = [2, -1]^\top, \boldsymbol{x}_3 = [3, 1]^\top, \tag{2.43}$$

and the following negatively labeled data points:

$$\boldsymbol{x}_4 = [1, 0]^\top, \boldsymbol{x}_5 = [0, 1]^\top, \boldsymbol{x}_6 = [0, -1]^\top. \tag{2.44}$$

   a. Are the two classes linear separable? Answer this question by visualizing their distribution in $\mathbb{R}^2$.
   b. Now, we are interested in designing a hard-margin linear SVM. What are the support vectors? Please answer this by inspection. You must give your reasoning.
   c. Using primal formulation, compute the closed form solution of the linear SVM classifier by hand calculation. You must show each step of your calculation. The inequality constraints may be simplified by exploiting the support vectors and KKT conditions.
   d. Using dual formulation, compute the closed form solution of the linear SVM classifier by hand calculation. You must show each step of your calculation. The inequality constraints may be simplified by exploiting the support vectors and KKT conditions.

4. Suppose we are given the following positively labeled data points:

$$\boldsymbol{x}_1 = [0.5, 0]^\top, \boldsymbol{x}_2 = [1.5, 1]^\top, \boldsymbol{x}_3 = [1.5, -1]^\top, \boldsymbol{x}_4 = [2, 0]^\top, \tag{2.45}$$

and the following negatively labeled data points:

$$\boldsymbol{x}_5 = [1, 0]^\top, \boldsymbol{x}_6 = [0, 1]^\top, \boldsymbol{x}_7 = [0, -1]^\top, \boldsymbol{x}_8 = [-1, 0]^\top. \tag{2.46}$$

   a. Are the two classes linearly separable? Answer this question by visualizing their distribution in $\mathbb{R}^2$.
   b. Now, we are interested in designing a soft-margin linear SVM. Using MATLAB, plot the decision boundaries for various choices of C.
   c. What do you observe when $C \to \infty$?

5. Suppose we are given the following positively labeled data points:

$$\boldsymbol{x}_1 = [3, 3]^\top, \boldsymbol{x}_2 = [3, -3]^\top, \boldsymbol{x}_3 = [-3, -3]^\top, \boldsymbol{x}_4 = [-3, 3]^\top, \tag{2.47}$$

and the following negatively labeled data points:

$$\boldsymbol{x}_5 = [1, 1]^\top, \boldsymbol{x}_6 = [1, -1]^\top, \boldsymbol{x}_7 = [-1, -1]^\top, \boldsymbol{x}_8 = [-1, 1]^\top. \tag{2.48}$$

a. Are the two classes linearly separable? Answer this question by visualizing their distribution in $\mathbb{R}^2$.
b. Find a feature mapping $\boldsymbol{\varphi} : \mathbb{R}^2 \mapsto F \subset \mathbb{R}^3$ so that two classes are linear separable in the feature space F. Show this by drawing data distribution in F.
c. What is the corresponding kernel?
d. What are the support vectors in F?
e. Using dual formulation, compute the closed form solution of a kernel SVM classifier by hand calculation. You must show each step of your calculation. The inequality constraints may be simplified by exploiting the support vectors and KKT conditions.

Chapter 3
Linear, Logistic, and Kernel Regression

3.1 Introduction

In machine learning, regression analysis refers to a process for estimating the relationships between dependent variables and independent variables. This method is mainly used to predict and find the cause-and-effect relationship between variables. For example, in a linear regression, a researcher tries to find the line that best fits the data according to a certain mathematical criterion (see Fig. 3.1a). Another important regression problem is the logistic regression. For example, in Fig. 3.1b, the dependent variables are binary properties such as yes or no for a given question, and the goal is to fit the binary data using continuously varying independent variables. It is easy to understand that this problem is closely related to the binary classification problem. For the case of Fig. 3.1c, the technical issue is a bit different from the other two. Here, the distribution cannot be regressed out by a linear line. Moreover, the dependent variable is not binary, but has continuous values. In fact, a better regression approach is to fit the data with a smoothly varying curve. In fact, this is directly related to a nonlinear regression problem.

Although regression analysis is a classical approach that can be dated back to the least squares method by Legendre in 1805 and by Gauss in 1809, regression analysis is still a key idea of the deep learning approaches, as will be discussed later. Therefore, we will visit the classical regression approach to discuss three specific forms of regression analysis: linear regression, logistic regression, and kernel regression. Later on, this overview will prove useful in understanding modern regression approaches using deep neural networks.

J. C. Ye, *Geometry of Deep Learning*, Mathematics in Industry 37,
https://doi.org/10.1007/978-981-16-6046-7_3

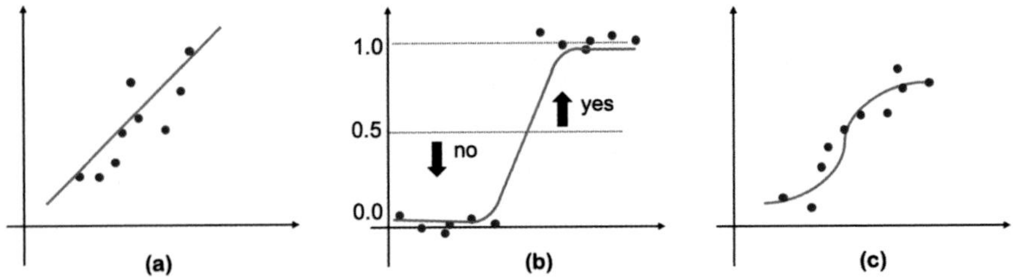

Fig. 3.1 Example of various regression problems. The x-axes are for the independent variables, and y-axes are for the dependent variables. (**a**) linear regression, (**b**) logistic regression, and (**c**) nonlinear regression using a polynomial kernel

3.2 Linear Regression

3.2.1 Ordinary Least Squares (OLS)

A linear regression uses a linear model as shown in Fig. 3.1a. More specifically, the dependent variable can be calculated from a linear combination of the input variables. It is also common to refer to a linear model as Ordinary Least Squares (OLS) linear regression or just Least Squares (LS) regression. For example, a simple linear regression model is given by

$$y_i = \beta_0 + \beta_1 x_i + \epsilon_i, \quad i = 1, \cdots, n \tag{3.1}$$

and the goal is to estimate the parameter set $\boldsymbol{\beta} = \{\beta_0, \beta_1\}$ from the training data $\{x_i, y_i\}_{i=1}^n$.

In general, a linear regression problem can be represented by

$$y_i = \langle \boldsymbol{x}_i, \boldsymbol{\beta} \rangle + \epsilon_i, \quad i = 1, \cdots, n, \tag{3.2}$$

where $(\boldsymbol{x}_i, y_i) \in \mathbb{R}^p \times \mathbb{R}$ is the i-th training data, and $\boldsymbol{\beta} \in \mathbb{R}^p$ is referred to as the regression coefficient. This can be represented in matrix form as

$$\boldsymbol{y} = \boldsymbol{X}^\top \boldsymbol{\beta} + \boldsymbol{\epsilon}, \tag{3.3}$$

where

$$\boldsymbol{y} := \begin{bmatrix} y_1 \\ \vdots \\ y_n \end{bmatrix}, \quad \boldsymbol{X} := \begin{bmatrix} \boldsymbol{x}_1 & \cdots & \boldsymbol{x}_n \end{bmatrix}, \quad \boldsymbol{\epsilon} := \begin{bmatrix} \epsilon_1 \\ \vdots \\ \epsilon_n \end{bmatrix}.$$

In this mathematical formulation, $\boldsymbol{x}_i$ corresponds to the independent variable, whereas y_i is the dependent variable.

Then, the regression analysis using l_2 loss or the mean squared error (MSE) loss can be done by

$$\min_{\boldsymbol{\beta}} \ell(\boldsymbol{\beta}), \quad \ell(\boldsymbol{\beta}) := \frac{1}{2}\|\boldsymbol{y} - \boldsymbol{X}^\top \boldsymbol{\beta}\|^2, \tag{3.4}$$

where the loss can be further expanded as

$$\begin{aligned}\ell(\boldsymbol{\beta}) &:= \frac{1}{2}\|\boldsymbol{y} - \boldsymbol{X}^\top \boldsymbol{\beta}\|^2 \\ &= \frac{1}{2}(\boldsymbol{y} - \boldsymbol{X}^\top \boldsymbol{\beta})^\top(\boldsymbol{y} - \boldsymbol{X}^\top \boldsymbol{\beta}) \\ &= \frac{1}{2}\left(\boldsymbol{y}^\top \boldsymbol{y} - \boldsymbol{y}^\top \boldsymbol{X}^\top \boldsymbol{\beta} - \boldsymbol{\beta}^\top \boldsymbol{X}\boldsymbol{y} + \boldsymbol{\beta}^\top \boldsymbol{X}\boldsymbol{X}^\top \boldsymbol{\beta}\right),\end{aligned}$$

The parameter that minimizes the MSE loss can be found by setting the gradient of the loss with respect to $\boldsymbol{\beta}$ to zero. To calculate the gradient for the vector-valued function, the following lemma is useful.

Lemma 3.1 *[5] Let $\boldsymbol{x}, \boldsymbol{a}$ and $\boldsymbol{B}$ denotes vectors and a matrix with appropriate sizes, respectively. Then, we have*

$$\frac{\partial \boldsymbol{x}^\top \boldsymbol{a}}{\partial \boldsymbol{x}} = \frac{\partial \boldsymbol{a}^\top \boldsymbol{x}}{\partial \boldsymbol{x}} = \boldsymbol{a}, \tag{3.5}$$

$$\frac{\partial \boldsymbol{x}^\top \boldsymbol{B}\boldsymbol{x}}{\partial \boldsymbol{x}} = (\boldsymbol{B} + \boldsymbol{B}^\top)\boldsymbol{x}. \tag{3.6}$$

Using Lemma 3.1, we have

$$\left.\frac{\partial \ell(\boldsymbol{\beta})}{\partial \boldsymbol{\beta}}\right|_{\boldsymbol{\beta}=\hat{\boldsymbol{\beta}}} = -\boldsymbol{X}\boldsymbol{y} + \boldsymbol{X}\boldsymbol{X}^\top \hat{\boldsymbol{\beta}} = \boldsymbol{0},$$

where $\hat{\boldsymbol{\beta}}$ is the minimizer. If $\boldsymbol{X}\boldsymbol{X}^\top$ is invertible, or $\boldsymbol{X}$ has the full row rank, then we have

$$\hat{\boldsymbol{\beta}} = \left(\boldsymbol{X}\boldsymbol{X}^\top\right)^{-1} \boldsymbol{X}\boldsymbol{y}. \tag{3.7}$$

The full rank condition is important for the existence of the matrix inverse, which will be revisited again in the ridge regression.

This regression setup is closely related to the *general linear model (GLM)*, which has been successfully used for statistical analysis. For example, GLM analysis is one of the main workhorses for the functional MRI data analysis [14]. The main idea of functional MRI is that multiple temporal frames of MR images of a brain are obtained during a given task (for example, motion tasks), and then the temporal

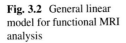

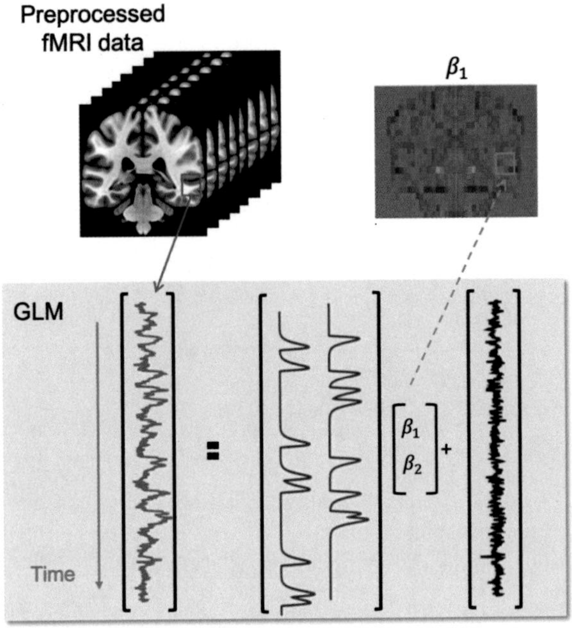

Fig. 3.2 General linear model for functional MRI analysis

variation of the MR values at each voxel location is analyzed to check whether its temporal variation is correlated with a given task. Here the temporal time series data $\boldsymbol{y}$ from one voxel is described as a linear combination of the model ($\boldsymbol{X}^\top$), which is often termed as the "design matrix", containing a set of regressors as in Fig. 3.2 representing the independent variable and the residuals (i.e., the errors), then the results are stored, displayed, and possibly analyzed further in the form of voxelwise maps as shown in the top right of Fig. 3.2 when $\boldsymbol{\beta} = [\beta_1, \beta_2]^\top$.

3.3 Logistic Regression

3.3.1 Logits and Linear Regression

Similar to the example in Fig. 3.1b, there are many important problems for which the dependent variable has limited values. For example, in binary logistic regression for analyzing smoking behavior, the dependent variable is a dummy variable: coded 0 (did not smoke) or 1 (did smoke). In another example, one is interested in fitting a linear model to the probability of the event. In this case, the dependent variable only takes values between 0 and 1. In this case, transforming the independent variables does not remedy all of the potential problems. Instead, the key idea of the logistic regression is transforming the dependent variable.

Specifically, we define the term *odds*:

$$\text{odds} = \frac{q}{1-q}, \tag{3.8}$$

where q is a probability in a range of 0–1. The odds have a range of 0–∞ with values greater than 1 associated with an event being more likely to occur than to not occur and values less than 1 associated with an event that is less likely to occur. Then, the term *logit* is defined as the log of the odds:

$$\text{logit} := \log(\text{odds}) = \log\left(\frac{q}{1-q}\right).$$

This transformation is useful because it creates a variable with a range from $-\infty$ to ∞ with zero associated with an event equally likely to occur and not occur. One of the important advantages of this transformation of the dependent variable is that it solves the problem we encountered in fitting a linear model to probabilities. If we transform our probabilities to logits, then the range of the logit is not restricted, so that we can apply a standard linear regression.

Specifically, using the logits transform, a linear regression model for the probability is given by

$$\log\left(\frac{q}{1-q}\right) = \beta_0 + \beta_1 x, \tag{3.9}$$

from which we have

$$\begin{aligned} q &= \frac{1}{1+e^{-(\beta_0+\beta_1 x)}} \\ &= \text{Sig}(\beta_0 + \beta_1 x), \end{aligned} \tag{3.10}$$

where $\text{Sig}(x)$ denotes the sigmoid function:

$$\text{Sig}(x) = \frac{1}{1+e^{-x}},$$

whose shape is shown in Fig. 3.3. It is remarkable that although the nonlinear transform is originally applied to the dependent variable for linear regression, the net result is the introduction of the nonlinearity after the linear term. In fact, this is closely related to the modern deep neural networks that have nonlinearities after the linear layers.

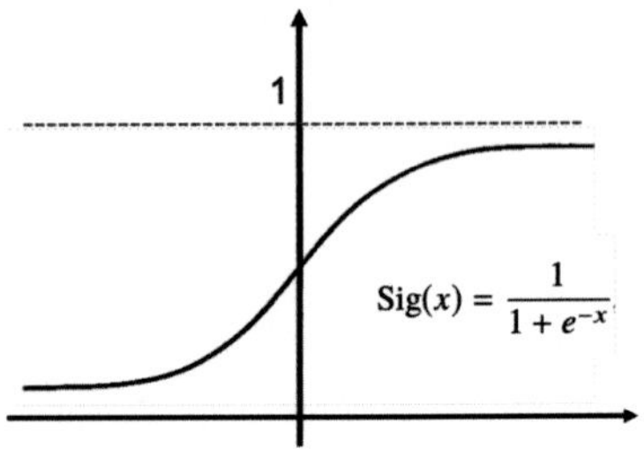

Fig. 3.3 Sigmoid function

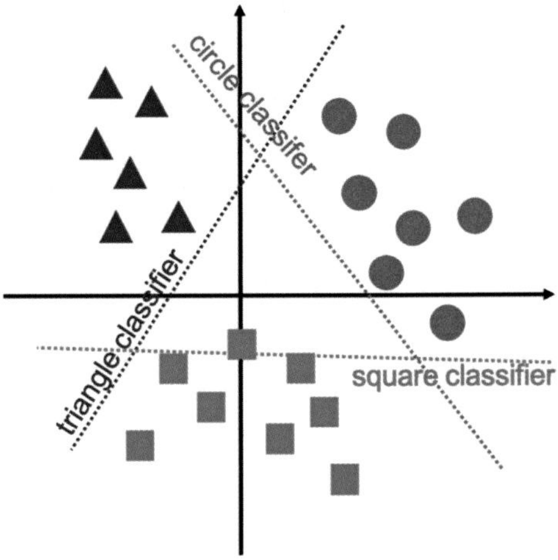

Fig. 3.4 Multi-class classification problem

3.3.2 Multiclass Classification Using Logistic Regression

In SVM, we mainly discussed the binary classification problem, in which a hyperplane is defined to separate two classes. Now, consider Fig. 3.4, where we want to define three hyperplanes that can split the data into multiple categories.

A direct extension of the SVM for the multiple class classifier design problem is to consider all the combinatorial combinations of the hyperplanes. More specifically, a data $\boldsymbol{x}_i$ can be on either side of the hyperplane so that given three hyperplanes in Fig. 3.4, one could design a classifier that can potentially classify $2^3 = 8$ classes. Although this approach may reduce the number of hyperplanes for a given number of classes c, one of the main technical difficulties of such extension of SVM is that we need to consider all combinatorial combinations of the constraint sets, which is difficult to implement.

A quick remedy for this multi-class classifier design problem is to use the logistic regression. More specifically, for given c-class categories, we define a probability vector $\boldsymbol{q} = [q_1, \cdots, q_c]^\top \in \mathbb{R}^c$, where $q_i \in [0, 1]$ denotes the probability that a data belongs to the class i. Then, by extending (3.9) to vector-valued probabilities for a given dependent variable $\boldsymbol{x} \in \mathbb{R}^p$, we have

$$\begin{bmatrix} \log\left(\frac{q_1}{1-q_1}\right) \\ \vdots \\ \log\left(\frac{q_c}{1-q_c}\right) \end{bmatrix} = \boldsymbol{W}^\top \boldsymbol{x} + \boldsymbol{b} \tag{3.11}$$

where $\boldsymbol{W} \in \mathbb{R}^{p\times c}$ denotes the matrix composed of c-normal vectors in the p-dimensional spaces, and $\boldsymbol{b} \in \mathbb{R}^c$ is the associated bias term. Then, we can easily see that the corresponding probability vector is given by

$$\boldsymbol{p} = \mathrm{Sig}(\boldsymbol{W}^\top \boldsymbol{x} + \boldsymbol{b}), \tag{3.12}$$

where $\mathrm{Sig}(\cdot)$ is an element-wise sigmoid function. Then, by ranking the magnitude of the probability, one could classify the data into the corresponding categories. In fact, this technique is a standard method in modern classifier design using deep neural networks. We will revisit this issue later.

3.4 Ridge Regression

Recall that the basic assumption for the linear regression solution in (3.7) is that $\boldsymbol{X}^\top$ has full column rank or $\boldsymbol{X}$ has the full row rank. However, when $\boldsymbol{X}^\top$ is high-dimensional, the columns of $\boldsymbol{X}^\top$ can be collinear, which in statistical terms refers to the event of two (or multiple) covariates being highly linearly related. Consequently, $\boldsymbol{X}^\top$ may not be of full column rank or close to not being the full column rank, and we cannot use the standard linear regression. To deal with this issue, the ridge regression is useful.

Specifically, the following regularized least squares problem is solved:

$$\min_{\boldsymbol{\beta}} \ell_{ridge}(\boldsymbol{\beta}),$$

where

$$\ell_{ridge}(\boldsymbol{\beta}) := \frac{1}{2}\|\boldsymbol{y} - \boldsymbol{X}^\top \boldsymbol{\beta}\|^2 + \frac{\lambda}{2}\|\boldsymbol{\beta}\|^2, \tag{3.13}$$

where $\lambda > 0$ is the regularization parameter. This type of regularization is often called the Tikhonov regularization. Using Lemma 3.1, we can easily show

$$\left.\frac{\partial \ell_{ridge}(\boldsymbol{\beta})}{\partial \boldsymbol{\beta}}\right|_{\boldsymbol{\beta}=\hat{\boldsymbol{\beta}}} = -\boldsymbol{X}\boldsymbol{y} + \boldsymbol{X}\boldsymbol{X}^\top \hat{\boldsymbol{\beta}} + \lambda\hat{\boldsymbol{\beta}} = \boldsymbol{0},$$

which leads to

$$\hat{\boldsymbol{\beta}} = \left(\boldsymbol{X}\boldsymbol{X}^\top + \lambda \boldsymbol{I}\right)^{-1} \boldsymbol{X}\boldsymbol{y}. \tag{3.14}$$

Using the following matrix inversion lemma [3],

$$(\boldsymbol{I} + \boldsymbol{U}\boldsymbol{C}\boldsymbol{V})^{-1} = \boldsymbol{I} - \boldsymbol{U}\left(\boldsymbol{C}^{-1} + \boldsymbol{V}\boldsymbol{U}\right)^{-1}\boldsymbol{V}, \tag{3.15}$$

Eq. (3.14) can also be equivalently written by

$$\begin{aligned}
\hat{\boldsymbol{\beta}} &= \left(\boldsymbol{X}\boldsymbol{X}^\top + \lambda \boldsymbol{I}\right)^{-1} \boldsymbol{X}\boldsymbol{y} \\
&= \frac{1}{\lambda}\left(\boldsymbol{X}\boldsymbol{X}^\top/\lambda + \boldsymbol{I}\right)^{-1} \boldsymbol{X}\boldsymbol{y} \\
&= \frac{1}{\lambda}\left\{\boldsymbol{I} - \boldsymbol{X}\left(\lambda\boldsymbol{I} + \boldsymbol{X}^\top\boldsymbol{X}\right)^{-1}\boldsymbol{X}^\top\right\}\boldsymbol{X}\boldsymbol{y} \\
&= \frac{1}{\lambda}\boldsymbol{X}\left\{\boldsymbol{I} - \left(\lambda\boldsymbol{I} + \boldsymbol{X}^\top\boldsymbol{X}\right)^{-1}\boldsymbol{X}^\top\boldsymbol{X}\right\}\boldsymbol{y} \\
&= \frac{1}{\lambda}\boldsymbol{X}\left(\lambda\boldsymbol{I} + \boldsymbol{X}^\top\boldsymbol{X}\right)^{-1}\left\{\left(\lambda\boldsymbol{I} + \boldsymbol{X}^\top\boldsymbol{X}\right) - \boldsymbol{X}^\top\boldsymbol{X}\right\}\boldsymbol{y} \\
&= \boldsymbol{X}\left(\boldsymbol{X}^\top\boldsymbol{X} + \lambda\boldsymbol{I}\right)^{-1}\boldsymbol{y}.
\end{aligned} \tag{3.16}$$

In particular, the expression in (3.16) is useful when $\boldsymbol{X}$ is a tall matrix, since the size of the matrix inversion is much smaller than that of (3.14). Even if this is not the case, the expression in (3.16) is extremely useful to derive the kernel ridge regression, which is the main topic in the next section.

3.5 Kernel Regression

Recall that a nonlinear kernel SVM was developed based on the observation that the nonlinear decision boundary in the original input space can be often represented as a linear boundary in the high-dimensional feature space. A similar idea can be used for regression. Specifically, the goal is to implement the linear regression in the high-dimensional feature space, but the net result is that the resulting regression becomes nonlinear in the original space (see Fig. 3.5).

In order to use a kernel trick similar to that used in the kernel SVM, let us revisit the linear regression problem in (3.2). Using the parameter estimation from the ridge

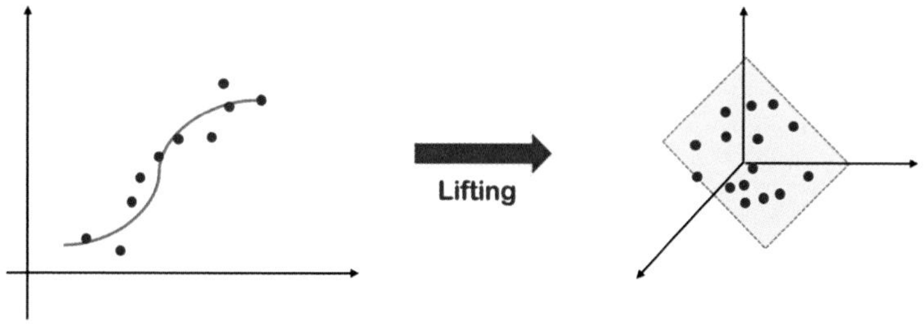

Fig. 3.5 Kernel regression concept

regression (3.16), the estimated function $\hat{f}(\boldsymbol{x})$ for a given independent variable $\boldsymbol{x} \in \mathbb{R}^p$ is given by

$$\begin{aligned}\hat{f}(\boldsymbol{x}) &:= \boldsymbol{x}^\top \hat{\boldsymbol{\beta}} \\ &= \boldsymbol{x}^\top \boldsymbol{X}(\boldsymbol{X}^\top \boldsymbol{X} + \lambda \boldsymbol{I})^{-1} \boldsymbol{y} \\ &= \left[\langle \boldsymbol{x}, \boldsymbol{x}_1 \rangle \cdots \langle \boldsymbol{x}, \boldsymbol{x}_n \rangle\right] \left(\begin{bmatrix} \langle \boldsymbol{x}_1, \boldsymbol{x}_1 \rangle & \cdots & \langle \boldsymbol{x}_1, \boldsymbol{x}_n \rangle \\ \vdots & \ddots & \vdots \\ \langle \boldsymbol{x}_n, \boldsymbol{x}_1 \rangle & \cdots & \langle \boldsymbol{x}_n, \boldsymbol{x}_n \rangle \end{bmatrix} + \lambda \boldsymbol{I} \right)^{-1} \boldsymbol{y},\end{aligned} \tag{3.17}$$

where we use

$$\boldsymbol{x}^\top \boldsymbol{X} = \left[\langle \boldsymbol{x}, \boldsymbol{x}_1 \rangle \cdots \langle \boldsymbol{x}, \boldsymbol{x}_n \rangle\right]$$

and

$$\boldsymbol{X}^\top \boldsymbol{X} = \begin{bmatrix} \boldsymbol{x}_1^\top \\ \vdots \\ \boldsymbol{x}_n^\top \end{bmatrix} \left[\boldsymbol{x}_1 \cdots \boldsymbol{x}_n\right] = \begin{bmatrix} \langle \boldsymbol{x}_1, \boldsymbol{x}_1 \rangle & \cdots & \langle \boldsymbol{x}_1, \boldsymbol{x}_n \rangle \\ \vdots & \ddots & \vdots \\ \langle \boldsymbol{x}_n, \boldsymbol{x}_1 \rangle & \cdots & \langle \boldsymbol{x}_n, \boldsymbol{x}_n \rangle . \end{bmatrix}$$

Since everything is represented by the inner product of the input vectors, we can now lift the data $\boldsymbol{x}$ to a feature space using $\boldsymbol{\varphi}(\boldsymbol{x})$ to compute the inner product in the high-dimensional feature space. Then, using the kernel trick, the inner product in the feature space can be replaced by the kernel:

$$\langle \boldsymbol{x}, \boldsymbol{x}_i \rangle \mapsto k(\boldsymbol{x}, \boldsymbol{x}_i) := \langle \boldsymbol{\varphi}(\boldsymbol{x}), \boldsymbol{\varphi}(\boldsymbol{x}_i) \rangle. \tag{3.18}$$

Accordingly, (3.17) can be extended to the feature space as:

$$\hat{f}(\boldsymbol{x}) = \left[k(\boldsymbol{x}, \boldsymbol{x}_1) \cdots k(\boldsymbol{x}, \boldsymbol{x}_n)\right] (\boldsymbol{K} + \lambda \boldsymbol{I})^{-1} \boldsymbol{y}, \tag{3.19}$$

where the $\boldsymbol{K} \in \mathbb{R}^{n \times n}$ is the kernel Gram matrix given by

$$\boldsymbol{K} := \begin{bmatrix} k(\boldsymbol{x}_1, \boldsymbol{x}_1) & \cdots & k(\boldsymbol{x}_1, \boldsymbol{x}_n) \\ \vdots & \ddots & \vdots \\ k(\boldsymbol{x}_n, \boldsymbol{x}_1) & \cdots & k(\boldsymbol{x}_n, \boldsymbol{x}_n) \end{bmatrix}. \tag{3.20}$$

Equivalently, (3.19) can be derived from the following regression problem with kernel:

$$y_i = \sum_{j=1}^{p} \alpha_j k(\boldsymbol{x}_i, \boldsymbol{x}_j) + \epsilon \tag{3.21}$$

which is a nonlinear extension of (3.2). Then, (3.19) is obtained using the following optimization problem:

$$\min_{\boldsymbol{\alpha} \in \mathbb{R}^p} \sum_{i=1}^{n} \left(y_i - \sum_{j=1}^{p} \alpha_j k(\boldsymbol{x}_i, \boldsymbol{x}_j) \right)^2 + \lambda \boldsymbol{\alpha}^\top \boldsymbol{K} \boldsymbol{\alpha}, \tag{3.22}$$

where $\boldsymbol{K}$ is the kernel Gram matrix in (3.20). This implies that the regularization term should be weighted by the kernel to take into account of the deformation in the feature space. More rigorous derivation of (3.22) is obtained from the so-called *representer theorem*, which is the topic of the next chapter.

Figure 3.6 shows the examples of linear regression and kernel regression using the polynomial and radial basis function (RBF) kernels. We can clearly see that nonlinear kernel regression follows the trend much better.

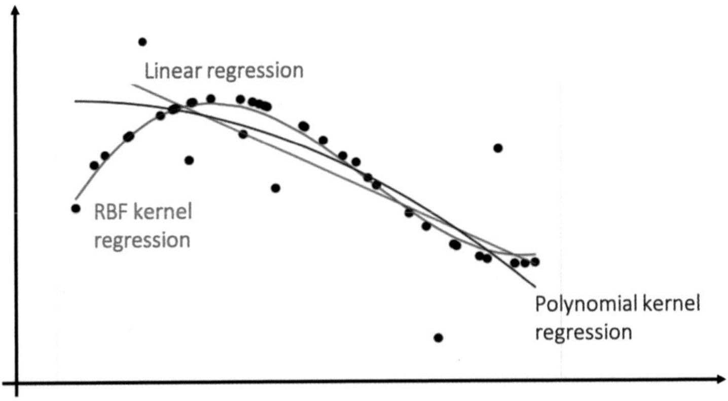

Fig. 3.6 Linear and nonlinear kernel regression

3.6 Bias–Variance Trade-off in Regression

In this section, we will discuss the important issue of the bias and variance trade-off in regression analysis.

Let $\{\boldsymbol{x}_i, \boldsymbol{y}_i\}_{i=1}^n$ denote the training data set, where $\boldsymbol{x}_i \in \mathbb{R}^p \subset \mathcal{X}$ is an independent variable and $\boldsymbol{y}_i \in \mathbb{R}^p \subset \mathcal{Y}$ is a dependent variable that has dependency on $\boldsymbol{x}_i$. The reason we use the boldface characters $\boldsymbol{x}_i$ and $\boldsymbol{y}_i$ is that they can be vectors. In regression analysis, the dependent variable is often represented as a functional relationship with respect to the independent variable:

$$\boldsymbol{y}_i = \boldsymbol{f}_\Theta(\boldsymbol{x}_i) + \boldsymbol{\epsilon}_i, \tag{3.23}$$

where $\boldsymbol{\epsilon}_i$ denotes an additive error term that may stand in for unmodeled parts, and $\boldsymbol{f}_{\boldsymbol{\Theta}}(\cdot)$ is a regression function (which can be possibly a nonlinear function) with the input variable $\boldsymbol{x}_i$ and parameterized by $\boldsymbol{\Theta}$. With a slight abuse of notation, we often use $\boldsymbol{f} := \boldsymbol{f}_{\boldsymbol{\Theta}}$ when the dependency on the parameter $\boldsymbol{\Theta}$ is obvious.

In (3.23), $\boldsymbol{\Theta}$ is the regression parameter set that should be estimated from the training data set. Usually, this parameter set is estimated by minimizing a loss. For example, one of the most popular loss functions is l_2 or the MSE loss, in which case the parameter estimation problem is given by

$$\min_{\boldsymbol{\Theta}} \frac{1}{2} \sum_{i=1}^{n} \|\boldsymbol{y}_i - \boldsymbol{f}_{\boldsymbol{\Theta}}(\boldsymbol{x}_i)\|^2. \tag{3.24}$$

Another popular tool that is often used in regression analysis is the regularization. In regularized regression analysis, an additional term is added to impose a constraint on the parameter. More specifically, the following optimization problem is solved to estimate the parameter $\boldsymbol{\Theta}$:

$$\min_{\boldsymbol{\Theta}} \frac{1}{2} \sum_{i=1}^{n} \|\boldsymbol{y}_i - \boldsymbol{f}_{\boldsymbol{\Theta}}(\boldsymbol{x}_i)\|^2 + \lambda R(\boldsymbol{\Theta}), \tag{3.25}$$

where $R(\boldsymbol{\Theta})$ and λ are often called the *regularization function* and *regularization parameter*, respectively.

With the estimated parameter $\hat{\boldsymbol{\Theta}}$, the estimated function $\hat{\boldsymbol{f}}$ is defined as

$$\hat{\boldsymbol{f}}(\boldsymbol{x}) := \boldsymbol{f}_{\hat{\boldsymbol{\Theta}}}(\boldsymbol{x}). \tag{3.26}$$

Suppose that the noise $\boldsymbol{\epsilon}$ is zero mean i.i.d. Gaussian with the variance σ^2. Then, the MSE error of the regression problem is given by

$$\begin{aligned} E\|\boldsymbol{y} - \hat{\boldsymbol{f}}\|^2 &= E\|\boldsymbol{f} + \boldsymbol{\epsilon} - \hat{\boldsymbol{f}}\|^2 \\ &= E\|\boldsymbol{f} + \boldsymbol{\epsilon} - \hat{\boldsymbol{f}} + E[\hat{\boldsymbol{f}}] - E[\hat{\boldsymbol{f}}]\|^2 \end{aligned}$$

$$= E\|f - E[\hat{f}]\|^2 + E\|\hat{f} - E[\hat{f}]\|^2 + E\|\epsilon\|^2$$
$$= \|f - E[\hat{f}]\|^2 + E\|\hat{f} - E[\hat{f}]\|^2 + E\|\epsilon\|^2$$
$$= \|\text{Bias}(\hat{f})\|^2 + \text{Var}(\hat{f}) + p\sigma^2, \tag{3.27}$$

where we use the following for the third equality:

$$E[\epsilon^\top(f - E[\hat{f}])] = 0,$$
$$E[\epsilon^\top(\hat{f} - E[\hat{f}])] = 0,$$
$$E[(\hat{f} - E[\hat{f}])^\top(f - E[\hat{f}])] = 0,$$

and the fourth equation comes from the fact that f and $E[\hat{f}]$ are deterministic. Equation (3.27) clearly shows that the MSE expression of the prediction error is composed of bias and variance components. This leads to the so-called bias–variance trade-off in regression problem, which can be explained in detail in the following example.

3.6.1 Examples

Here, we will investigate the bias and variance trade-off for the linear regression problem, where the regression function is given by

$$f(\boldsymbol{x}) = \langle \boldsymbol{x}, \boldsymbol{\beta} \rangle = \boldsymbol{x}^\top \boldsymbol{\beta}. \tag{3.28}$$

By defining the expectation operation $E[\cdot]$, the bias and variance of the OLS in (3.7) can be computed as follows:

$$\begin{aligned}\text{Bias}(\widehat{f}) &:= \boldsymbol{x}^\top\boldsymbol{\beta} - E[\boldsymbol{x}^\top\hat{\boldsymbol{\beta}}]\\ &= \boldsymbol{x}^\top\boldsymbol{\beta} - \boldsymbol{x}^\top E[(\boldsymbol{X}\boldsymbol{X}^\top)^{-1}\boldsymbol{X}\boldsymbol{y}]\\ &= \boldsymbol{x}^\top\boldsymbol{\beta} - \boldsymbol{x}^\top(\boldsymbol{X}\boldsymbol{X}^\top)^{-1}\boldsymbol{X}E[\boldsymbol{y}]\\ &= \boldsymbol{x}^\top\boldsymbol{\beta} - \boldsymbol{x}^\top(\boldsymbol{X}\boldsymbol{X}^\top)^{-1}\boldsymbol{X}\boldsymbol{X}^\top\boldsymbol{\beta} = \boldsymbol{0},\end{aligned}$$

since $E[\boldsymbol{y}] = E[\boldsymbol{X}^\top\boldsymbol{\beta} + \boldsymbol{\epsilon}] = \boldsymbol{X}^\top\boldsymbol{\beta} + E[\boldsymbol{\epsilon}] = \boldsymbol{X}^\top\boldsymbol{\beta}$. Since the bias is zero, $\hat{f}$ is often called an unbiased estimator. Similarly, the covariance can be computed by

$$\begin{aligned}\text{Var}(\hat{f}) &:= E\left[\boldsymbol{x}^\top(\hat{\boldsymbol{\beta}} - \boldsymbol{\beta})(\hat{\boldsymbol{\beta}} - \boldsymbol{\beta})^\top\boldsymbol{x}\right]\\ &= E\left[\boldsymbol{x}^\top(\boldsymbol{X}\boldsymbol{X}^\top)^{-1}\boldsymbol{X}\boldsymbol{\epsilon}\boldsymbol{\epsilon}^\top\boldsymbol{X}^\top(\boldsymbol{X}\boldsymbol{X}^\top)^{-1}\boldsymbol{x}\right]\end{aligned}$$

$$= \boldsymbol{x}^\top (\boldsymbol{X}\boldsymbol{X}^\top)^{-1} \boldsymbol{X} E\left[\boldsymbol{\epsilon}\boldsymbol{\epsilon}^\top\right] \boldsymbol{X}^\top (\boldsymbol{X}\boldsymbol{X}^\top)^{-1} \boldsymbol{x}$$
$$= \sigma^2 \boldsymbol{x}^\top (\boldsymbol{X}\boldsymbol{X}^\top)^{-1} \boldsymbol{x}.$$

On the other hand, the bias and covariance of the ridge regression in (3.14) are given by

$$\begin{aligned}
\text{Bias}(\hat{\boldsymbol{f}}) &:= \boldsymbol{x}^\top \boldsymbol{\beta} - E[\boldsymbol{x}^\top (\boldsymbol{X}\boldsymbol{X}^\top + \lambda \boldsymbol{I})^{-1} \boldsymbol{X}\boldsymbol{y}] \\
&= \boldsymbol{x}^\top \left(\boldsymbol{I} - (\boldsymbol{X}\boldsymbol{X}^\top + \lambda \boldsymbol{I})^{-1} \boldsymbol{X}\boldsymbol{X}^\top \right) \boldsymbol{\beta} \\
&= \lambda \boldsymbol{x}^\top (\boldsymbol{X}\boldsymbol{X}^\top + \lambda \boldsymbol{I})^{-1} \boldsymbol{\beta},
\end{aligned}$$

and

$$\begin{aligned}
\text{Var}(\hat{\boldsymbol{f}}) &= E\left[\boldsymbol{x}^\top (\boldsymbol{X}\boldsymbol{X}^\top + \lambda \boldsymbol{I})^{-1} \boldsymbol{X}\boldsymbol{\epsilon}\boldsymbol{\epsilon}^\top \boldsymbol{X}^\top (\boldsymbol{X}\boldsymbol{X}^\top + \lambda \boldsymbol{I})^{-1} \boldsymbol{x}\right] \\
&= \sigma^2 \boldsymbol{x}^\top (\boldsymbol{X}\boldsymbol{X}^\top + \lambda \boldsymbol{I})^{-1} \boldsymbol{X}\boldsymbol{X}^\top (\boldsymbol{X}\boldsymbol{X}^\top + \lambda \boldsymbol{I})^{-1} \boldsymbol{x}, \qquad (3.29)
\end{aligned}$$

where we use $E\left[\boldsymbol{\epsilon}\boldsymbol{\epsilon}^\top\right] = \sigma^2 \boldsymbol{I}$.

Accordingly, we can see that as λ becomes larger, the variance decreases and the bias increases as shown in Fig. 3.7. This implies that the bias–variance trade-off of a ridge regression depends on the regularization parameter. One could find the optimal parameter λ^* that leads to the minimal total prediction error which gives the best bias–variance trade-off. The search for this optimal hyperparameter is one of the important research topics in classical ridge regression problems.

Fig. 3.7 Bias-variance trade-off in ridge regression

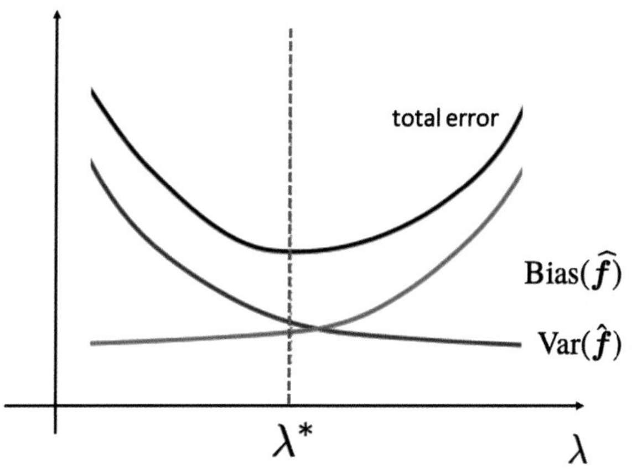

3.7 Exercises

1. Prove the matrix inversion lemma in Eq. (3.15).
2. The blood pressures, y (mmHg), and the ages, x years, of 7 patients are shown in the following table:

Patient id	1	2	3	4	5	6	7
x	42	70	45	30	55	25	57
y (mmHg)	98	130	121	88	182	80	125

   a. Obtain the OLS estimate of blood pressure with respect to age.
   b. Plot the regression line on the scatter plots.

3. A mechanic part is tested under various temperature conditions. The table below summarizes observational data on the part for 10 trials, where the all other experimental conditions are same except for the temperature (shown as degrees). Damaged represents the number of damaged parts, and Undamaged represents the number of parts that were not damaged.

Trial id	1	2	3	4	5	6	7	8	9	10
Temperature	53	57	58	63	66	67	67	67	68	69
Damaged	5	1	1	1	0	0	0	0	0	1
Undamaged	7	6	5	6	8	8	7	6	5	6

   a. Write down the logistic regression model.
   b. What is the estimated failure probability for a given temperature T?

4. Show that the ridge regression in (3.14) is equivalent to the linear regression with the following augmented dependent and independent variables:

$$\tilde{y} = \begin{bmatrix} \boldsymbol{y} \\ \sqrt{\lambda}\boldsymbol{I} \end{bmatrix}, \quad \tilde{\boldsymbol{X}} = \left[\boldsymbol{X}\ \sqrt{\lambda}\boldsymbol{I}\right],$$

   where $\boldsymbol{I}$ is the $p \times p$ identity matrix.

5. Consider the regression problem in the following table, where x is the independent variable and y is the dependent variable.

x	11	22	32	41	55	67	78	89	100	50	71	91
y	2330	2750	2309	2500	2100	1120	1010	1640	1931	1705	1751	2002

a. Perform the linear regression. What is the remaining residual error?
b. Consider the following Gaussian kernel:

$$K(x, x_i) = \frac{1}{h\sqrt{2\pi}} \exp\left(-\frac{1}{2}\left(\frac{x - x_i}{h}\right)^2\right).$$

c. Perform the kernel regression with $h = 5$, 10 and 15. What do you observe?

6. By directly solving (3.22), derive the kernel regression in (3.17).
7. Show that the variance of the kernel regression in (3.29) increases with decreasing regularization parameter λ.

Chapter 4
Reproducing Kernel Hilbert Space, Representer Theorem

4.1 Introduction

One of the key concepts in machine learning is the feature space, which is often referred to as the *latent space*. A feature space is usually a higher or lower-dimensional space than the original one where the input data lie (which is often referred to as the *ambient space*). Recall that in the kernel SVM, by lifting the data to a higher-dimensional feature space, one can find a linear classifier that can separate two different classes of samples (see Fig. 4.1a). Similarly, in kernel regression, rather than searching for nonlinear functions that can fit the data in the ambient space, the main idea is to compute a linear regressor in a higher-dimensional feature space as shown in Fig. 4.1b. On the other hand, in the principal component analysis (PCA), the input signals are projected on a lower-dimensional feature space using singular vector decomposition (see Fig. 4.1c).

In this section, we formally define a feature space that has good mathematical properties. Here, the "good" mathematical properties refer to the well-defined structure such as existence of the inner product, the completeness, reproducing properties, etc. In fact, the feature space with these properties is often called the *reproducing kernel Hilbert space (RKHS)* [11]. Although the RKHS is only a small subset of the Hilbert space, its mathematical properties are highly versatile, which makes the algorithm development simpler.

The RKHS theory has wide applications, including complex analysis, harmonic analysis, and quantum mechanics. Reproducing kernel Hilbert spaces are particularly important in the field of machine learning theory because of the celebrated representer theorem [11, 15] which states that every function in an RKHS that minimizes an empirical risk functional can be written as a linear combination of the kernel function evaluated at training samples. Indeed, the representer theorem has played a key role in classical machine learning problems, since it provides a means to reduce infinite dimensional optimization problems to tractable finite-dimensional ones.

J. C. Ye, *Geometry of Deep Learning*, Mathematics in Industry 37,
https://doi.org/10.1007/978-981-16-6046-7_4

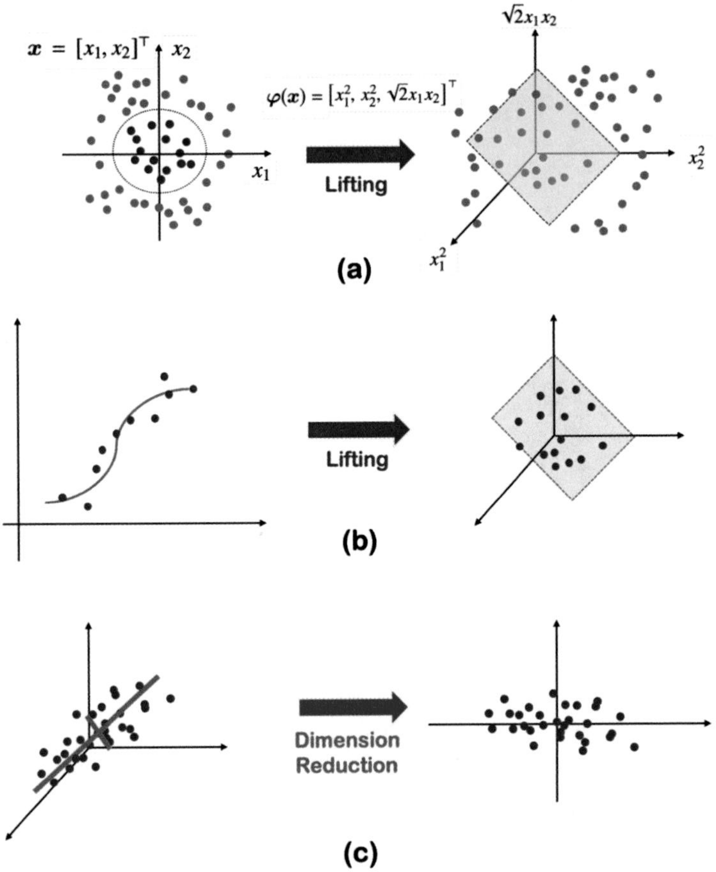

Fig. 4.1 Example of feature space embedding in (**a**) kernel SVM, (**b**) kernel regression, and (**c**) principle component analysis

In this chapter, we review the RKHS theory and the representer theorem. Then, we revisit the classifier and regression problems to show how kernel SVM and regression can be derived from the representer theorem. Then, we discuss the limitation of the kernel machines. Later we will show how these limitations of kernel machines can be largely overcome by modern deep learning approaches.

4.2 Reproducing Kernel Hilbert Space (RKHS)

As the theory of the RKHS originates from core mathematics, the rigorous definition is very abstract, which is often difficult to understand for students working on machine learning applications. Therefore, this section tries to explain the concept

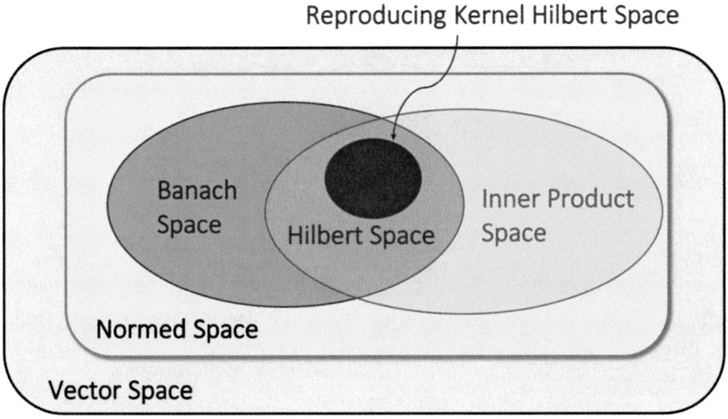

Fig. 4.2 RKHS, Hilbert space, Banach space, and vector space

from a more machine learning perspective so that students can understand why the RKHS theory has been the main workhorse in the classical machine learning theory.

Before diving into details, the readers are reminded that the RKHS is only a subset of the Hilbert space as shown in Fig. 4.2, i.e. the Hilbert space is more general than the RKHS. For the formal definition of the Hilbert space, please refer to Chap. 1.

4.2.1 Feature Map and Kernels

Here we start with the formal definition of a kernel:

Definition 4.1 Let $\mathcal{X}$ be a non-empty set. A function $k : \mathcal{X} \times \mathcal{X} \mapsto \mathbb{R}$ is called *a kernel* if there exists a Hilbert space $\mathcal{H}$ and a feature map $\boldsymbol{\phi} : \mathcal{X} \mapsto \mathcal{H}$ such that $\forall\, \boldsymbol{x}, \boldsymbol{x}' \in \mathcal{X}$:

$$k(\boldsymbol{x}, \boldsymbol{x}') := \langle \boldsymbol{\phi}(\boldsymbol{x}), \boldsymbol{\phi}(\boldsymbol{x}') \rangle_{\mathcal{H}}. \tag{4.1}$$

For example, a feature mapping we used to explain the kernel SVM was

$$\boldsymbol{\phi}(\boldsymbol{x}) = [\phi_1, \phi_2, \phi_3]^\top = \left[x_1^2 \; x_2^2 \; \sqrt{2}x_1x_2\right]^\top, \tag{4.2}$$

where $\mathcal{X} = \mathbb{R}^2$ (see Fig. 4.1a). We also showed that the corresponding kernel is given by

$$\begin{aligned} k(\boldsymbol{x}, \boldsymbol{y}) &= \langle \boldsymbol{\phi}(\boldsymbol{x}), \boldsymbol{\phi}(\boldsymbol{y}) \rangle \\ &= x_1^2 y_1^2 + x_2^2 y_2^2 + 2x_1 x_2 y_1 y_2 \\ &= (\langle \boldsymbol{x}, \boldsymbol{y} \rangle)^2, \end{aligned}$$

for all $\boldsymbol{x} = [x_1\ x_2]^\top$, $\boldsymbol{y} = [y_1\ y_2]^\top \in \mathbb{R}^2$, which corresponds to a polynomial kernel with degree 2. Note also that the feature space can be infinite-dimensional, such as $l^2(\mathbb{Z})$. In this case, using the definition of the inner product in $l^2(\mathbb{Z})$ (see (1.5)), the kernel is defined as

$$k(\boldsymbol{x}, \boldsymbol{x}') = \sum_{l=-\infty}^{\infty} \phi_l(\boldsymbol{x}) \phi_l(\boldsymbol{x}'),$$

where $\boldsymbol{\phi} = \{\phi_l\}_{l=-\infty}^{\infty} \in \mathcal{H}$.

Here it is important to emphasize that there exist almost no conditions on $\mathcal{X}$, i.e. $\mathcal{X}$ does not need an inner product, etc. On the other hand, the feature space $\mathcal{H}$ should be a Hilbert space. This implies that the feature map imposes a mathematical structure to the data set which does not necessarily have mathematical structures. This is an important machine learning apparatus as it provides a versatile tool to set the mathematical structure for all data in practice. For example, the bag-of-words (BOW) kernel [16] used for document classification is such an example that imposes a mathematical structure for an unstructured data such as documentations (see Fig. 4.3).

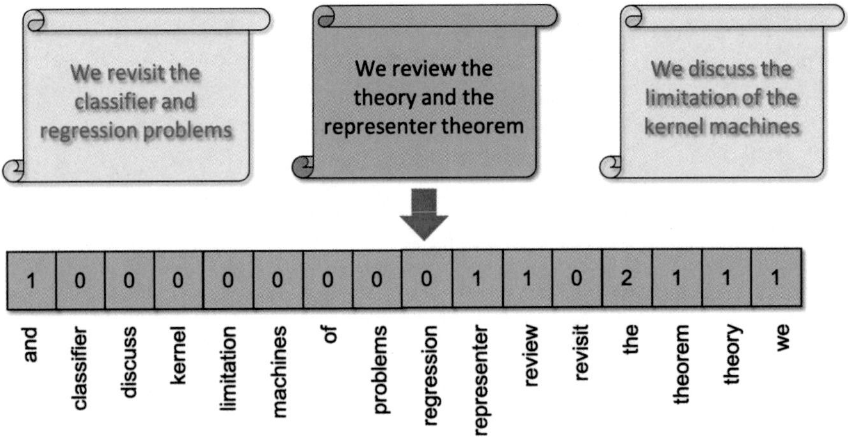

Fig. 4.3 Bag-of-words embedding to the feature space

Example (Bag-of-Words Kernel)
Suppose that the l-th element of the feature mapping $\boldsymbol{\phi}(\boldsymbol{x})$ for a document $\boldsymbol{x}$ denotes the number of l-th words (from a dictionary) appearing in the document $\boldsymbol{x}$. If we want to classify documents by their word counts, we can use the kernel $k(\boldsymbol{x}, \boldsymbol{y}) = \langle \boldsymbol{\phi}(\boldsymbol{x}), \boldsymbol{\phi}(\boldsymbol{y}) \rangle$.

In the kernel SVM and/or kernel regression, the optimization problem for the design of a classifier and/or regressor is formulated using kernels without ever using the feature map. Then, if we are given a function of two arguments, $k(\boldsymbol{x}, \boldsymbol{x}')$, how can we determine if it is a valid kernel? To answer this question, we need to check whether there exists a valid feature map. For this, the concept of the *positive definiteness* is important.

Definition 4.2 A symmetric function $k : \mathcal{X} \times \mathcal{X} \mapsto \mathbb{R}$ is positive definite if $\forall n \geq 1$, $\forall (a_1, \cdots, a_n) \in \mathbb{R}^n$, $\forall (\boldsymbol{x}_1, \cdots, \boldsymbol{x}_n) \in \mathcal{X}^n$

$$\sum_{i=1}^{n} \sum_{j=1}^{n} a_i a_j k(\boldsymbol{x}_i, \boldsymbol{x}_j) \geq 0. \tag{4.3}$$

Although this condition is both necessary and sufficient, the forward direction is more intuitive in understanding why the kernel function should be positive definite. More specifically, if we define the kernel as in (4.1), we have

$$\begin{aligned} \sum_{i=1}^{n} \sum_{j=1}^{n} a_i a_j k(\boldsymbol{x}_i, \boldsymbol{x}_j) &= \sum_{i=1}^{n} \sum_{j=1}^{n} a_i a_j \langle \boldsymbol{\phi}(\boldsymbol{x}_i), \boldsymbol{\phi}(\boldsymbol{x}_j) \rangle_{\mathcal{H}} \\ &= \left\| \sum_{i=1}^{n} a_i \boldsymbol{\phi}(\boldsymbol{x}_i) \right\|_{\mathcal{H}}^2 \geq 0. \end{aligned}$$

Therefore, existence of the feature mapping guarantees the positive definiteness of the kernel.

4.2.2 Definition of RKHS

With the definition of kernels and feature mapping, we are now ready to define the reproducing kernel Hilbert space. Toward this goal, let us revisit the feature mapping we used to explain the kernel SVM:

$$\boldsymbol{\phi}(\boldsymbol{x}) = [\phi_1, \phi_2, \phi_2]^\top = \left[x_1^2 \; x_2^2 \; \sqrt{2} x_1 x_2\right]^\top.$$

Suppose we define a function $f : \mathcal{X} \mapsto \mathbb{R}$ via a feature maps:

$$\begin{aligned} f(\boldsymbol{x}) &= \sum_{l=1}^{3} f_l \phi_l(\boldsymbol{x}) \\ &= f_1 x_1^2 + f_2 x_2^2 + f_3 \left(\sqrt{2} x_1 x_2\right). \end{aligned}$$

In terms of feature space coordinates, f is represented by $f(\cdot)$:

$$f = f(\cdot) := \begin{bmatrix} f_1 & f_2 & f_3 \end{bmatrix}^\top,$$

so that $f(\boldsymbol{x})$ can be represented as an inner product:

$$f(\boldsymbol{x}) = \langle f(\cdot), \boldsymbol{\phi}(\boldsymbol{x}) \rangle_{\mathcal{H}}, \tag{4.4}$$

where the feature map $\boldsymbol{\phi}(\boldsymbol{x})$ is often called *the point evaluation function at* $\boldsymbol{x}$ in the RKHS literature.

Now, the key ingredient of the RKHS is that rather than considering all of the Hilbert space $\mathcal{H}$, we consider its subset $\mathcal{H}_\phi$ (recall Fig. 4.2) that is generated by the evaluation function $\boldsymbol{\phi}$. More specifically, for all $f(\cdot) \in \mathcal{H}_\phi$ there exists a set $\{\boldsymbol{x}_i\}_{i=1}^n$, $\boldsymbol{x}_i \in \mathcal{X}$ such that

$$f(\cdot) = \sum_{i=1}^{n} \alpha_i \boldsymbol{\phi}(\boldsymbol{x}_i). \tag{4.5}$$

This is equivalent to saying that $\mathcal{H}_\phi$ is a linear span of $\{\boldsymbol{\phi}(\boldsymbol{x}) : \boldsymbol{x} \in \mathcal{X}\}$. Then, by plugging (4.5) into (4.4), we have

$$\begin{aligned} f(\boldsymbol{x}) &= \langle f(\cdot), \boldsymbol{\phi}(\boldsymbol{x}) \rangle_{\mathcal{H}} \\ &= \sum_{i=1}^{n} \alpha_i \langle \boldsymbol{\phi}(\boldsymbol{x}_i), \boldsymbol{\phi}(\boldsymbol{x}) \rangle_{\mathcal{H}} \\ &= \sum_{i=1}^{n} \alpha_i k(\boldsymbol{x}_i, \boldsymbol{x}). \end{aligned} \tag{4.6}$$

As a special case, we can easily see that the coordinate of a kernel in the feature space, $k(\boldsymbol{x}', \cdot)$ for a given $\boldsymbol{x}' \in \mathcal{X}$, lives in an RKHS $\mathcal{H}_\phi$, since we have

$$k(\boldsymbol{x}', \boldsymbol{x}) = \langle k(\boldsymbol{x}', \cdot), \boldsymbol{\phi}(\boldsymbol{x}) \rangle_{\mathcal{H}} = \langle \boldsymbol{\phi}(\boldsymbol{x}'), \boldsymbol{\phi}(\boldsymbol{x}) \rangle, \tag{4.7}$$

where the last equality comes from the definition of a kernel. Therefore, we can see that

$$k(\boldsymbol{x}', \cdot) = \boldsymbol{\phi}(\boldsymbol{x}'), \tag{4.8}$$

which corresponds to (4.5) with $n = 1$. Accordingly, we can write a kernel in terms of a inner product in the underlying Hilbert space:

$$k(\boldsymbol{x}, \boldsymbol{x}') = \langle k(\boldsymbol{x}, \cdot), k(\boldsymbol{x}', \cdot)\rangle_{\mathcal{H}}. \tag{4.9}$$

Furthermore, we can write (4.4) as follows:

$$f(\boldsymbol{x}) = \langle f(\cdot), k(\boldsymbol{x}, \cdot)\rangle_{\mathcal{H}}, \tag{4.10}$$

which is known as the *reproducing property* [11].

As such, for all $f(\cdot), g(\cdot) \in \mathcal{H}_\phi$ we can show that there exist $\{\alpha_i\}_{i=1}^r$ and $\{\beta_i\}_{i=1}^s$ such that $f(\cdot) = \sum_{i=1}^r \alpha_i k(\boldsymbol{x}_i, \cdot)$ and $g(\cdot) = \sum_{i=1}^s \beta_i k(\boldsymbol{x}_i, \cdot)$, since $\boldsymbol{\phi}(\boldsymbol{x}) = k(\boldsymbol{x}, \cdot)$. Therefore, we often interchangeably use $\mathcal{H}_k$ to denote $\mathcal{H}_\phi$ if the kernel $k(\boldsymbol{x}, \boldsymbol{x}')$ is specified. This leads to the explicit representation of their inner product:

$$\langle f, g\rangle_{\mathcal{H}} = \sum_{i=1}^r \sum_{j=1}^s \alpha_i \beta_j \langle k(\boldsymbol{x}_i, \cdot), k(\boldsymbol{x}_i', \cdot)\rangle \tag{4.11}$$

$$= \sum_{i=1}^r \sum_{j=1}^s \alpha_i \beta_j k(\boldsymbol{x}_i, \boldsymbol{x}_j'). \tag{4.12}$$

The induced norm is then defined by

$$\|f\|_{\mathcal{H}} = \sqrt{\langle f, f\rangle_{\mathcal{H}}} = \sum_{i=1}^r \sum_{j=1}^r \alpha_i \alpha_j k(\boldsymbol{x}_i, \boldsymbol{x}_j'). \tag{4.13}$$

By summarizing these findings, we are ready to provide an intuitive definition of RKHS.

Definition 4.3 Let $k : \mathcal{X} \times \mathcal{X} \mapsto \mathbb{R}$ be a positive definite kernel. The RKHS, $\mathcal{H}_k$, generated by the kernel k, is a linear span of $\{k(\boldsymbol{x}, \cdot) : \boldsymbol{x} \in \mathcal{X}\}$ equipped with the inner product

$$\langle f, g\rangle_{\mathcal{H}} = \sum_{i=1}^r \sum_{j=1}^s \alpha_i \beta_j k(\boldsymbol{x}_i, \boldsymbol{x}_j'), \tag{4.14}$$

where $f(\cdot) = \sum_{i=1}^r \alpha_i k(\boldsymbol{x}_i, \cdot)$ and $g(\cdot) = \sum_{i=1}^s \beta_i k(\boldsymbol{x}_i', \cdot)$.

From the (classical) machine learning perspective, the most important reason to use the RKHS is Eq. (4.5), which states that the feature map of the target function can be represented as a *linear* span of $\{k(\boldsymbol{x}, \cdot) : \boldsymbol{x} \in \mathcal{X}\}$ or, equivalently, $\{\boldsymbol{\phi}(\boldsymbol{x}) : \boldsymbol{x} \in \mathcal{X}\}$. This implies that as long as we have a sufficient number of training data, we can estimate the target function by estimating their feature space coordinates.

In fact, one of the important breakthroughs of the modern neural network approach is to relax the assumption that the feature map of the target function should be represented as a linear span. This issue will be discussed in detail later.

4.3 Representer Theorem

Given the definition of kernels and the RKHS, the representer theorem is a simple consequence. Recall that in machine learning problems, the loss is defined as the error energy between the actual target and the estimated one. For example, in the linear regression problem, the MSE loss for the given training data $\{\boldsymbol{x}_i, y_i\}_{i=1}^n$ is defined by

$$\ell_2\left(\{\boldsymbol{x}_i, y_i, f(\boldsymbol{x}_i)\}_{i=1}^n\right) = \sum_{i=1}^n \|y_i - f(\boldsymbol{x}_i)\|^2, \tag{4.15}$$

where

$$f(\boldsymbol{x}_i) = \langle \boldsymbol{x}_i, \boldsymbol{\beta} \rangle,$$

with $\boldsymbol{\beta}$ being the unknown parameter to estimate. In the soft-margin SVM, the loss is given by the hinge loss:

$$\ell_{hinge}\left(\{\boldsymbol{x}_i, y_i, f(\boldsymbol{x}_i)\}_{i=1}^n\right) = \sum_{i=1}^n \max\{0, 1 - y_i f(\boldsymbol{x}_i)\}, \tag{4.16}$$

where

$$f(\boldsymbol{x}_i) = \langle \boldsymbol{w}, \boldsymbol{x}_i \rangle + b,$$

with $\boldsymbol{w}$ and b denoting the parameters to estimate. For the general loss function, the celebrated representer theorem is given as follows:

Theorem 4.1 *[11, 15] Consider a positive definite real-valued kernel* $k : \mathcal{X} \times \mathcal{X} \mapsto \mathbb{R}$ *on a non-empty set* $\mathcal{X}$ *with the corresponding RKHS* $\mathcal{H}_k$. *Let there be given training data set* $\{\boldsymbol{x}_i, y_i\}_{i=1}^n$ *with* $\boldsymbol{x}_i \in \mathcal{X}$ *and* $y_i \in \mathbb{R}$ *and a strictly increasing real-valued regularization function* $R : [0, \infty) \mapsto \mathbb{R}$. *Then, for arbitrary loss function*

$\ell\left(\{\boldsymbol{x}_i, y_i, f(\boldsymbol{x}_i)\}_{i=1}^n\right)$, *any minimizer for the following optimization problem:*

$$f^* = \arg\min_{f \in \mathcal{H}_k} \ell\left(\{\boldsymbol{x}_i, y_i, f(\boldsymbol{x}_i)\}_{i=1}^n\right) + R(\|f\|_{\mathcal{H}}) \tag{4.17}$$

admits a representation of the form

$$f^*(\cdot) = \sum_{i=1}^n \alpha_i k(\boldsymbol{x}_i, \cdot) = \sum_{i=1}^n \alpha_i \boldsymbol{\phi}(\boldsymbol{x}_i) \tag{4.18}$$

for some $\alpha_i \in \mathbb{R}, i = 1, \cdots, n$*; or it is equivalently represented by*

$$f^*(\boldsymbol{x}) = \sum_{i=1}^n \alpha_i k(\boldsymbol{x}_i, \boldsymbol{x}). \tag{4.19}$$

The proof of the representer theorem can easily be found in the standard machine learning textbook [11], so we do not revisit it here. Instead, we briefly touch upon the main idea of the proof, since it also highlights the limitations of kernel machines. Specifically, the feature space coordinate of the minimizer f^*, denoted by $f^*(\cdot)$, should be represented by the linear combination of the feature maps from the training data $\{\boldsymbol{\phi}(\boldsymbol{x}_i)\}_{i=1}^n$ and its orthogonal complement. But when we perform the point evaluation with $\{\boldsymbol{\phi}(\boldsymbol{x}_i)\}_{i=1}^n$ using the inner product during the training phase, the contribution from the orthogonal complement disappears, which leads to the final form in (4.18).

4.4 Application of Representer Theorem

In this section, we revisit the kernel SVM and regression to show how the representer theorem can simplify the derivation.

4.4.1 Kernel Ridge Regression

Recall that the ridge regression was given by the following optimization problem:

$$\min_{\boldsymbol{\beta}} \sum_{i=1}^n \|y_i - \langle \boldsymbol{x}_i, \boldsymbol{\beta}\rangle\|^2 + \lambda\|\boldsymbol{\beta}\|^2.$$

By extending this in nonparameteric form, the kernel ridge regression is given by the following minimization problem:

$$\min_{f\in\mathcal{H}_k}\sum_{i=1}^{n}\|y_i-f(\boldsymbol{x}_i)\|^2+\lambda\|f\|_{\mathcal{H}}^2, \tag{4.20}$$

where $\mathcal{H}_k$ is the RKHS with the positive definite kernel k. From Theorem 4.1, we know that the minimizer should have the form

$$f(\cdot)=\sum_{j=1}^{n}\alpha_j\boldsymbol{\phi}(\boldsymbol{x}_j). \tag{4.21}$$

Using (4.4), the MSE loss becomes

$$\begin{aligned}\sum_{i=1}^{n}\|y_i-f(\boldsymbol{x}_i)\|^2&=\sum_{i=1}^{n}\|y_i-\langle f(\cdot),\boldsymbol{\phi}(\boldsymbol{x}_i)\rangle\|^2\\&=\sum_{i=1}^{n}\|y_i-\sum_{j=1}^{n}\alpha_j\langle\boldsymbol{\phi}(\boldsymbol{x}_j),\boldsymbol{\phi}(\boldsymbol{x}_i)\rangle\|^2\\&=\sum_{i=1}^{n}\|y_i-\sum_{j=1}^{n}\alpha_jk(\boldsymbol{x}_j,\boldsymbol{x}_i)\|^2\\&=\|\boldsymbol{y}-\boldsymbol{K}\boldsymbol{\alpha}\|^2,\end{aligned}$$

where $\boldsymbol{K}\in\mathbb{R}^{n\times n}$ denotes the kernel Gram matrix given by

$$\boldsymbol{K}=\begin{bmatrix}k(\boldsymbol{x}_1,\boldsymbol{x}_i)&\cdots&k(\boldsymbol{x}_1,\boldsymbol{x}_n)\\\vdots&\ddots&\vdots\\k(\boldsymbol{x}_n,\boldsymbol{x}_1)&\cdots&k(\boldsymbol{x}_n,\boldsymbol{x}_n)\end{bmatrix} \tag{4.22}$$

and

$$\boldsymbol{y}=\begin{bmatrix}y_1&\cdots&y_n\end{bmatrix}^\top,\quad\boldsymbol{\alpha}=\begin{bmatrix}\alpha_1&\cdots&\alpha_n\end{bmatrix}^\top. \tag{4.23}$$

Similarly, the regularization term becomes

$$\begin{aligned}\|f\|_{\mathcal{H}}^2&=\langle f(\cdot),f(\cdot)\rangle\\&=\sum_{i=1}^{n}\sum_{j=1}^{n}\alpha_i\alpha_j\langle\boldsymbol{\phi}(\boldsymbol{x}_i),\boldsymbol{\phi}(\boldsymbol{x}_j)\rangle\end{aligned}$$

$$= \sum_{i=1}^{n} \sum_{j=1}^{n} \alpha_i \alpha_j k(\boldsymbol{x}_i, \boldsymbol{x}_j)$$
$$= \boldsymbol{\alpha}^{\top} \boldsymbol{K} \boldsymbol{\alpha}.$$

Therefore, (4.20) can be equivalently represented by the finite dimensional optimization problem:

$$\hat{\boldsymbol{\alpha}} := \arg \min_{\boldsymbol{\alpha} \in \mathbb{R}^n} \|\boldsymbol{y} - \boldsymbol{K}\boldsymbol{\alpha}\|^2 + \lambda \boldsymbol{\alpha}^{\top} \boldsymbol{K} \boldsymbol{\alpha}. \tag{4.24}$$

The problem is convex; so using the first order necessary condition, we have

$$(\boldsymbol{K}^2 + \lambda \boldsymbol{K}) \hat{\boldsymbol{\alpha}} = \boldsymbol{K} \boldsymbol{y}.$$

where we use $\boldsymbol{K}^{\top} = \boldsymbol{K}$ due to the symmetry of the Gram matrix. If $\boldsymbol{K}$ is invertible (which is usually the case for the standard choice of kernels), we have

$$\hat{\boldsymbol{\alpha}} = (\boldsymbol{K} + \lambda \boldsymbol{I})^{-1} \boldsymbol{y}.$$

Finally, using (4.4) and (4.21) we have

$$f^*(\boldsymbol{x}) = \langle f(\cdot), \boldsymbol{\phi}(\boldsymbol{x}) \rangle$$
$$= \sum_{i=1}^{n} \alpha_i \langle \boldsymbol{\phi}(\boldsymbol{x}_i), \boldsymbol{\phi}(\boldsymbol{x}) \rangle$$
$$= \left[k(\boldsymbol{x}_1, \boldsymbol{x}) \cdots k(\boldsymbol{x}_n, \boldsymbol{x}) \right] (\boldsymbol{K} + \lambda \boldsymbol{I})^{-1} \boldsymbol{y},$$

which is what we obtained before.

4.4.2 Kernel SVM

Recall that the soft-margin SVM formulation (without bias) can be represented by

$$\min_{\boldsymbol{w}} \frac{1}{2} \|\boldsymbol{w}\|^2 + C \sum_{i=1}^{n} \ell_{hinge} \left(y_i, \langle \boldsymbol{w}, \boldsymbol{x}_i \rangle \right), \tag{4.25}$$

where ℓ_{hinge} is the hinge loss

$$\ell_{hinge}(y, \hat{y}) = \max\{0, 1 - y\hat{y}\}. \tag{4.26}$$

This problem can be solved using the representer theorem. Specifically, an extended formulation of (4.25) in the RKHS is given by

$$\min_{f \in \mathcal{H}_k} \frac{1}{2}\|f\|_{\mathcal{H}}^2 + C \sum_{i=1}^{n} \ell_{hinge}\left(y_i, f(\boldsymbol{x}_i)\right), \tag{4.27}$$

whose minimizer f has the following coordinate in the feature space:

$$f(\cdot) = \sum_{j=1}^{n} \alpha_j k(\boldsymbol{x}_j, \cdot). \tag{4.28}$$

Using this, the hinge loss term becomes

$$\ell_{hinge}\left(y_i, f(\boldsymbol{x}_i)\right) = \max\{0, 1 - y_i \sum_{j=1}^{n} \alpha_j k(\boldsymbol{x}_j, \boldsymbol{x}_i)\}. \tag{4.29}$$

Similarly, the regularization term becomes

$$\|f\|_{\mathcal{H}}^2 = \boldsymbol{\alpha}^\top \boldsymbol{K}\boldsymbol{\alpha},$$

where $\boldsymbol{K}$ is the kernel Gram matrix in (4.22). Now, (4.27) can be represented in an constrained form

$$\begin{aligned} \min_{\boldsymbol{\alpha},\boldsymbol{\xi}} \quad & \frac{1}{2}\boldsymbol{\alpha}^\top \boldsymbol{K}\boldsymbol{\alpha} + C\sum_{i=1}^{n} \xi_i \\ \text{subject to } & 1 - y_i \sum_{j=1}^{n} \alpha_j k(\boldsymbol{x}_j, \boldsymbol{x}_i) \leq \xi_i, \\ & \xi_i \geq 0, \quad \forall i. \end{aligned} \tag{4.30}$$

For the given primal problem in (4.30), the corresponding Lagrangian dual is given by

$$\begin{aligned} \max_{\boldsymbol{\lambda},\boldsymbol{\gamma}} \quad & g(\boldsymbol{\lambda}, \boldsymbol{\gamma}) \\ \text{subject to} \quad & \boldsymbol{\lambda} \geq 0, \ \boldsymbol{\gamma} \geq 0, \end{aligned} \tag{4.31}$$

$$g(\boldsymbol{\lambda}, \boldsymbol{\gamma}) = \min_{\boldsymbol{\alpha}, \boldsymbol{\xi}} \left\{ \frac{1}{2}\boldsymbol{\alpha}^\top \boldsymbol{K}\boldsymbol{\alpha} + C\sum_{i=1}^n \xi_i \right. \tag{4.32}$$

$$\left. + \sum_{i=1}^n \lambda_i (1 - y_i \sum_{j=1}^n \alpha_j k(\boldsymbol{x}_j, \boldsymbol{x}_i) - \xi_i) - \sum_{i=1}^n \gamma_i \xi_i \right\},$$

which can be further simplified as

$$g(\boldsymbol{\lambda}, \boldsymbol{\gamma}) = \min_{\boldsymbol{\alpha}, \boldsymbol{\xi}} \left\{ \frac{1}{2}\boldsymbol{\alpha}^\top \boldsymbol{K}\boldsymbol{\alpha} + \sum_{i=1}^n \lambda_i (1 - \xi_i) + (C - \gamma_i)\xi_i - \boldsymbol{r}^\top \boldsymbol{K}\boldsymbol{\alpha} \right\}, \tag{4.33}$$

where

$$\boldsymbol{r} = \begin{bmatrix} y_1\lambda_1 & \cdots & y_n\lambda_n \end{bmatrix}^\top .$$

The first-order optimality conditions with respect to $\boldsymbol{\alpha}$ and $\boldsymbol{\xi}$ lead to the following equations:

$$\boldsymbol{K}\boldsymbol{\alpha} = \boldsymbol{K}\boldsymbol{r} \Longrightarrow \boldsymbol{\alpha} = \boldsymbol{r} \tag{4.34}$$

and

$$\lambda_i + \gamma_i = C. \tag{4.35}$$

By plugging (4.34) and (4.35) into Eq. (4.32), we have

$$g(\boldsymbol{\lambda}, \boldsymbol{\gamma}) = \sum_{i=1}^n \lambda_i - \frac{1}{2}\sum_{i=1}^n \sum_{j=1}^n \lambda_i \lambda_j y_i y_j k(\boldsymbol{k}_i, \boldsymbol{k}_j)$$

where $0 \leq \lambda_i \leq C$ and the classifier is given by

$$f(\boldsymbol{x}) = \sum_{j=1}^n y_j \lambda_j k(\boldsymbol{x}_j, \boldsymbol{x}), \tag{4.36}$$

which is equivalent to the kernel SVM we derived before.

4.5 Pros and Cons of Kernel Machines

The kernel machine has many important advantages that deserve further discussion. This approach is based on the beautiful theory of the RKHS, which leads to the closed form solution in designing classifiers and regressors thanks to the representer theorem. Therefore, the classical research issue is not about the machine learning algorithm itself, but rather to find the feature space embedding that can effectively represent the data in the ambient space.

Having said this, there are several limitations associated with the classical kernel machines. First, the reason that enables a closed form solution in terms of the representer theorem is the assumption that the feature space forms an RKHS. This implies that the mapping from the feature space to the final function is assumed to be linear. This approach is somewhat unbalanced given that only the mapping from the ambient space to feature space is nonlinear, whereas the feature space representation is linear. Moreover, as discussed before, the RKHS is only a subset of underlying Hilbert space; therefore, restricting feature space within the RKHS severely reduces available function class from the underlying Hilbert space (see Fig. 4.2). As such, it limits the flexibility of the learning algorithm and resulting expressiveness.

Finally, the feature mapping and the associated kernel in the classical machine learning approach are primarily selected in a top-down manner based on human intuition or mathematical modeling that has no space that can be automatically learned from the data. In fact, the learning part of the kernel machine is for the linear weighting parameters in the representer (i.e. α_i's in (4.18)), whereas the feature map itself is deterministic once the kernel is selected in a top-down manner. This significantly limits the capability of learning. Later, we will investigate how this limitation of the kernel machine can be mitigated by modern deep learning approaches.

4.6 Exercises

1. Show that the following kernels are positive definite.

   a. Cosine kernel: $k(x, y) = \cos(x - y)$ for $\forall x, y \in \mathbb{R}$.
   b. Polynomial kernel with degree exactly p:

   $$k(\boldsymbol{x}, \boldsymbol{y}) = (\boldsymbol{x}^\top \boldsymbol{y})^p.$$

   c. Polynomial kernel with degree up to p:

   $$k(\boldsymbol{x}, \boldsymbol{y}) = (\boldsymbol{x}^\top \boldsymbol{y} + 1)^p.$$

   d. Radial basis function kernel with width σ:

   $$k(\boldsymbol{x}, \boldsymbol{y}) = \exp(-\|\boldsymbol{x} - \boldsymbol{y}\|^2/(2\sigma^2)).$$

   e. Sigmoid kernel:

   $$\tanh(\eta \boldsymbol{x}^\top \boldsymbol{y} + \nu).$$

2. Let k_1 and k_2 be two positive definite kernels on a set $\mathcal{X}$, and α, β two positive scalars. Show that $\alpha k_1 + \beta k_2$ is positive definite.
3. Let k_1 be a positive definite kernel on a set $\mathcal{X}$. Then, for any polynomial $p(\cdot)$ with non-negative coefficients, show that the following is also a positive definite kernel on a set $\mathcal{X}$:

$$k(x, y) = p(k_1(x, y)), \quad x, y \in \mathcal{X}.$$

4. Let $\{\mathcal{X}_i\}_{i=1}^{p}$ be a sequence of sets and k_i be a collection of corresponding positive definite functions on $\mathcal{X}_i$. Then, show that

$$k\left(x_1, \cdots, x_p; y_1, \cdots, y_p\right) = k_1(x_1, y_1) \cdots k_p(x_p, y_p), \quad x_i, y_i \in \mathcal{X}_i, \forall i$$

is a kernel on the space $\mathcal{X} := \mathcal{X}_1 \times \cdots \mathcal{X}_p$.
5. Let $X_0 \subset X$, then the restriction of k to $X_0 \times X_0$ is also a reproducing kernel.
6. Let k be a valid kernel on $\mathcal{X}$. Is the following normalized function a valid positive definite kernel?

$$k_{norm}(x, y) = \begin{cases} 0, & \text{if } k(x, x) = 0 \text{ or } k(y, y) = 0 \\ \frac{k(x,y)}{\sqrt{k(x,x)}\sqrt{k(y,y)}}, & \text{otherwise} \end{cases}, \quad \forall x, y \in \mathcal{X}.$$

7. Consider a normalized kernel k such that $k(x, x) = 1$ for all $x \in \mathcal{X}$. Define a pseudo-metric on $\mathcal{X}$ as

$$d_{\mathcal{X}}(x, y) = \|k(x, \cdot) - k(y, \cdot)\|_{\mathcal{H}}. \tag{4.37}$$

   a. Show that

$$d_{\mathcal{X}}(x, y) = 2(1 - k(x, y)).$$

   b. Show that $d_{\mathcal{X}}(x, y)$ is not a metric. Which property of the metric does it violate?

8. Define the mean of the feature space

$$\boldsymbol{\mu}_\phi = \frac{1}{n}\sum \boldsymbol{\phi}(\boldsymbol{x}_i).$$

   a. Show that

$$\|\boldsymbol{\mu}_\phi\|_{\mathcal{H}}^2 = \frac{1}{n^2}\sum_{i=1}^{n}\sum_{j=1}^{n} k(\boldsymbol{x}_i, \boldsymbol{x}_j).$$

b. Show that

$$\sigma_{\phi}^{2} := \frac{1}{n}\sum_{i=1}^{n} \|\boldsymbol{\phi}(\boldsymbol{x}_i) - \boldsymbol{\mu}_{\phi}\|_{\mathcal{H}}^{2} = \frac{1}{n}\mathrm{Tr}(\boldsymbol{K}) - \|\boldsymbol{\mu}_{\phi}\|_{\mathcal{H}}^{2},$$

where $\mathrm{Tr}(\cdot)$ denotes the matrix trace, and $\boldsymbol{K}$ is the kernel Gram matrix

$$\boldsymbol{K} = \begin{bmatrix} k(\boldsymbol{x}_1, \boldsymbol{x}_i) & \cdots & k(\boldsymbol{x}_1, \boldsymbol{x}_n) \\ \vdots & \ddots & \vdots \\ k(\boldsymbol{x}_n, \boldsymbol{x}_1) & \cdots & k(\boldsymbol{x}_n, \boldsymbol{x}_n) \end{bmatrix}.$$

9. The kernel SVM formulation in (4.27) is often called the 1-SVM. In this problem, we are interested in obtaining the 2-SVM, which is defined by

$$\min_{f\in\mathcal{H}_k} \frac{1}{2}\|f\|_{\mathcal{H}}^{2} + C\sum_{i=1}^{n} \ell_{hinge}^{2}\left(y_i, f(\boldsymbol{x}_i)\right),$$

where ℓ_{hinge}^{2} is the *square* hinge loss:

$$\ell_{hinge}^{2}(y, \hat{y}) = \left(\max\{0, 1 - y\hat{y}\}\right)^{2}.$$

Write the primal and dual problems associated with the 2-SVM, and compare the result with the 1-SVM.

10. Consider the following kernel regression problem:

$$\min_{f\in\mathcal{H}_k} \frac{1}{2}\|f\|_{\mathcal{H}}^{2} + C\sum_{i=1}^{n} \ell_{logit}\left(y_i, f(\boldsymbol{x}_i)\right),$$

where ℓ_{logit} is the logistic regression loss:

$$\ell_{logit}(y, \hat{y}) = \log(1 + e^{-y\hat{y}}).$$

Write the dual problems and find the solution as simply as possible.

Part II
Building Blocks of Deep Learning

"I get very excited when we discover a way of making neural networks better and when that's closely related to how the brain works."

– Geoffrey Hinton

Chapter 5
Biological Neural Networks

5.1 Introduction

A biological neural network is composed of a group of connected neurons. A single neuron may be connected to many other neurons and the total number of neurons and connections in a network may be significantly high. One of the amazing aspects of biological neural networks is that when the neurons are connected to each other, higher-level intelligence, which cannot be observed from a single neuron, emerges. The exact mechanism of the emergence of intelligence from the neuronal network has been an intense research topic for neuroscientists, biologists, and engineers, and is not yet fully understood. In fact, computational modeling and mathematical analysis of biological neural networks are integral parts of the neuroscience discipline called computational neuroscience, which is also closely related to the artificial neural network community. The main assumption in this discipline is that through the computational modeling the probable working mechanism of the biological network can be unveiled. Moreover, understanding the working principles of biological neuronal networks has been believed to open the horizon to designing high-performance artificial neuronal networks.

Therefore, in this chapter, we will review the basic neurobiology regarding individual neurons and their networks, and introduce some interesting neuro-scientific discoveries that have inspired artificial neural networks. However, these introductory materials are by no means extensive, so interested readers are advised to read standard textbooks in neuroscience [17–19].

J. C. Ye, *Geometry of Deep Learning*, Mathematics in Industry 37,
https://doi.org/10.1007/978-981-16-6046-7_5

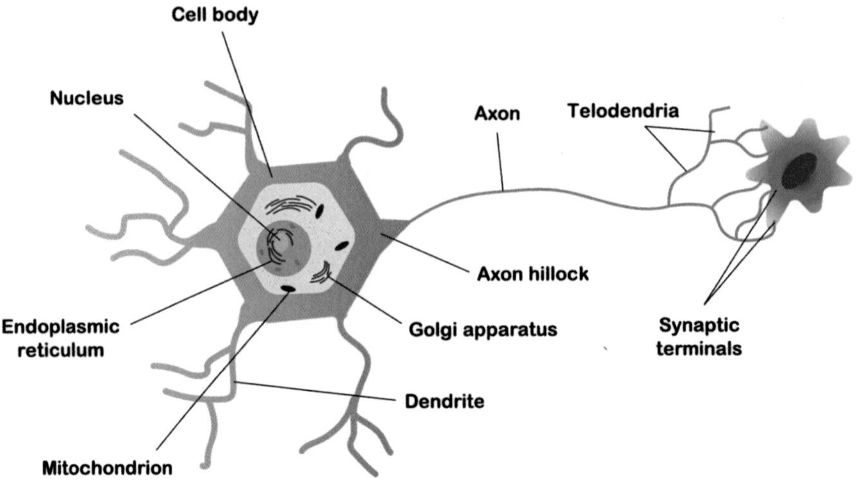

Fig. 5.1 Anatomy of neurons

5.2 Neurons

5.2.1 Anatomy of Neurons

A typical neuron consists of a cell body (soma), dendrites, and a single axon (see Fig. 5.1). The axon and dendrites are filaments that extrude from the cell body. Dendrites typically branch heavily and extend a few hundred micrometers from the soma. The axon leaves the soma at the *axon hillock,* and moves up to 1 m in humans or more in other species. The end branches of an axon are called telodendria. At the extreme tip of the axon's branches are synaptic terminals, where the neuron can transmit a signal to another cell via the synapse.

The endoplasmic reticulum (ER) in the soma performs many general functions, including folding protein molecules and transporting synthesized proteins in vesicles to the Golgi apparatus. Proteins synthesized in the ER are packaged into vesicles, which then fuse with the Golgi apparatus. These cargo proteins are modified in the Golgi apparatus and destined for secretion via exocytosis or for use in the cell as shown in Fig. 5.2.

5.2.2 Signal Transmission Mechanism

Neurons specialize in forwarding signals to individual target cells via synapses. At a synapse, the membrane of the presynaptic neuron comes into close proximity to the membrane of the postsynaptic cell (see Fig. 5.3). Although there are electric synapses where the presynaptic and postsynaptic neurons are directly fused together

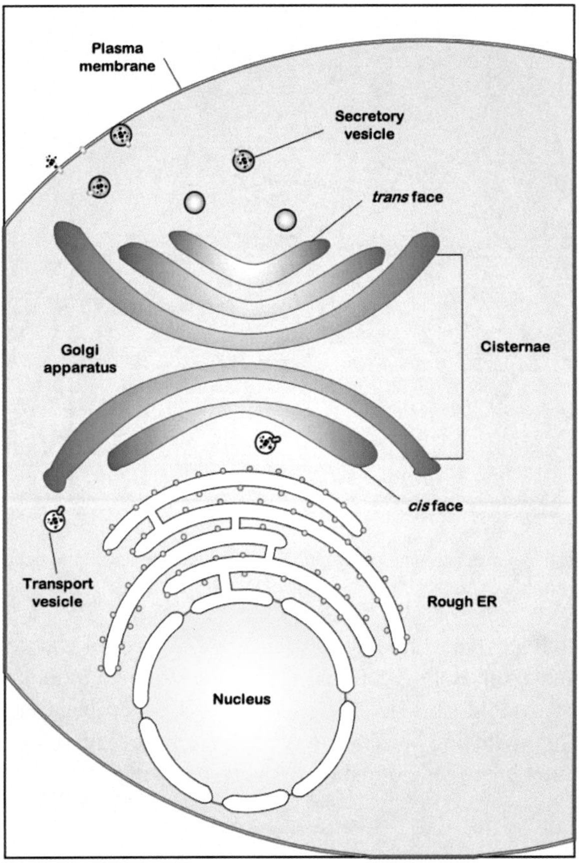

Fig. 5.2 ER and Golgi apparatus for protein synthesis and transport

for fast electric signal transmission [18, 19], chemical synapses, which transmit the action potential via neurotransmitters, are the most common and are of great interest for artificial neural networks.

As shown in Fig. 5.3, in a chemical synapse, electrical activity in the presynaptic neuron is converted into the release of neurotransmitters that bind to receptors located in the membrane of the postsynaptic cell. The neurotransmitters are usually packaged in a synaptic vesicle, as shown in Fig. 5.3. Therefore, the amount of the actual neurotransmitter at the postsynaptic terminal is an integer multiple of the number of neurotransmitters in each vesicle, so this phenomenon is often referred to as *quantal release*. The release is regulated by a voltage-dependent calcium channel. The released neurotransmitter then binds to the receptors on the postsynaptic dendrites, which can trigger an electrical response that can produce excitatory postsynaptic potentials (EPSPs) or inhibitory postsynaptic potentials (IPSPs).

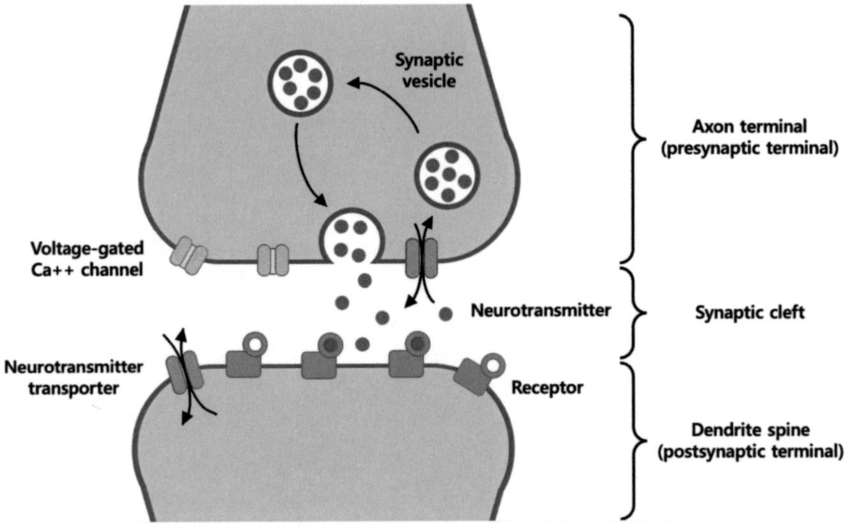

Fig. 5.3 Chemical synapse between presynaptic terminal and postsynaptic dendrite

The axon hillock (see Fig. 5.1) is a specialized part of the cell body that is connected to the axon. Both IPSPs and EPSPs are summed in the axon hillock and once a trigger threshold is exceeded, an action potential propagates through the rest of the axon. This switching behavior of the axon hillock plays a very important role in the information processing of neural networks, as will be discussed in detail later in Chap. 6.

5.2.3 Synaptic Plasticity

Synaptic plasticity is the ability of synapses to strengthen or weaken over time as their activity increases or decreases. In fact, synaptic plasticity is one of the important neurochemical foundations of learning and memory that is often mimicked by artificial neural networks.

Two of the best studied forms of the synaptic plasticity in the neuronal cell are long-term potentiation (LTP) and long-term depression (LTD). Specifically, LTP is a sustained strengthening of the synapses based on recent patterns of activity. These are patterns of synaptic activity that cause a long-lasting increase in signal transmission between two neurons. The opposite of LTP is long-term depression (LTD), which leads to a long-lasting decrease in synaptic strength.

In contrast to the artificial neural network, in which the synaptic plasticity changes are usually modeled by simple weight changes, the synaptic plastic change in biological neurons often results from the change in the number of neurotransmitter receptors located on a synapse. For example, as shown in Fig. 5.4,

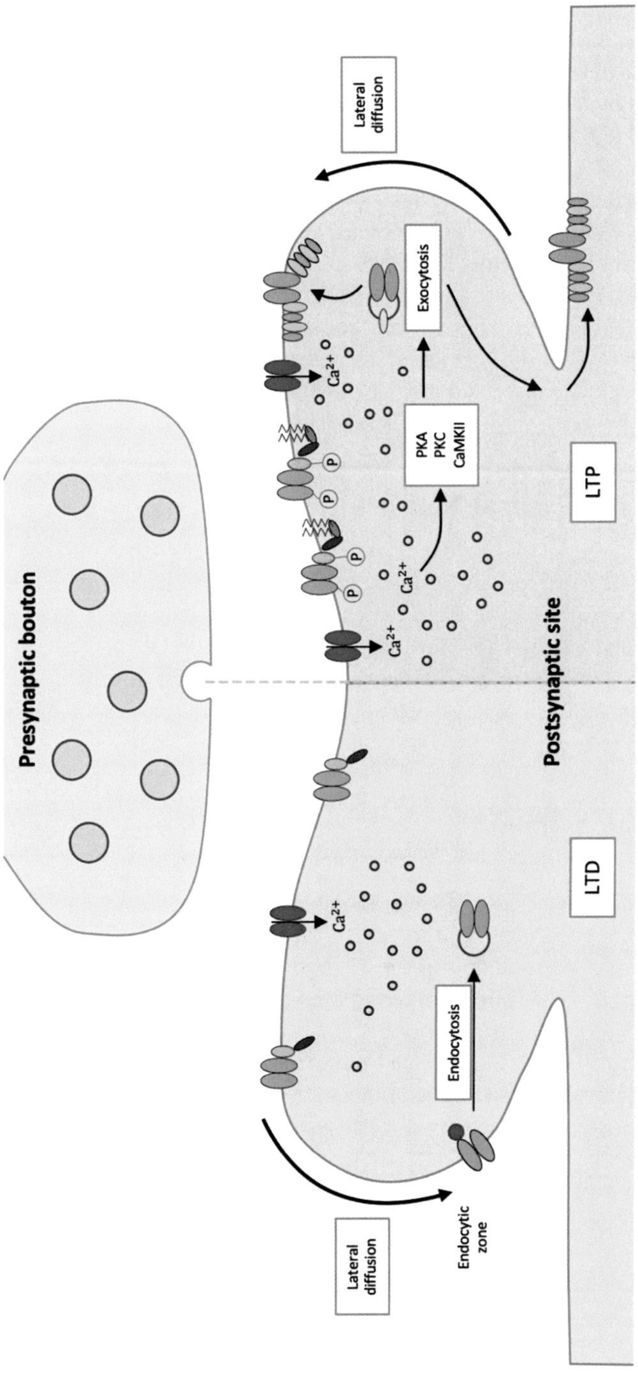

Fig. 5.4 Biological mechanism of LTP and LTD

during the LTP additional receptors are fused to the membrane by exocytosis, which are then moved to the postsynaptic dendrite by lateral diffusion within the membrane. On the other hand, in the case of LTD, some of the redundant receptors are moved into the endocytosis region by lateral diffusion within the membrane, and then absorbed by the cell via endocytosis.

Because of the dynamics of learning and synaptic plasticity, it becomes clear that the trafficking of these receptors is an important mechanism to meet the demand and supply of the receptors at various synaptic locations in the neurons. There are various mechanisms that are being intensively researched by neurobiologists. For example, assembled receptors leave the endoplasmic reticulum (ER) and reach the neural surface via the Golgi network. Packets of nascent receptors are transported along microtubule tracks from the cell body to synaptic sites through microtubule networks. Figure 5.5 shows critical steps in receptor assembly, transport, intracellular trafficking, slow release and insertion at synapses.

5.3 Biological Neural Network

One of the most mysterious features of the brain is the emergence of higher-level information processing from the connections of neurons. To understand this emergent property, one of the most extensively studied biological neural networks is the visual system. Therefore, in this section we review the information processing in the visual system.

5.3.1 Visual System

The visual system is a part of the central nervous system that enables organisms to process visual detail as eyesight. It detects and interprets information from visible light to create a representation of the environment. The visual system performs a number of complex tasks, from capturing light to identifying and categorizing visual objects.

As shown in Fig. 5.6, the reflected light from objects shines on the retina. The retina uses photoreceptors to convert this image into electrical impulses. The optic nerve then carries these impulses through the optic canal. Upon reaching the optic chiasm, the nerve fibers decussate (left becomes right). Most of the optic nerve fibers terminate in the lateral geniculate nucleus (LGN). The LGN forwards the impulses to V1 of the visual cortex. The LGN also sends some fibers to V2 and V3. V1 performs edge detection to understand spatial organization. V1 also creates a bottom-up saliency map to guide attention.

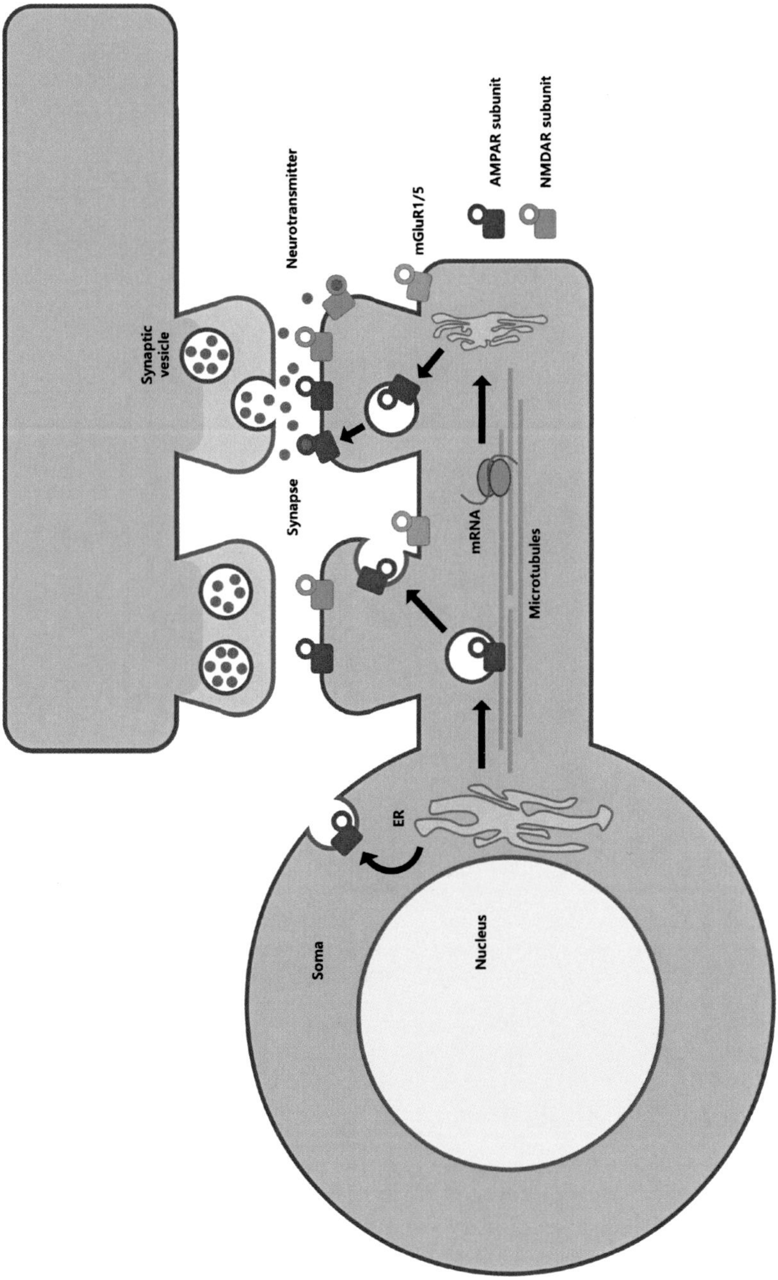

Fig. 5.5 Receptor trafficking for synaptic plasticity

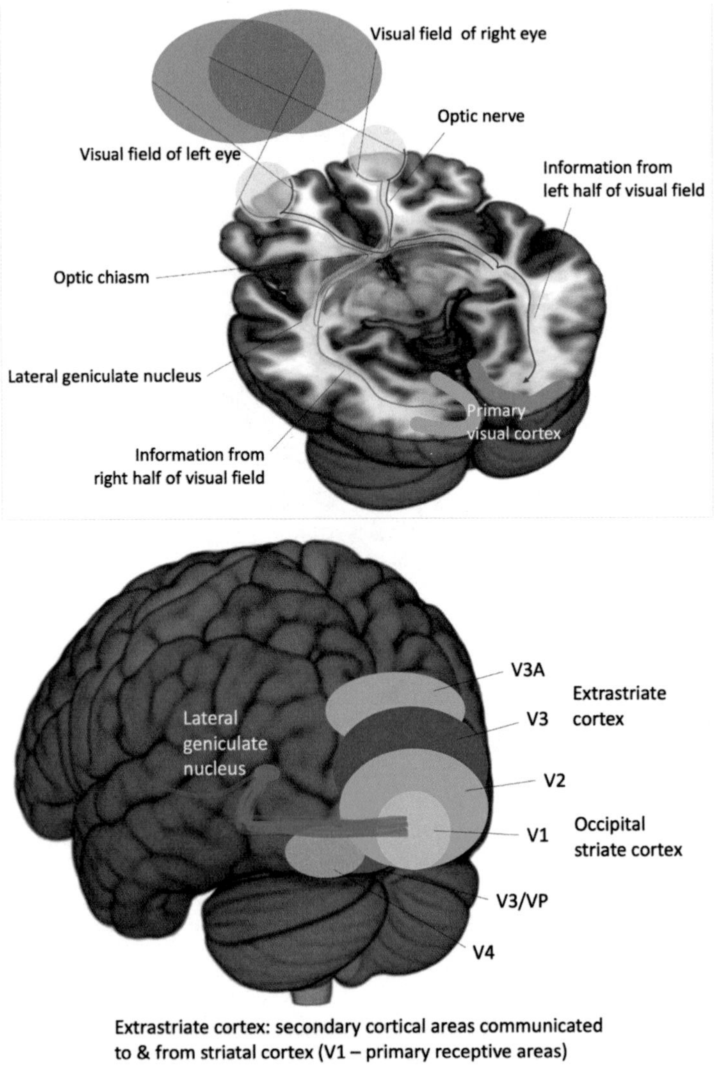

Fig. 5.6 Anatomy of visual system and information processing

5.3.2 *Hubel and Wiesel Model*

One of the most important discoveries of Hubel and Wiesel [20] is the hierarchical visual information flow in the primary visual cortex. Specifically, by examining the primary visual cortex of cats, Hubel and Wiesel found two classes of functional cells in the primary visual cortex: simple cells and complex cells. More specifically, simple cells at V1 L4 respond best to edge-like stimuli with a certain orientation,

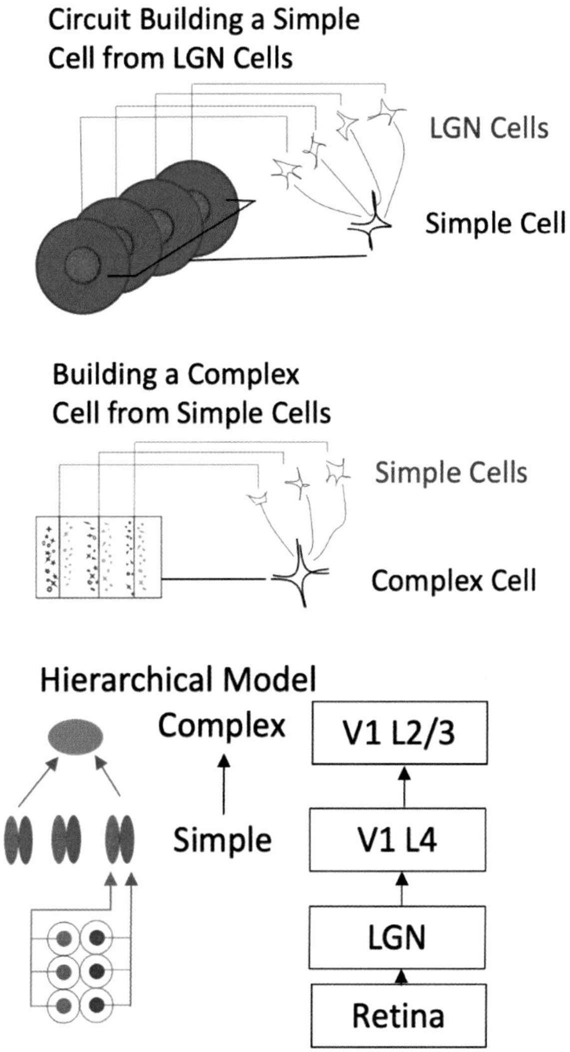

Fig. 5.7 Hubel and Wiesel model for primary visual cortex

position and phase within their relatively small receptive fields (Fig. 5.7). They realized that such a response of the simple cells could be obtained by pooling the activity of a small set of input cells with the same receptive field that is observed in LGN cells. They also observed that complex cells at V1 L2/L3, although selective for oriented bars and edges too, tend to have larger receptive fields and have some tolerance with regard to the exact position within their receptive fields. Hubel and Wiesel found that position tolerance at the complex cell level could be obtained by grouping simple cells at the level below with the same preferred orientation but slightly different positions. As will be discussed later, the operation of pooling LGN cells with the same receptive field is similar to the convolution operation, which

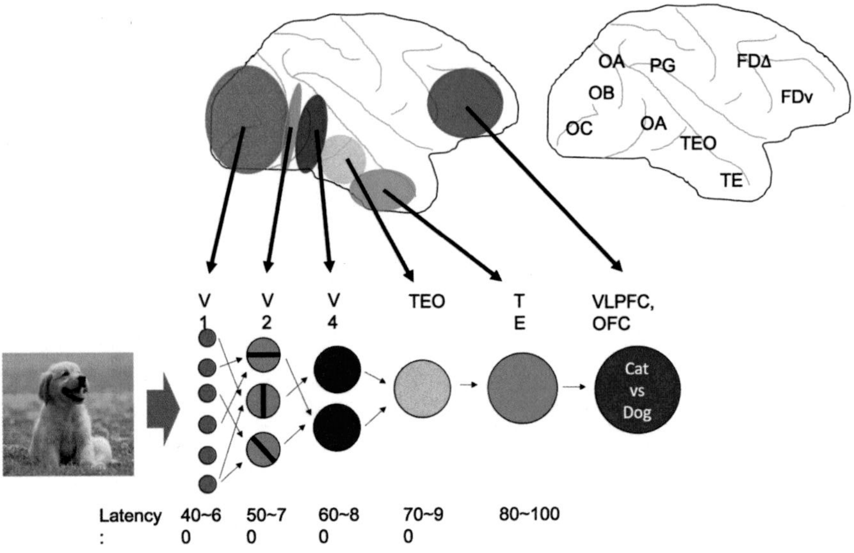

Fig. 5.8 Hierarchical models of visual information processing

inspired Yann LeCun to invent the convolutional neural network for handwritten zip code identification [21].

The extension of these ideas from the primary visual cortex to higher areas of the visual cortex led to a class of object recognition models, the feedforward hierarchical models [22]. Specifically, as shown in Fig. 5.8, as we go from V1 to TE, the size of the receptive field increases and the latency for the response increases. This implies that there is a neuronal connection along this path, which forms a neuronal hierarchy. A more surprising finding is that as we go along this pathway, neurons become sensitive to more complex inputs that are not sensitive to transforms.

5.3.3 Jennifer Aniston Cell

An extreme form or surprising example of this information processing hierarchy can be found in the discovery of the so-called "Jennifer Aniston Cell" [23], which represents a complex but specific concept or object. For those who do not know Jennifer Aniston, she was one of the most popular American actresses of the 1990s, having starred in America's favorite sitcom, *Friends*.

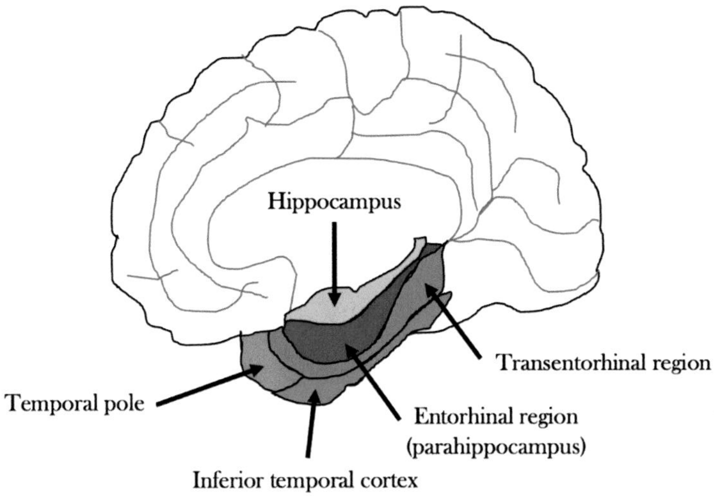

Fig. 5.9 The anatomical location of the medial temporal lobe

The study involved eight epilepsy patients who were temporarily implanted with a single cell recording device to monitor the activity of brain cells in the medial temporal lobe (MTL). The medial temporal lobe contains a system of anatomically related structures that are essential for declarative memory (conscious memory for facts and events). The system consists of the hippocampal region (Cornu Ammonis (CA) fields, dentate gyrus, and subicular complex) and the adjacent perirhinal, entorhinal, and parahippocampal cortices (see Fig. 5.9).

During the single cell recording, the authors in [23] noticed a strange pattern on the medial temporal lobe (MTL) of the brain in one of their participants. Every time the patient saw a picture of Jennifer Aniston, a specific neuron in the brain fired. They tried to show the words "Jennifer Aniston," and again it would fire. They tried other ways to summon Jennifer Aniston in other ways, and each time it fired. The conclusion was inevitable: for this particular person, there was a single neuron that embodied the concept of Jennifer Aniston.

The experiment showed that individual neurons in the MTL respond to the faces of certain people. The researchers say that these types of cell are involved in sophisticated aspects of visual processing, such as identifying a person, rather than just a simple shape. This observation leads to a fundamental question: can a single neuron embody a single concept? Although this issue will be investigated thoroughly throughout the book, the short answer is "no" because it is not the single neuron in isolation, but a neuron from a densely connected neural network that can extract the high-level concept.

5.4 Exercises

1. Explain the role of the following structure in a neuron:
    a. Soma
    b. Dendrite
    c. ER
    d. Golgi apparatus
    e. Axon hillock
    f. Synapse
2. It is important to have a sense for the relative orders of magnitude of cellular components. Please specify each physical parameter for a synapse.
    a. Vesicle diameter
    b. Synapse width
    c. Vesicles released per active zone per action potential
    d. Synaptic cleft width
3. Explain the differences between electrical and chemical synapses.
4. Explain the different types of neurotransmitters and their roles.
5. Explain the differences between ionotropic receptors and metabotropic receptors.
6. Explain the mechanism of LTD and LTP.
7. What is the role of the neurotransmitter trafficking?
8. Explain the visual information processing step by step.
9. Explain why the Hubel and Wiesel model implies the convolutional processing in the visual cortex.
10. What is the main observation from the Jennifer Aniston cell?

Chapter 6
Artificial Neural Networks and Backpropagation

6.1 Introduction

Inspired by the biological neural network, here we discuss its mathematical abstraction known as the artificial neural network (ANN). Although efforts have been made to model all aspects of the biological neuron using a mathematical model, all of them may not be necessary: rather, there are some key aspects that should not be neglected when modeling a neuron. This includes the weight adaptation and the nonlinearity. In fact, without them, we cannot expect any learning behavior.

In this chapter, we first describe a mathematical model for a single neuron, and explain its multilayer realization using a feedforward neural network. We then discuss standard methods of updating weight, often referred to as neural network training. One of the most important parts of neural network training is gradient computation, so the rest of this chapter discusses the main weight update techniques known as *backpropagation* in detail.

6.2 Artificial Neural Networks

6.2.1 Notation

Since the mathematical description of an artificial neural network involves several indices for neuron, layers, training sample, etc., here we would like to summarize them for reference so that they can be used in the rest of the chapter.

First, each training data set is usually represented as bold face lower case letters with the index n: for example, the following are used to indicate the n-th training-data-related variables:

$$\boldsymbol{x}_n, \boldsymbol{y}_n, \{\boldsymbol{x}_n, \boldsymbol{y}_n\}_{n=1}^N, \boldsymbol{o}_n, \boldsymbol{g}_n.$$

J. C. Ye, *Geometry of Deep Learning*, Mathematics in Industry 37,
https://doi.org/10.1007/978-981-16-6046-7_6

Second, with a slight abuse of notation, the subscript i and j for the light face lower-case letters denotes the i-th and j-th element of a vector: for example, o_i is the i-th element of the vector $\boldsymbol{o} \in \mathbb{R}^d$:

$$o_i = [\boldsymbol{o}]_i, \quad \text{or} \quad \boldsymbol{o} = \begin{bmatrix} o_1 \cdots o_d \end{bmatrix}^\top.$$

Similarly, the double index ij indicates the (i, j) element of a matrix: for example, w_{ij} is the (i, j)-th element of a matrix $\boldsymbol{W} \in \mathbb{R}^{p \times q}$:

$$w_{ij} = [\boldsymbol{W}]_{i,j} \quad \text{or} \quad \boldsymbol{W} = \begin{bmatrix} w_{11} & \cdots & w_{1q} \\ \vdots & \ddots & \vdots \\ w_{p1} & \cdots & w_{pq} \end{bmatrix}.$$

This index notation is often used to refer to the i-th or j-th neuron in each layer of a neural network. To avoid potential confusion, if we refer to the i-th element of the n-th training data vector $\boldsymbol{x}_n$ is referred to as $(\boldsymbol{x}_n)_i$. Next, to denote the l-th layer, the following superscript notation is used:

$$\boldsymbol{g}^{(l)}, \boldsymbol{W}^{(l)}, \boldsymbol{b}^{(l)}, d^{(l)}.$$

Accordingly, by combining the training index n, for example $\boldsymbol{g}_n^{(l)}$ refers to the l-th layer $\boldsymbol{g}$ vector for the n-th training data. Finally, the t-th update using an optimizer such as the stochastic gradient method can be denoted by $[t]$: for example,

$$\boldsymbol{\Theta}[t], \boldsymbol{V}[t]$$

refer to the t-th update of the parameter map $\boldsymbol{\Theta}$ and $\boldsymbol{V}$, respectively.

6.2.2 Modeling a Single Neuron

Consider a typical biological neuron in Fig. 6.1 and its mathematical diagram in Fig. 6.2. Let $o_j, j = 1, \cdots, d$ denote the presynaptic potential from the j-th dendric synapse. For mathematical simplicity, we assume that the potential occurs synchronously, and arrives simultaneously at the axon hillock. At the axon hillock, they are summed together, and fires an action potential if the summed signal is greater than the specific threshold value. This process can be mathematically modeled as

$$net_i = \sigma\left(\sum_{j=1}^{d} w_{ij} o_j + b_i\right), \tag{6.1}$$

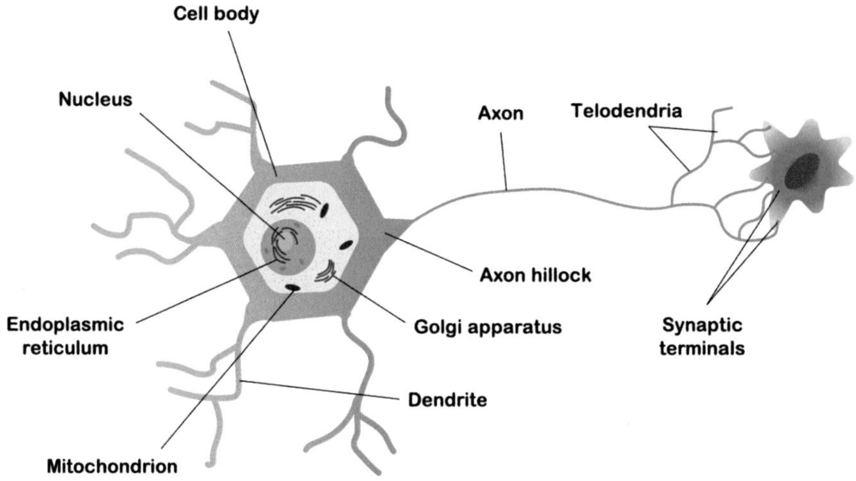

Fig. 6.1 Anatomy of neurons

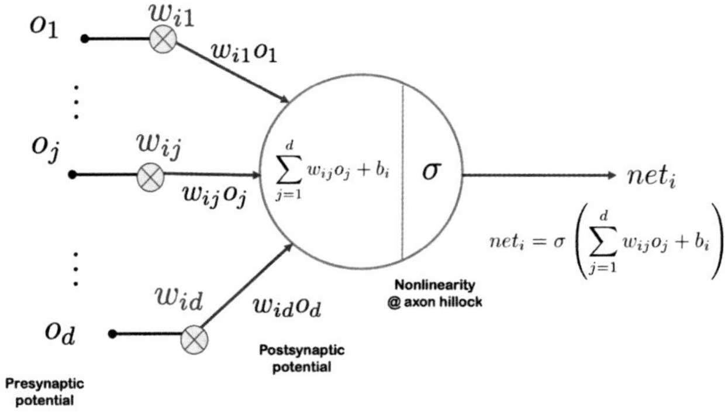

Fig. 6.2 A mathematical model of a single neuron

where net_i denotes the action potential arriving at the i-th synaptic terminal of the telodendria, and b_i is the bias term for the nonlinearity $\sigma(\cdot)$ at the axon hillock. Note that the w_{ij} is the weight parameter determined by the synaptic plasticity, and the positive values imply that $w_{ij}o_j$ are the excitatory postsynaptic potentials (EPSPs), whereas the negative weights correspond to the inhibitory postsynaptic potentials (IPSPs).

In artificial neural networks (ANNs), the nonlinearity $\sigma(\cdot)$ in (6.1) is modeled in various ways as shown in Fig. 6.3. This nonlinearity is often called the *activation function.* Nonlinearity may be perhaps the most important feature of neural networks, since learning and adaptation never happen without nonlinearity. The mathematical proof of this argument is somewhat complicated, so the discussion will be deferred to later.

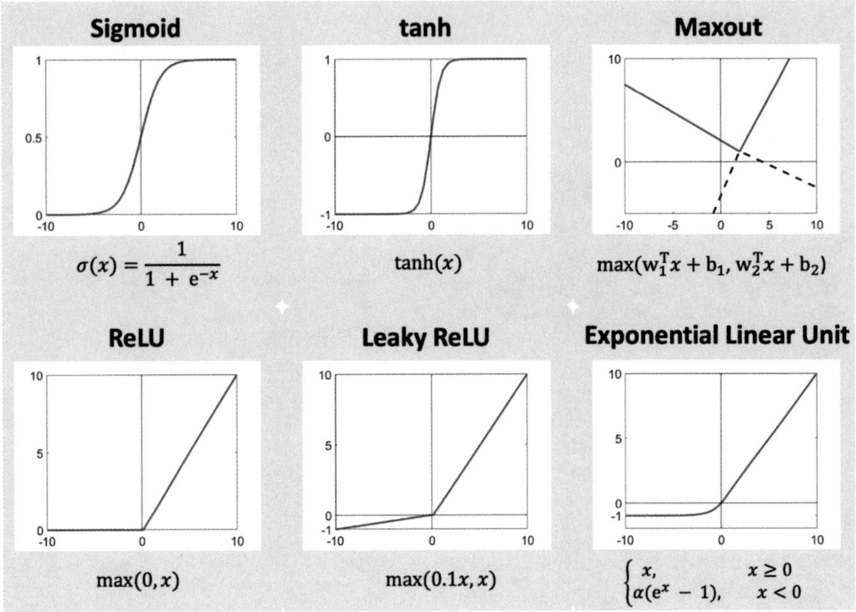

Fig. 6.3 Various forms of activation functions

Among the various forms of the activation functions, one of the most successful ones in modern deep learning is the rectified linear unit (ReLU), which is defined as [24]

$$\sigma(x) = \text{ReLU}(x) := \max\{0, x\}. \tag{6.2}$$

The ReLU activation function is called *active* when the output is nonzero. It is believed that the non-vanishing gradient in the positive range contributed to the success of modern deep learning. Specifically, we have

$$\frac{\partial \text{ReLU}(x)}{\partial c} = \begin{cases} 1, & \text{if } x > 0 \\ 0, & \text{otherwise} \end{cases}, \tag{6.3}$$

which shows that the gradient is always 1 whenever the ReLU is active. Note that we set the gradient 0 at $x = 0$ by convention, since the ReLU is not differentiable at $x = 0$.

In evaluating the activation function $\sigma(x)$, the gain function, which refers to the input/output ratio, is also useful:

$$\gamma(x) := \frac{\sigma(x)}{x}, \quad x \neq 0. \tag{6.4}$$

For example, the ReLU satisfies the following important property:

$$\gamma(x)=\frac{\partial\sigma(x)}{\partial x}=\begin{cases}1, & \text{if } x>0\\ 0, & \text{otherwise}\end{cases}, \tag{6.5}$$

which will be used later in analyzing the backpropagation algorithm.

There is an additional advantage of using the ReLU compared to other nonlinearities. As will be explained in detail later, the ReLU divides the input and feature space into two disjoint sets, i.e. active and inactive areas, resulting in a piecewise linear approximation of a nonlinear mapping onto the partitioned geometry. Accordingly, a neural network within each partition can be viewed as locally linear, even though the overall map is highly nonlinear. This is the geometric picture of a deep neural network that we would like to highlight for readers in this book.

6.2.3 Feedforward Multilayer ANN

Biological neural networks are composed of multiple neurons that are connected to each other. This connection can have complicated topology, such as recurrent connection, asynchronous connection, inter-neurons, etc.

One of the most simple forms of the neural network connection is the multi-layer feedforward neural network as shown in Figs. 6.4 and 6.8. Specifically, let $o_j^{(l-1)}$ denote the j-th output of the $(l-1)$-th layer neuron, which is given as the j-th dendrite presynaptic potential input for the l-th layer neuron, and $w_{ij}^{(l)}$ corresponds to the synaptic weights at the l-th layer. Then, by extending the model in (6.1) we have

$$o_i^{(l)}=\sigma\left(\sum_{j=1}^{d^{(l)}} w_{ij}^{(l)}o_j^{(l-1)}+b_i^{(l)}\right), \tag{6.6}$$

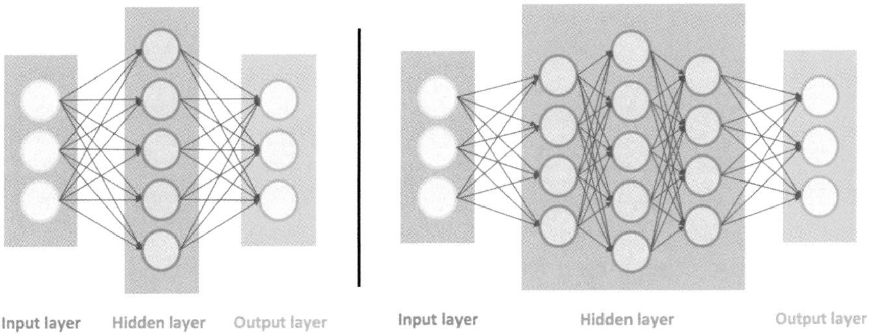

Fig. 6.4 Examples of multilayer feedforward neural networks

for $i = 1, \cdots, d^{(l)}$, where $d^{(l)}$ denotes the number of dendrites of the l-th layer neuron. This can be represented in a matrix form

$$\boldsymbol{o}^{(l)} = \boldsymbol{\sigma}\left(\boldsymbol{W}^{(l)}\boldsymbol{o}^{(l-1)} + \boldsymbol{b}^{(l)}\right), \tag{6.7}$$

where $\boldsymbol{W}^{(l)} \in \mathbb{R}^{d^{(l)} \times d^{(l-1)}}$ is the weight matrix whose (i, j) elements are given by $w_{ij}^{(l)}$, $\boldsymbol{\sigma}(\cdot)$ denotes the nonlinearity $\sigma(\cdot)$ applied for each elements of the vector, and

$$\boldsymbol{o}^{(l)} = \left[o_1^{(l)} \cdots o_{d^{(l)}}^{(l)}\right]^\top \in \mathbb{R}^{d^{(l)}}, \tag{6.8}$$

$$\boldsymbol{b}^{(l)} = \left[b_1^{(l)} \cdots b_{d^{(l)}}^{(l)}\right]^\top \in \mathbb{R}^{d^{(l)}}. \tag{6.9}$$

Another way to simplify the multilayer representation is using the hidden nodes from linear layers in between. Specifically, an L-layer feedforward neural network can be represented recursively using the hidden node $\boldsymbol{g}^{(l)}$ by

$$\boldsymbol{o}^{(l)} = \boldsymbol{\sigma}(\boldsymbol{g}^{(l)}), \quad \boldsymbol{g}^{(l)} = \boldsymbol{W}^{(l)}\boldsymbol{o}^{(l-1)} + \boldsymbol{b}^{(l)}, \tag{6.10}$$

for $l = 1, \cdots, L$.

6.3 Artificial Neural Network Training

6.3.1 *Problem Formulation*

For given training data $\{\boldsymbol{x}_n, \boldsymbol{y}_n\}_{n=1}^N$, a neural network training problem can be then formulated as follows:

$$\hat{\boldsymbol{\Theta}} = \arg\min_{\boldsymbol{\Theta}} c(\boldsymbol{\Theta}), \tag{6.11}$$

where the cost function is given by

$$c(\boldsymbol{\Theta}) := \sum_{n=1}^N \ell\left(\boldsymbol{y}_n, \boldsymbol{f}_{\boldsymbol{\Theta}}(\boldsymbol{x}_n)\right). \tag{6.12}$$

Here, $\ell(\cdot, \cdot)$ denotes a loss function, and $\boldsymbol{f}_{\boldsymbol{\Theta}}(\boldsymbol{x}_n)$ is a regression function with the input $\boldsymbol{x}_n$, which is parameterized by the parameter set $\boldsymbol{\Theta}$.

For the case of an L-layer feedforward neural network, the regression function $\boldsymbol{f}_\Theta(\boldsymbol{x}_n)$ in (6.12) can be represented by

$$\boldsymbol{f}_\Theta(\boldsymbol{x}_n) := \left(\boldsymbol{\sigma} \circ \boldsymbol{g}^{(L)} \circ \boldsymbol{\sigma} \circ \boldsymbol{g}^{(L-1)} \cdots \circ \boldsymbol{g}^{(1)}\right)(\boldsymbol{x}_n), \tag{6.13}$$

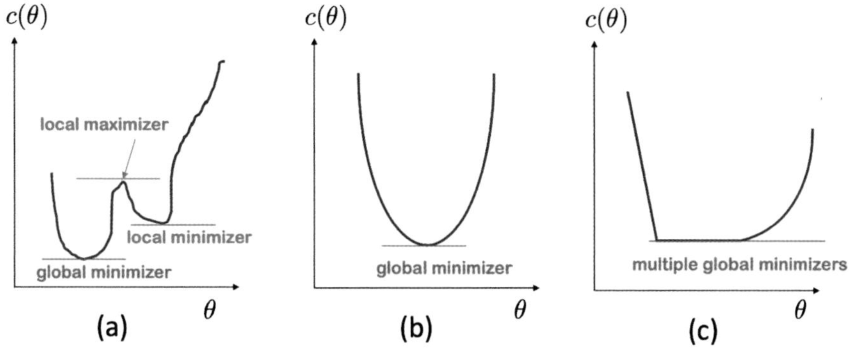

Fig. 6.5 Examples of cost functions for a 1-D optimization problem: (**a**) both local and global minimizers exists, (**b**) only a single global minimizer exist, (**c**) multiple global minimizers exist

where the parameter set $\boldsymbol{\Theta}$ is composed of the synaptic weight and bias for each layer:

$$\boldsymbol{\Theta} = \begin{bmatrix} \boldsymbol{W}^{(1)}, & \boldsymbol{b}^{(1)} \\ \vdots & \vdots \\ \boldsymbol{W}^{(L)}, & \boldsymbol{b}^{(L)} \end{bmatrix}. \tag{6.14}$$

As discussed before for kernel machines in Chap. 4, the formulation in (6.11) is so general that it covers classification, regression, etc., by simply changing the loss function (for example, l_2 loss for the regression, and the hinge loss for the classification). Unfortunately, in contrast to the kernel machines, one of the main difficulties in the neural network training is that the cost function $c(\boldsymbol{\Theta})$ is not convex, and indeed there exist many local minimizers (see Fig. 6.5). Therefore, the neural network training critically depends on the choice of optimization algorithm, initialization, step size, etc.

6.3.2 *Optimizers*

In view of the parameterized neural network in (6.13), the key question is how the minimizers for the optimization problem (6.11) can be found. As already mentioned, the main technical challenge of this minimization problem is that there are many local minimizers, as shown in Fig. 6.5a. Another tricky issue is that sometimes there are many global minimizers, as shown in Fig. 6.5c. Although all the global minimizers can be equally good in the training phase, each global minimizer may have different generalization performance in the test phase. This issue is important and will be discussed later. Furthermore, different global minimizers can be achieved depending on the specific choice of an optimizer, which is often called the *implicit bias* or *inductive bias* of an optimization algorithm. This topic will also be discussed later.

One of the most important observations in designing optimization algorithms is that the following first-order necessary condition (FONC) holds at local minimizers.

Lemma 6.1 *Let $c : \mathbb{R}^p \mapsto \mathbb{R}$ be a differentiable function. If $\boldsymbol{\Theta}^*$ is a local minimizer, then*

$$\left.\frac{\partial c}{\partial \boldsymbol{\Theta}}\right|_{\boldsymbol{\Theta}=\boldsymbol{\Theta}^*} = \mathbf{0}. \tag{6.15}$$

Indeed, various optimization algorithms exploit the FONC, and the main difference between them is the way they avoid the local minimum and provide fast convergence. In the following, we start with the discussion of the classical gradient descent method and its stochastic extension called the *stochastic gradient descent (SGD)*, after which various improvements will be discussed.

6.3.2.1 Gradient Descent

For the given training data $\{\boldsymbol{x}_n, \boldsymbol{y}_n\}_{n=1}^N$, the gradient of the cost function in (6.12) is given by

$$\begin{aligned}\frac{\partial c}{\partial \boldsymbol{\Theta}}(\boldsymbol{\Theta}) &= \frac{\partial \left(\sum_{n=1}^N \ell\left(\boldsymbol{y}_n, \boldsymbol{f}_{\boldsymbol{\Theta}}(\boldsymbol{x}_n)\right)\right)}{\partial \boldsymbol{\Theta}} \\ &= \sum_{n=1}^N \frac{\partial \ell}{\partial \boldsymbol{\Theta}}\left(\boldsymbol{y}_n, \boldsymbol{f}_{\boldsymbol{\Theta}}(\boldsymbol{x}_n)\right),\end{aligned} \tag{6.16}$$

which is equal to the sum of the gradient at each of the training data. Since the gradient is the steep direction for the increasing cost function, the steep descent algorithm is to update the parameter in its opposite direction:

$$\begin{aligned}\boldsymbol{\Theta}[t+1] &= \boldsymbol{\Theta}[t] - \eta \left.\frac{\partial c}{\partial \boldsymbol{\Theta}}(\boldsymbol{\Theta})\right|_{\boldsymbol{\Theta}=\boldsymbol{\Theta}[t]} \\ &= \boldsymbol{\Theta}[t] - \eta \sum_{n=1}^N \left.\frac{\partial \ell}{\partial \boldsymbol{\Theta}}\left(\boldsymbol{y}_n, \boldsymbol{f}_{\boldsymbol{\Theta}}(\boldsymbol{x}_n)\right)\right|_{\boldsymbol{\Theta}=\boldsymbol{\Theta}[t]},\end{aligned} \tag{6.17}$$

where $\eta > 0$ denotes the step size and $\boldsymbol{\Theta}[t]$ is the t-th update of the parameter $\boldsymbol{\Theta}$. Figure 6.6a illustrates why gradient descent is a good way to minimize the cost for the convex optimization problem. As the gradient of the cost points toward the uphill direction of the cost, the parameter update should be in its negative direction.

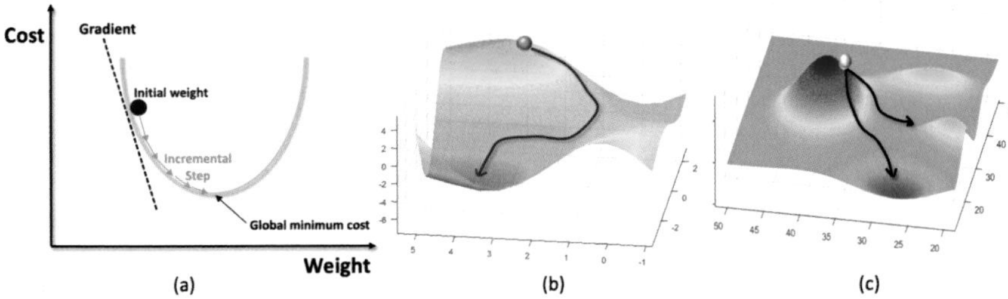

Fig. 6.6 Steepest gradient descent example: (**a**) convex cases, where steepest descent succeeds, (**b**) non-convex case, where the steepest descent cannot go uphill, (**c**) steepest gradient leads to different local minimizers depending on the initialization

After a small step, a new gradient is computed and a new search direction is found. By iterating the procedure, we can achieve the global minimum.

One of the downsides of the gradient descent method is that when the gradient becomes zero at a local minimizers at t^*, the update equation in (6.17) make the iteration stuck in the local minimizers, i.e.:

$$\boldsymbol{\Theta}[t+1] = \boldsymbol{\Theta}[t], \quad t \geq t^*. \tag{6.18}$$

For example, Fig. 6.6b,c show the potential limitation of the gradient descent. For the case of Fig. 6.6b, during the path toward the global minimum, there exists uphill directions, which cannot be overcome by the gradient methods. On the other hand, Fig. 6.6c shows that depending on the initialization, different local minimizers can be found by the gradient descent due to the different intermediate path. In fact, the situations in Fig. 6.6b,c are more likely situations in neural network training, since the optimization problem is highly non-convex due to the cascaded connection of nonlinearities. In addition, despite using the same initialization, the optimizer can converge to a completely different solution depending on the step size or certain optimization algorithms. In fact, algorithmic bias is a major research topic in modern deep learning, often referred to as *inductive bias*.

This can be another reason why neural network training is difficult and depends heavily on who is training the model. For example, even if multiple students are given the exact same training set, network architecture, GPU, etc., it is usually observed that some students are successfully training the neural network and others are not. The main reason for such a difference is usually due to their commitment and self-confidence, which leads to different optimization algorithms with different inductive biases. Successful students usually try different initializations, optimizers, different learning rates, etc. until the model works, while unsuccessful students usually stick to the parameters all the time without trying to carefully change them. Instead, they often claim that the failure is not their fault, but because of the wrong model they started with. If the training problem were convex, then regardless of the inductive bias they have in training, all students could be successful. Unfortunately, neural network training is highly non-convex, so it is highly dependent on the student's inductive bias. The good news is that once students learn how to make a model work, the intuition they gain from such experiences usually works for training more complicated neural networks.

Indeed, advances in algorithms to optimize deep neural networks can be viewed as overcoming operator dependency. The following describes the various methods of systematically reducing the operator-dependent inductive bias for training neural networks, although the same problem still exists, albeit in a reduced manner, due to the non-convexity of the problems.

6.3.2.2 Stochastic Gradient Descent (SGD) Method

We say that the update equations in (6.17) are based on full gradients, since at each iteration we need to compute the gradient with respect to the whole data set. However, if n is large, computational cost for the gradient calculation is quite heavy. Moreover, by using the full gradient, it is difficult to avoid the local minimizer, since the gradient descent direction is always toward the lower cost value.

To address the problem, the SGD algorithm uses an easily computable estimate of the gradient using a small subset of training data. Although it is a bit noisy, this noisy gradient can even be helpful in avoiding local minimizers. For example, let $I[t] \subset \{1, \cdots, N\}$ denote a random subset of the index set $\{1, \cdots, N\}$ at the t-th update. Then, our estimate of the full gradient at the t-th iteration is given by

$$\left.\frac{\partial c}{\partial \boldsymbol{\Theta}}(\boldsymbol{\Theta})\right|_{\boldsymbol{\Theta}=\boldsymbol{\Theta}[t]} \simeq \frac{N}{|I[t]|} \sum_{i \in I[t]} \left.\frac{\partial \ell}{\partial \boldsymbol{\Theta}}\big(\boldsymbol{y}_n, \boldsymbol{f}_{\boldsymbol{\Theta}}(\boldsymbol{x}_n)\big)\right|_{\boldsymbol{\Theta}=\boldsymbol{\Theta}[t]}, \tag{6.19}$$

where $|I[t]|$ denotes the number of elements in $I[t]$. As the SGD utilizes a small random subset of the original training data set (i.e. $|I[t]| \ll N$) in calculating the gradient, the computational complexity for each update is much smaller than the original gradient descent method. Moreover, it is not exactly the same as the true gradient direction so that the resulting noise can provide a means to escape from the local minimizers.

6.3.2.3 Momentum Method

Another way to overcome the local minimum is to take into account the previous updates as additional terms to avoid getting stuck in local minima. Specifically, a desirable update equation may be written as

$$\boldsymbol{\Theta}[t+1] = \boldsymbol{\Theta}[t] - \eta \sum_{s=1}^{t} \beta^{t-s} \frac{\partial c}{\partial \boldsymbol{\Theta}}(\boldsymbol{\Theta}[s]) \tag{6.20}$$

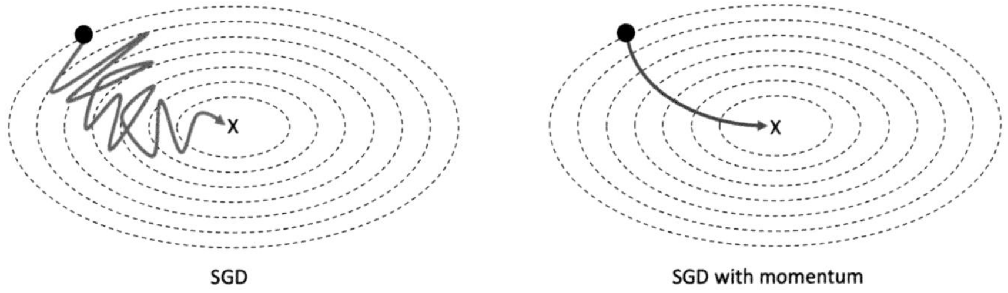

Fig. 6.7 Example trajectory of update in (**a**) stochastic gradient, (**b**) SGD with momentum method

for an appropriate forgetting factor $0 < \beta < 1$. This implies that the contribution from the past gradient is gradually reduced in calculating the current update direction. However, the main limitation of using (6.20) is that all the history of the past gradients should be saved, which requires huge GPU memory. Instead, the following recursive formulation is mostly used which provide the equivalent representation:

$$\begin{aligned} \boldsymbol{V}[t] &= \beta(\boldsymbol{\Theta}[t] - \boldsymbol{\Theta}[t-1]) - \eta \frac{\partial c}{\partial \boldsymbol{\Theta}}(\boldsymbol{\Theta}[t]), \\ \boldsymbol{\Theta}[t+1] &= \boldsymbol{\Theta}[t] + \boldsymbol{V}[t]. \end{aligned} \tag{6.21}$$

This type of method is called the *momentum method*, and is particularly useful when it is combined with the SGD. The example update trajectory of the SGD with momentum is shown in Fig. 6.7b. Compared to the fluctuating path, the momentum method provides a smoothed solution path thanks to the averaging effects from the past gradient, which results in fast convergence.

6.3.2.4 Other Variations

In neural networks, several other variants of the optimizers are often used, among which ADAGrad [25], RMSprop [26], and Adam [27] are most popular. The main ideas of these variants is that instead of using the fixed step size η for all elements of the gradient, an element-wise adaptive step size is used. For example, for the case of the steepest descent in (6.17), we use the following update equation:

$$\boldsymbol{\Theta}[t+1] = \boldsymbol{\Theta}[t] - \boldsymbol{\Upsilon}[t] \odot \frac{\partial c}{\partial \boldsymbol{\Theta}}(\boldsymbol{\Theta}[t]), \tag{6.22}$$

where $\boldsymbol{\Upsilon}[t]$ is a matrix with the step size and $\odot$ is the element-wise multiplication. In fact, the main difference in these algorithms is how to update the matrix $\boldsymbol{\Upsilon}[t]$ at each iteration. For more details for specific update rules, see the original papers [25–27].

6.4 The Backpropagation Algorithm

In the previous section, various optimization algorithms for neural network training were discussed based on the assumption that the gradient $\frac{\partial c}{\partial \boldsymbol{\Theta}}(\boldsymbol{\Theta}[t])$ is computed. However, given the complicated nonlinear nature of the feedforward neural network, the computation of the gradient is not trivial.

In machine learning, backpropagation (backprop, or BP) [28] is a standard way of calculating the gradient in training feedforward neural networks, by providing an explicit and computationally efficient way of computing the gradient. The term backpropagation and its general use in neural networks were originally derived in Rumelhart, Hinton and Williams [28]. Their main idea is that although the multilayer neural network is composed of complicated connections of neurons with a large number of unknown weights, the recursive structure of the multilayer neural network in (6.10) lends itself to computationally efficient optimization methods.

6.4.1 Derivation of the Backpropagation Algorithm

The following lemma, which was previously introduced in Chap. 1, is useful in deriving the BP algorithm:

Lemma 6.2 *Let $\boldsymbol{A} \in \mathbb{R}^{m \times n}$ and $\boldsymbol{x} \in \mathbb{R}^n$. Then, we have*

$$\frac{\partial \boldsymbol{A}\boldsymbol{x}}{\partial \mathrm{VEC}(\boldsymbol{A})} = \boldsymbol{x} \otimes \boldsymbol{I}_m. \tag{6.23}$$

Lemma 6.3 *For the vectors $\boldsymbol{x} \in \mathbb{R}^m$, $\boldsymbol{y} \in \mathbb{R}^n$, we have*

$$\mathrm{VEC}(\boldsymbol{x}\boldsymbol{y}^\top) = (\boldsymbol{y} \otimes \boldsymbol{I}_m)\boldsymbol{x}, \tag{6.24}$$

where $\boldsymbol{I}_m$ denotes the $m \times m$ identity matrix.

For the derivation of the backpropagation algorithm, we tentatively assume that the bias terms are zero, i.e. $\boldsymbol{b}^{(l)} = \boldsymbol{0},\ l = 1, \cdots, L$. In this case, the neural network parameter $\boldsymbol{\Theta}$ in (6.14) can be simplified as

$$\boldsymbol{\Theta} = \begin{bmatrix} \boldsymbol{W}^{(1)} \\ \vdots \\ \boldsymbol{W}^{(L)} \end{bmatrix}, \tag{6.25}$$

where $\boldsymbol{W}^{(l)} \in \mathbb{R}^{d^{(l)} \times d^{(l-1)}}$. Using the denominator layout as explained in Chap. 1, we have

$$\frac{\partial c}{\partial \boldsymbol{\Theta}} = \begin{bmatrix} \frac{\partial c}{\partial \boldsymbol{W}^{(1)}} \\ \vdots \\ \frac{\partial c}{\partial \boldsymbol{W}^{(L)}} \end{bmatrix}, \tag{6.26}$$

so that the weight at the l-th layer can be updated with the increment:

$$\Delta\boldsymbol{\Theta} = \begin{bmatrix} \Delta\boldsymbol{W}^{(1)} \\ \vdots \\ \Delta\boldsymbol{W}^{(L)} \end{bmatrix}, \quad \text{where} \quad \Delta\boldsymbol{W}^{(l)} = -\eta \frac{\partial c}{\partial \boldsymbol{W}^{(l)}}. \tag{6.27}$$

Therefore, $\partial c / \partial \boldsymbol{W}^{(l)}$ should be specified. More specifically, for a given training data set $\{\boldsymbol{x}_n, \boldsymbol{y}_n\}_{n=1}^N$, recall that the cost function $c(\boldsymbol{\Theta})$ in (6.12) is given by

$$c(\boldsymbol{\Theta}) = \sum_{n=1}^{N} \ell\left(\boldsymbol{y}_n, \boldsymbol{f}_{\Theta}(\boldsymbol{x}_n)\right), \tag{6.28}$$

where $\boldsymbol{f}_{\Theta}(\boldsymbol{x}_n)$ is defined in (6.13). Now define the l-th layer variable with respect to the n-th training data:

$$\boldsymbol{o}_n^{(l)} = \boldsymbol{\sigma}(\boldsymbol{g}_n^{(l)}), \quad \boldsymbol{g}_n^{(l)} = \boldsymbol{W}^{(l)} \boldsymbol{o}_n^{(l-1)}, \tag{6.29}$$

for $l = 1, \cdots, L$, with the initialization

$$\boldsymbol{o}_n^{(0)} := \boldsymbol{x}_n, \tag{6.30}$$

where the bias is assumed zero. Then, we have

$$\boldsymbol{o}_n^{(L)} = \boldsymbol{f}_{\Theta}(\boldsymbol{x}_n),$$

Using the chain rule for the denominator convention (see Eq. (1.40))

$$\frac{\partial c(\boldsymbol{g}(\boldsymbol{u}))}{\partial \boldsymbol{x}} = \frac{\partial \boldsymbol{u}}{\partial \boldsymbol{x}} \frac{\partial \boldsymbol{g}(\boldsymbol{u})}{\partial \boldsymbol{u}} \frac{\partial c(\boldsymbol{g})}{\partial \boldsymbol{g}} \tag{6.31}$$

we have

$$\frac{\partial c}{\partial \mathrm{VEC}(\boldsymbol{W}^{(l)})} = \sum_{n=1}^{N} \frac{\partial \boldsymbol{g}_n^{(l)}}{\partial \mathrm{VEC}(\boldsymbol{W}^{(l)})} \frac{\partial \ell\left(\boldsymbol{y}_n, \boldsymbol{o}_n^{(L)}\right)}{\partial \boldsymbol{g}_n^{(l)}}.$$

Furthermore, Lemma 6.2 informs us

$$\frac{\partial \boldsymbol{g}_n^{(l)}}{\partial \mathrm{VEC}(\boldsymbol{W}^{(l)})} = \boldsymbol{o}_n^{(l-1)} \otimes \boldsymbol{I}_{d^{(l)}}. \tag{6.32}$$

We further define the term:

$$\boldsymbol{\delta}_n^{(l)} := \frac{\partial \ell\left(\boldsymbol{y}_n, \boldsymbol{o}_n^{(L)}\right)}{\partial \boldsymbol{g}_n^{(l)}}, \tag{6.33}$$

which can be calculated using the chain rule (6.31) as follows:

$$\begin{aligned}\boldsymbol{\delta}_n^{(l)} &= \frac{\partial \boldsymbol{o}_n^{(l)}}{\partial \boldsymbol{g}_n^{(l)}} \frac{\partial \boldsymbol{g}_n^{(l+1)}}{\partial \boldsymbol{o}_n^{(l)}} \cdots \frac{\partial \boldsymbol{o}_n^{(L)}}{\partial \boldsymbol{g}_n^{(L)}} \frac{\partial \ell\left(\boldsymbol{y}_n, \boldsymbol{o}_n^{(L)}\right)}{\partial \boldsymbol{o}_n^{(L)}} \\ &= \boldsymbol{\Lambda}_n^{(l)} \boldsymbol{W}^{(l+1)\top} \boldsymbol{\Lambda}_n^{(l+1)} \boldsymbol{W}^{(l+2)\top} \ldots \boldsymbol{W}^{(L)\top} \boldsymbol{\Lambda}_n^{(L)} \boldsymbol{\epsilon}_n \end{aligned} \tag{6.34}$$

for $l = 1, \cdots, L$, and the error term $\boldsymbol{\epsilon}_n$ is computed by

$$\boldsymbol{\epsilon}_n = \frac{\partial \ell\left(\boldsymbol{y}_n, \boldsymbol{o}_n^{(L)}\right)}{\partial \boldsymbol{o}_n^{(L)}}.$$

In (6.34), we use

$$\boldsymbol{\Lambda}_n^{(l)} := \frac{\partial \boldsymbol{o}_n^{(l)}}{\partial \boldsymbol{g}_n^{(l)}} = \frac{\partial \boldsymbol{\sigma}\left(\boldsymbol{g}_n^{(l)}\right)}{\partial \boldsymbol{g}_n^{(l)}} \in \mathbb{R}^{d^{(l)} \times d^{(l)}}, \tag{6.35}$$

which is calculated using the denominator layout as explained in Chap. 1, and

$$\frac{\partial \boldsymbol{g}_n^{(l+1)}}{\partial \boldsymbol{o}_n^{(l)}} = \frac{\partial \boldsymbol{W}^{(l+1)} \boldsymbol{o}_n^{(l)}}{\partial \boldsymbol{o}_n^{(l)}} = \boldsymbol{W}^{(l+1)\top}, \tag{6.36}$$

which is obtained using the denominator convention (see (1.41) in Chap. 1). Accordingly, we have

$$\begin{aligned}\frac{\partial c}{\partial \textsc{Vec}(\boldsymbol{W}^{(l)})} &= \sum_{n=1}^{N} \frac{\partial \boldsymbol{g}_n^{(l)}}{\partial \textsc{Vec}(\boldsymbol{W}^{(l)})} \frac{\partial \ell\left(\boldsymbol{y}_n, \boldsymbol{o}_n^{(L)}\right)}{\partial \boldsymbol{g}_n^{(l)}} \\ &= \sum_{n=1}^{N} \left(\boldsymbol{o}_n^{(l-1)} \otimes \boldsymbol{I}_{d^{(l)}}\right) \boldsymbol{\delta}_n^{(l)} \\ &= \sum_{n=1}^{N} \textsc{Vec}\left(\boldsymbol{\delta}_n^{(l)} \boldsymbol{o}_n^{(l-1)\top}\right),\end{aligned}$$

where we use (6.32) and (6.33) for the second equality, and Lemma 6.3 for the last equality. Finally, we have the following derivative of the cost with respect to $\boldsymbol{W}^{(l)}$:

$$\begin{aligned}\frac{\partial c}{\partial \boldsymbol{W}^{(l)}} &= \textsc{UnVec}\left(\frac{\partial c}{\partial \textsc{Vec}(\boldsymbol{W}^{(l)})}\right) \\ &= \textsc{UnVec}\left(\sum_{n=1}^{N} \textsc{Vec}\left(\boldsymbol{\delta}_n^{(l)} \boldsymbol{o}_n^{(l-1)\top}\right)\right) \\ &= \sum_{n=1}^{N} \boldsymbol{\delta}_n^{(l)} \boldsymbol{o}_n^{(l-1)\top},\end{aligned}$$

where we use the linearity of $\textsc{UnVec}(\cdot)$ operator for the last equality. Therefore, the weight update increment is given by

$$\begin{aligned}\Delta \boldsymbol{W}^{(l)} &= -\eta \frac{\partial c}{\partial \boldsymbol{W}^{(l)}} \\ &= -\eta \sum_{n=1}^{N} \boldsymbol{\delta}_n^{(l)} \boldsymbol{o}_n^{(l-1)\top}.\end{aligned} \tag{6.37}$$

6.4.2 *Geometrical Interpretation of BP Algorithm*

This weight update scheme in (6.37) is the key in BP. Not only is the final form of the weight update in (6.37) very concise, but it also has a very important geometric meaning, which deserves further discussion. In particular, the update is totally determined by the outer product of the two terms $\boldsymbol{\delta}_n^{(l)}$ and $\boldsymbol{o}_n^{(l-1)}$, i.e. $\boldsymbol{\delta}_n^{(l)} \boldsymbol{o}_n^{(l-1)\top}$. Why are these terms so important? This is the main discussion point in this section.

First, recall that $\boldsymbol{o}_n^{(l-1)}$ is the $(l-1)$-th layer neural network output given by (6.29). Since this term is calculated in the forward path of the neural network, it is nothing but the *forward-propagated input* to the l-th layer neuron. Second, recall that

$$\boldsymbol{\epsilon}_n = \frac{\partial \ell\left(\boldsymbol{y}_n, \boldsymbol{o}_n^{(L)}\right)}{\partial \boldsymbol{o}_n^{(L)}}.$$

If we use the l_2 loss, this term becomes

$$\begin{aligned}\boldsymbol{\epsilon}_n &= \frac{\partial\left(\frac{1}{2}\|\boldsymbol{y}_n - \boldsymbol{o}_n^{(L)}\|^2\right)}{\partial \boldsymbol{o}_n^{(L)}} \\ &= \boldsymbol{o}_n^{(L)} - \boldsymbol{y}_n,\end{aligned}$$

which is indeed the estimation error of the neural network output. Since we have

$$\boldsymbol{\delta}_n^{(l)} = \boldsymbol{\Lambda}_n^{(l)} \boldsymbol{W}^{(l+1)\top} \boldsymbol{\Lambda}_n^{(l+1)} \boldsymbol{W}^{(l+2)\top} \cdots \boldsymbol{W}^{(L)\top} \boldsymbol{\Lambda}_n^{(L)} \boldsymbol{\epsilon}_n \,, \tag{6.38}$$

this implies that $\boldsymbol{\delta}_n^{(l)}$ is indeed *the backward-propagated estimation error* down to the l-th layer. Therefore, we can find that the weight update is determined by the outer product of the forward-propagated input and backward-propagated estimation error.

In terms of calculation, the forward and backward terms $\boldsymbol{o}_n^{(l)}$ and $\boldsymbol{\delta}_n^{(l)}$ can be efficiently calculated using recursive formulae. More specifically, we have

$$\boldsymbol{o}_n^{(l-1)} = \boldsymbol{\sigma}\left(\boldsymbol{W}^{(l-1)} \boldsymbol{o}_n^{(l-2)}\right), \tag{6.39}$$

$$\boldsymbol{\delta}_n^{(l)} = \boldsymbol{\Lambda}_n^{(l)} \boldsymbol{W}^{(l+1)\top} \boldsymbol{\delta}_n^{(l+1)}, \tag{6.40}$$

with the initialization by

$$\boldsymbol{o}_n^{(0)} = \boldsymbol{x}_n, \quad \boldsymbol{\delta}_n^{(L)} = \boldsymbol{\epsilon}_n. \tag{6.41}$$

The geometric interpretation and recursive formulae are illustrated in Fig. 6.8.

6.4.3 Variational Interpretation of BP Algorithm

The variational principle is a scientific principle used within the *calculus of variations* [29], which develops general methods for finding functions that minimize the value of quantities that depend upon those functions. The calculus of variations

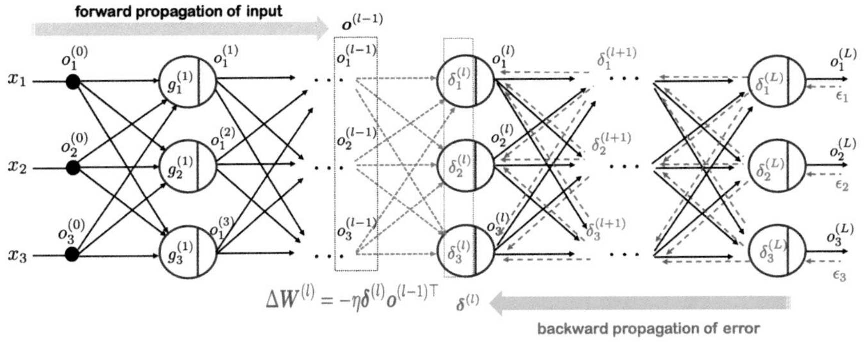

Fig. 6.8 Geometry of backpropagation

is a field of mathematical analysis pioneered by Isaac Newton, which uses variations to reduce the energy function [29].

Given the incremental variation in (6.37), we are therefore interested in finding whether it indeed reduces the energy function. For this, let us consider a simplified form of the loss function with l_2 loss with $N = 1$. In the following, we show that for the case of neural networks with ReLU activation functions, the BP algorithm is indeed equivalent to the variational approach.

More specifically, let the baseline energy function, which refers to the cost function before the perturbation, be given by

$$\ell(\boldsymbol{y}, \boldsymbol{o}^{(L)}) = \frac{1}{2}\|\boldsymbol{y} - \boldsymbol{o}^{(L)}\|^2, \tag{6.42}$$

where the subscript n for the training data index is neglected here for simplicity and

$$\boldsymbol{o}^{(L)} := \boldsymbol{\sigma}\left(\boldsymbol{W}^{(L)}\boldsymbol{o}^{(L-1)}\right), \tag{6.43}$$

One of the important observations is that for the case of the ReLU, (6.43) can be represented by

$$\boldsymbol{o}^{(L)} := \boldsymbol{\Gamma}^{(L)}\boldsymbol{g}^{(L)}, \quad \text{where} \quad \boldsymbol{g}^{(L)} = \boldsymbol{W}^{(L)}\boldsymbol{o}^{(L-1)}, \tag{6.44}$$

where $\boldsymbol{\Gamma}^{(L)} \in \mathbb{R}^{d^{(L)} \times d^{(L)}}$ is a diagonal matrix with 0 and 1 values given by

$$\boldsymbol{\Gamma}^{(L)} = \begin{bmatrix} \gamma_1 & \cdots & 0 & \cdots & 0 \\ \vdots & \ddots & \vdots & \ddots & \vdots \\ 0 & \cdots & \gamma_j & \cdots & 0 \\ \vdots & \ddots & \vdots & \ddots & \vdots \\ 0 & \cdots & 0 & \cdots & \gamma_{d^{(L)}} \end{bmatrix}, \tag{6.45}$$

where

$$\gamma_j = \gamma\left([g^{(L)}]_j\right), \tag{6.46}$$

where $[g^{(L)}]_j$ denotes the j-th element of the vector $g^{(L)}$ and $\gamma(\cdot)$ is defined in (6.4). Thanks to (6.5), we have

$$\boldsymbol{\Gamma}^{(l)} = \boldsymbol{\Lambda}^{(l)}, \quad l = 1, \cdots, L, \tag{6.47}$$

where $\boldsymbol{\Lambda}^{(l)}$ is defined as the derivative of the activation function in (6.35). Therefore, using the recursive formula, we have

$$\boldsymbol{o}^{(L)} = \boldsymbol{\Lambda}^{(L)} \boldsymbol{W}^{(L)} \cdots \boldsymbol{\Lambda}^{(l)} \boldsymbol{W}^{(l)} \boldsymbol{o}^{(l-1)}. \tag{6.48}$$

Using this, we now investigate whether the cost decreases with the perturbed weight

$$\Delta \boldsymbol{W}^{(l)} = -\eta \boldsymbol{\delta}^{(l)} \boldsymbol{o}^{(l-1)\top}. \tag{6.49}$$

When the step size η is sufficiently small, then the ReLU activation patterns from $\boldsymbol{W}^{(l)} + \Delta \boldsymbol{W}^{(l)}$ do not change from those by $\boldsymbol{W}^{(l)}$ (this issue will be discussed later), so that the new cost function value is given by

$$\widehat{\ell}(\boldsymbol{y}, \boldsymbol{o}^{(L)}) := \|\boldsymbol{y} - \boldsymbol{\Lambda}^{(L)} \boldsymbol{W}^{(L)} \cdots \boldsymbol{\Lambda}^{(l)} (\boldsymbol{W}^{(l)} + \Delta \boldsymbol{W}^{(l)}) \boldsymbol{o}^{(l-1)}\|^2.$$

Recall that we have

$$\begin{aligned}\boldsymbol{\delta}^{(L)} &= \boldsymbol{o}^{(L)} - \boldsymbol{y} \\ &= \boldsymbol{\Lambda}^{(L)} \boldsymbol{W}^{(L)} \cdots \boldsymbol{\Lambda}^{(l)} \boldsymbol{W}^{(l)} \boldsymbol{o}^{(l-1)} - \boldsymbol{y}.\end{aligned}$$

Accordingly, we have

$$\begin{aligned}\widehat{\ell}(\boldsymbol{y}, \boldsymbol{o}^{(L)}) &= \| -\boldsymbol{\delta}^{(L)} - \boldsymbol{\Lambda}^{(L)} \boldsymbol{W}^{(L)} \cdots \boldsymbol{\Lambda}^{(l)} \Delta \boldsymbol{W}^{(l)} \boldsymbol{o}^{(l-1)} \|^2 \\ &= \| -\boldsymbol{\delta}^{(L)} + \eta \boldsymbol{\Lambda}^{(L)} \boldsymbol{W}^{(L)} \cdots \boldsymbol{\Lambda}^{(l)} \boldsymbol{\delta}^{(l)} \boldsymbol{o}^{(l-1)\top} \boldsymbol{o}^{(l-1)} \|^2 \\ &= \left\| \left(\boldsymbol{I} - \eta \|\boldsymbol{o}^{(l-1)}\|^2 \boldsymbol{M}^{(l)} \right) \boldsymbol{\delta}^{(L)} \right\|^2,\end{aligned} \tag{6.50}$$

where we use $\|\boldsymbol{o}^{(l-1)}\|^2 = \boldsymbol{o}^{(l-1)\top} \boldsymbol{o}^{(l-1)}$ and

$$\boldsymbol{M}^{(l)} = \boldsymbol{\Lambda}^{(L)} \boldsymbol{W}^{(L)} \cdots \boldsymbol{W}^{(l+1)} \boldsymbol{\Lambda}^{(l)} \boldsymbol{\Lambda}^{(l)} \boldsymbol{W}^{(l+1)\top} \cdots \boldsymbol{W}^{(L)\top} \boldsymbol{\Lambda}^{(L)},$$

which comes from (6.38). Now, we can easily see that for all $\boldsymbol{x} \in \mathbb{R}^{d^{(L)}}$ we have

$$\boldsymbol{x}^\top \boldsymbol{M}^{(l)} \boldsymbol{x} = \|\boldsymbol{\Lambda}^{(l)} \boldsymbol{W}^{(l+1)\top} \cdots \boldsymbol{W}^{(L)\top} \boldsymbol{\Lambda}^{(L)} \boldsymbol{x}\|^2 \geq 0, \tag{6.51}$$

so that $\boldsymbol{M}^{(l)}$ is positive semidefinite, i.e. its eigenvalues are non-negative. Furthermore, we have

$$\left\| \left(\boldsymbol{I} - \eta \|\boldsymbol{o}^{(l-1)}\|^2 \boldsymbol{M}^{(l)}\right) \boldsymbol{\delta}^{(L)} \right\|^2 \leq \lambda_{\max}^2 \left(\boldsymbol{I} - \eta \|\boldsymbol{o}^{(l-1)}\|^2 \boldsymbol{M}^{(l)}\right) \times \|\boldsymbol{\delta}^{(L)}\|^2, \tag{6.52}$$

where $\lambda_{\max}(\boldsymbol{A})$ denotes the largest eigenvalue of $\boldsymbol{A}$. In addition, we have

$$\lambda_{\max}^2 \left(\boldsymbol{I} - \eta \|\boldsymbol{o}^{(l-1)}\|^2 \boldsymbol{M}^{(l)}\right) = \left(1 - \eta \|\boldsymbol{o}^{(l-1)}\|^2 \lambda_{\max}\left(\boldsymbol{M}^{(l)}\right)\right)^2.$$

Therefore, if the largest eigenvalue satisfies

$$0 \leq \lambda_{\max}\left(\boldsymbol{M}^{(l)}\right) \leq \frac{2}{\eta \|\boldsymbol{o}^{(l-1)}\|^2}, \tag{6.53}$$

we can show

$$\widehat{\ell}(\boldsymbol{y}, \boldsymbol{o}^{(L)}) \leq \|\boldsymbol{\delta}^{(l)}\|^2 = \ell(\boldsymbol{y}, \boldsymbol{o}^{(L)}),$$

so the cost function value decreases with the perturbation.

It is important to emphasize that this strong convergence result is due to the unique property of the ReLU in (6.47), which is never satisfied with other activation functions. This may be another reason for the success of the ReLU in modern deep learning. Having said this, care should be taken since this argument is true only for sufficiently small step size η, so that the ReLU activation patterns after the perturbation do not change. In fact, this may be another reason to choose an appropriate step size in the optimization algorithm.

6.4.4 Local Variational Formulation

Another way of understanding BP is via propagation of the cost function. As shown in Fig. 6.8, after the forward and backward propagation of the input and error, respectively, the resulting optimization problem for the weight update at the l-th layer is given by

$$\min_{\boldsymbol{W}} \| - \boldsymbol{\delta}^{(l)} - \boldsymbol{W} \boldsymbol{o}^{(l-1)}\|^2. \tag{6.54}$$

Note that we have a minus sign in front of $\boldsymbol{\delta}^{(l)}$ inspired by its global counterpart in (6.50). By inspection, we can easily see that the optimal solution for (6.54) is given by

$$\boldsymbol{W}^* = -\frac{1}{\|\boldsymbol{o}^{(l-1)}\|^2}\boldsymbol{\delta}^{(l)}\boldsymbol{o}^{(l-1)\top}, \tag{6.55}$$

since plugging (6.55) in (6.54) makes the cost function zero. Therefore, the optimal search direction for the weight update should be given by

$$\Delta \boldsymbol{W}^{(l)} = -\eta\boldsymbol{\delta}^{(l)}\boldsymbol{o}^{(l-1)\top}, \tag{6.56}$$

which is equivalent to (6.49). The take-away message here is that as long as we can obtain the back-propagated error and the forward-propagated input, we can obtain a local variational formulation, which can be solved by any means.

6.5 Exercises

1. Derive the general form of the activation function $\sigma(x)$ that satisfies the following differential equation:

$$\frac{\sigma(x)}{x} = \frac{\partial\sigma(x)}{\partial x}$$

2. Show that (6.21) is equivalent to (6.20).
3. Recall that L-layer feedforward neural network can be represented recursively by

$$\boldsymbol{o}^{(l)} = \boldsymbol{\sigma}(\boldsymbol{g}^{(l)}), \quad \boldsymbol{g}^{(l)} = \boldsymbol{W}^{(l)}\boldsymbol{o}^{(l-1)} + \boldsymbol{b}^{(l)}, \tag{6.57}$$

for $l = 1, \cdots, L$. When the training data size is 1, the weight update is given by

$$\Delta \boldsymbol{W}^{(l)} = -\gamma\boldsymbol{\delta}^{(l)}\boldsymbol{o}^{(l-1)\top}, \tag{6.58}$$

where $\gamma > 0$ is the step size and

$$\boldsymbol{\delta}^{(l)} := \frac{\partial\ell\left(\boldsymbol{y}, \boldsymbol{o}^{(L)}\right)}{\partial\boldsymbol{g}^{(l)}}. \tag{6.59}$$

a. Derive the update equation similar to (6.58) for the bias term, i.e. $\Delta\boldsymbol{b}^{(l)}$.
b. Suppose the weight matrix $\boldsymbol{W}^{(l)}, l =, \cdots, L$ is a diagonal matrix. Draw the network connection architecture similarly to Fig. 6.8. Then, derive the

backprop algorithm for the diagonal term of the weight matrix, assuming that the bias is zero. You must use the chain rule to derive this.

4. Let a two-layer ReLU neural network $\boldsymbol{f}_{\boldsymbol{\Theta}}$ have an input and output dimension for each layer in $\mathbb{R}^2$, i.e. $\boldsymbol{f}_{\boldsymbol{\Theta}} : \boldsymbol{x} \in \mathbb{R}^2 \mapsto \boldsymbol{f}_{\boldsymbol{\Theta}}(\boldsymbol{x}) \in \mathbb{R}^2$. Suppose that the parameter $\boldsymbol{\Theta}$ of the network is composed of weight and bias:

$$\boldsymbol{\Theta} = \left\{ \boldsymbol{W}^{(1)}, \boldsymbol{W}^{(2)}, \boldsymbol{b}^{(1)}, \boldsymbol{b}^{(2)} \right\}, \tag{6.60}$$

which are initialized as follows:

$$\boldsymbol{W}^{(1)} = \boldsymbol{W}^{(2)} = \begin{bmatrix} 1 & -1 \\ 0 & 1 \end{bmatrix}, \quad \boldsymbol{b}^{(1)} = \boldsymbol{b}^{(2)} = \begin{bmatrix} 1 \\ 0 \end{bmatrix}. \tag{6.61}$$

Then, for a given l_2 loss function

$$\ell(\boldsymbol{\Theta}) = \frac{1}{2}\|\boldsymbol{y} - \boldsymbol{f}(\boldsymbol{x})\|^2 \tag{6.62}$$

and a training data

$$\boldsymbol{x} = [1, -1]^\top, \quad \boldsymbol{y} = [1, 0]^\top, \tag{6.63}$$

compute the weight and bias update for the first two iterations of the backpropagation algorithm. It is suggested that the unit step size, i.e. $\gamma = 1$, be used.

5. We are now interested in extending (6.54) for the training data composed of N samples.

   a. Show that the following equality holds for the local variation formulation:

$$\min_{\boldsymbol{W}} \sum_{n=1}^{N} \| - \boldsymbol{\delta}_n^{(l)} - \boldsymbol{W}\boldsymbol{o}_n^{(l-1)}\|^2 = \min_{\boldsymbol{W}} \| - \boldsymbol{\Delta}^{(l)} - \boldsymbol{W}\boldsymbol{O}^{(l-1)}\|_F^2, \tag{6.64}$$

   where $\|\cdot\|_F$ denotes the Frobenious norm and

$$\boldsymbol{\Delta}^{(l)} = \left[\boldsymbol{\delta}_1^{(l)} \cdots \boldsymbol{\delta}_N^{(l)}\right], \quad \boldsymbol{O}^{(l-1)} = \left[\boldsymbol{o}_1^{(l-1)} \cdots \boldsymbol{o}_N^{(l-1)}\right].$$

   b. Show that there exists a step size $\gamma > 0$ such that the weight perturbation

$$\Delta \boldsymbol{W}^{(l)} = -\gamma \sum_{n=1}^{N} \boldsymbol{\delta}_n^{(l)} \boldsymbol{o}_n^{(l-1)\top}$$

   reduces the cost value in (6.64).

6. Suppose that our activation function is sigmoid. Derive the BP algorithm for the L-layer neural network. What is the main difference of the BP algorithm compared to the network with a ReLU? Is this an advantage or disadvantage? Answer this question in terms of variational perspective.
7. Now we are interested in extending the model in (6.6) to a convolutional neural network model

$$o_i^{(l)} = \sigma\left(\sum_{j=1}^{d^{(l)}} h_{i-j}^{(l)} o_j^{(l-1)} + b_i^{(l)}\right), \tag{6.65}$$

for $i = 1, \cdots, d^{(l)}$, where $h_i^{(l)}$ is the i-th element of the filter $\boldsymbol{h}^{(l)} = [h_1^{(l)}, \cdots, h_p^{(l)}]^\top$.

a. If we want to represent this convolutional neural network in a matrix form,

$$\boldsymbol{o}^{(l)} = \boldsymbol{\sigma}\left(\boldsymbol{W}^{(l)}\boldsymbol{o}^{(l-1)} + \boldsymbol{b}^{(l)}\right), \tag{6.66}$$

what is the corresponding weight matrix $\boldsymbol{W}^{(l)}$? Please show the structure of $\boldsymbol{W}^{(l)}$ explicitly in terms of $\boldsymbol{h}^{(l)}$ elements.

b. Derive the backpropagation algorithm for the filter update $\Delta \boldsymbol{h}^{(l)}$.

Chapter 7
Convolutional Neural Networks

7.1 Introduction

A convolutional neural network (CNN, or ConvNet) is a class of deep neural networks, widely used for analyzing and processing images. Multilayer perceptrons, which we discussed in the previous chapter, usually require fully connected networks, where each neuron in one layer is connected to all neurons in the next layer. Unfortunately, this type of connections inescapably increases the number of weights. In CNNs, the number of weights can be significantly reduced using their shared-weights architecture originated from translation invariant characteristics of the convolution.

A convolutional neural network was first developed by Yann LeCun for hand-written zip code identification [21], inspired by the famous experiments by Hubel and Wiesel for a cat's primary visual cortex [20]. Recall that Hubel and Wiesel found that simple cells in the primary visual cortex of a cat respond best to edge-like stimuli at a particular orientation, position, and phase within their relatively small receptive fields. Yann LeCun realized that the aggregation of LGN (lateral geniculate nucleus) cells with the same receptive field is similar to the convolution operation, which led him to construct a neural network as the cascaded applications of convolution, nonlinearity, and image subsampling, followed by fully connected layers that determine linear hyperplanes in the feature space for the classification tasks. The resulting network architecture, shown in Fig. 7.1, is called LeNet [21].

While the algorithm worked, training to learn 10 digits required 3 days! Many factors contributed to the slow speed, including the vanishing gradient problem, which will be discussed later. Therefore, simpler models that use task-specific handcrafted features such as support vector machines (SVMs) or kernel machines [11] were popular choices in the 1990s and 2000s, because of the artificial neural network's (ANN) computational cost and a lack of understanding of its working mechanism. In fact, the lack of understanding of the ANN has been the main criticism of many contemporary scientists, including the famous Vladmir Vapnik,

J. C. Ye, *Geometry of Deep Learning*, Mathematics in Industry 37,
https://doi.org/10.1007/978-981-16-6046-7_7

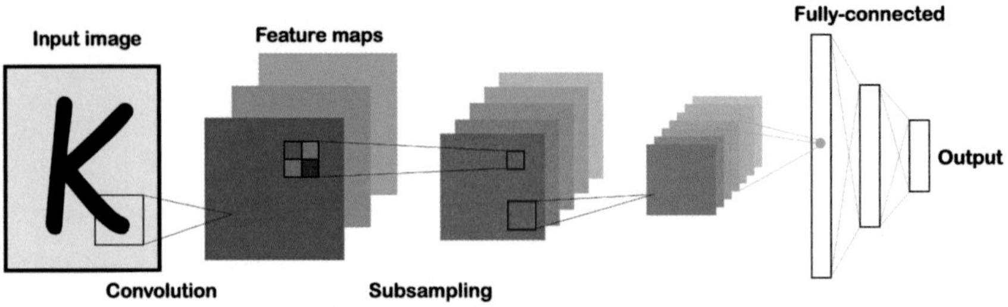

Fig. 7.1 LeNet: the first CNN proposed by Yann LeCun for zip code identificiation [21]

the inventor of the SVM. In the preface of his classical book entitled *The Nature of Statistical Learning Theory* [10], Vapnik expressed his concern saying that "Among artificial intelligence researchers the *hardliners* had considerable influence (it is precisely they who declared that complex theories do not work, simple algorithms do)".

Ironically, the advent of the SVM and kernel machines has led to a long period of decline in neural network research, often referred to as the "AI winter". During the AI winter, the neural network researchers were largely considered pseudo-scientists and even had difficulty in securing research funding. Although there have been several notable publications on neural networks during the AI winter, the revival of convolutional neural network research, up to the level of general public acceptance, has had to wait until the series of deep neural network breakthroughs at the ILSVRC (ImageNet Large Scale Visual Recognition Competition).

In the following section, we give a brief overview of the history of modern CNN research that has contributed to the revival of research on neural networks.

7.2 History of Modern CNNs

7.2.1 AlexNet

ImageNet is a large visual database designed for use in visual object recognition software research [8]. ImageNet contains more than 20,000 categories, consisting of several hundred images. Since 2010, the ImageNet project has an annual software contest, the ImageNet Large Scale Visual Recognition Challenge (ILSVRC) [7], where software programs compete to correctly classify and detect objects and scenes. Around 2011, a good ILSVRC classification error rate, which was based on classical machine learning approaches, was about 27%.

In the 2012 ImageNet Challenge, Krizhevsky et al. [9] proposed a CNN architectures, shown in Fig. 7.2, which is now known as AlexNet. The AlexNet architecture is composed of five convolution layers and three fully connected layers. In fact, the basic components of AlexNet were nearly the same as those of LeNet by

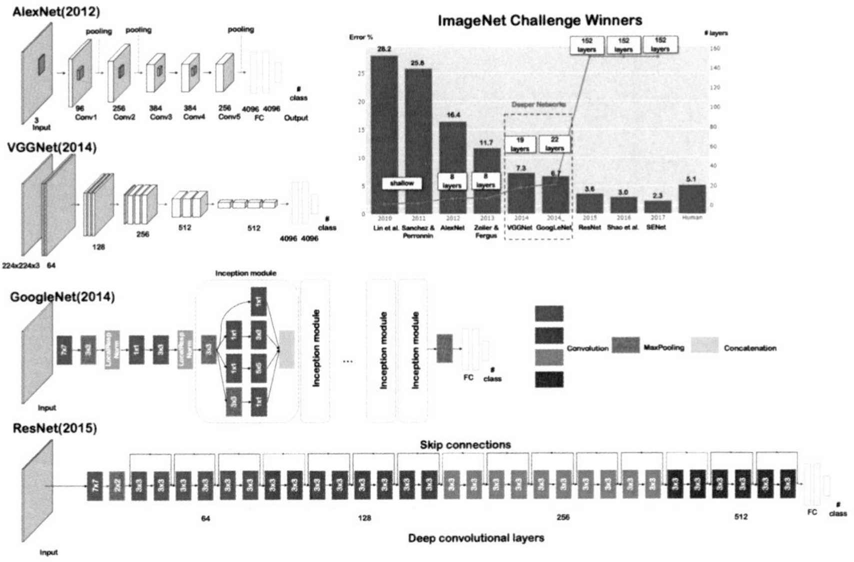

Fig. 7.2 The ImageNet challenges and the CNN winners that have completely changed the landscape of artificial intelligence

Yann LeCun [21], except the new nonlinearity using the rectified linear unit (ReLU). AlexNet got a Top-5 error rate (rate of not finding the true label of a given image among its top 5 predictions) of 15.3%. The next best result in the challenge, which was based on the classical kernel machines, trailed far behind (26.2%).

In fact, the celebrated victory of AlexNet declared the start of a "new era" in data science, as witnessed by more than 75k citations according to Google Scholar as of January 2021. With the introduction of AlexNet, the world was no longer the same, and all the subsequent winners at the ImageNet challenges were deep neural networks, and nowadays CNN surpasses the human observers in ImageNet classification. In the following, we introduce several subsequent CNN architectures which have made significant contributions in deep learning research.

7.2.2 *GoogLeNet*

GoogLeNet [30] was the winner at the 2014 ILSVRC (see Fig. 7.2). As the name "GoogLeNet" indicates, it is from Google, but one may wonder why it is not written as "GoogleNet". This is because the researchers of "GoogLeNet" tried to pay tribute to Yann LeCun's LeNet [21] by containing the word "LeNet".

The network architecture is quite different from AlexNet due to the so-called inception module[30], shown in Fig. 7.3. Specifically, at each inception module, there exist different sizes/types of convolutions for the same input and stack-

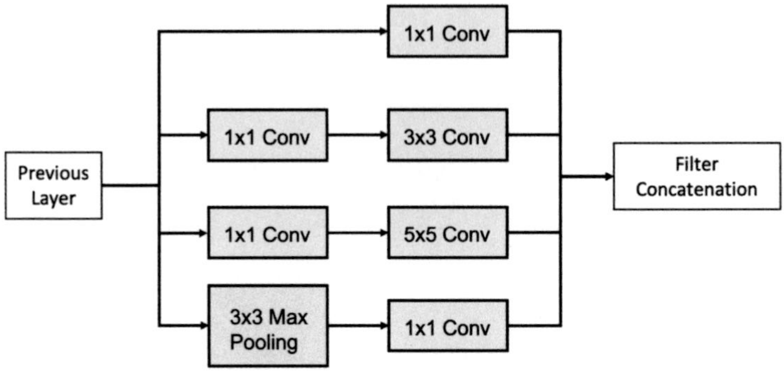

Fig. 7.3 Inception module in GoogLeNet

ing all the outputs. This idea was inspired by the famous 2010 science fiction film *Inception*, in which Leonardo DiCaprio starred. In the film, the renowned director Christoper Nolan wanted to explore "the idea of people sharing a dream space…That gives you the ability to access somebody's unconscious mind." The key concept which GoogLeNet borrowed from the film was the "dream within a dream" strategy, which led to the "network within a network" strategy that improves the overall performance.

7.2.3 VGGNet

VGGNet [31] was invented by the VGG (Visual Geometry Group) from University of Oxford for the 2014 ILSVRC (see Fig. 7.2). Although VGGNet was not the winner of the 2014 ILSVRC (GoogLeNet was the winner at that time, and the VGGNet came second), VGGNet has made a prolonged impact in the machine learning community due to its modular and simple architecture, yet resulting in a significant performance improvement over AlexNet [9]. In fact, the pretrained VGGNet model captures many important image features; therefore, it is still widely used for various purposes such as perceptual loss [32], etc. Later we will use VGGNet to visualize CNNs.

As shown in Fig. 7.2, VGGNet is composed of multiple layers of convolution, max pooling, the ReLU, followed by fully connected layers and softmax. One of the most important observations of VGGNet is that it achieves an improvement over AlexNet by replacing large kernel-sized filters with multiple 3×3 kernel-sized filters. As will be shown later, for a given receptive field size, cascaded application of a smaller size kernel followed by the ReLU makes the neural network more expressive than one with a larger kernel size. This is why VGGNet provided significantly improved performance over AlexNet despite its simple structure.

7.2.4 *ResNet*

In the history of ILSVRC, the Residual Network (ResNet) [33] is considered another masterpiece, as shown in its citation record of more than 68k as of January 2020.

Since the representation power of a deep neural network increases with the network depth, there has been strong research interest in increasing the network depth. For example, AlexNet [9] from 2012 LSVRC had only five convolutional layers, while the VGG network [31] and GoogLeNet [30] from 2014 LSVRC had 19 and 22 layers, respectively. However, people soon realized that a deeper neural network is hard to train. This is because of the vanishing gradient problem, where the gradient can be easily back-propagated to layers closer to the output, but is difficult to be back-propagated far from the output layer since the repeated multiplication may make the gradient so small. As discussed in the previous chapter, the ReLU nonlinearity partly mitigates the problem, since the forward and backward propagation are symmetric, but still the deep neural network turns out to be difficult to train due to an unfavorable *optimization landscape* [34]; this issue will be reviewed later.

As shown in Fig. 7.2, there exist bypass (or skip) connections in the ResNet, representing an identity mapping. The bypass connection was proposed to promote the gradient back-propagation. Thanks to the skip connection, ResNet makes it possible to train up to hundreds or even thousands of layers, achieving a significant performance improvement. Recent researches reveals that the bypass connection also improves the forward propagation, making the representation more expressive [35]. Furthermore, its optimization landscape can be significantly improved thanks to bypass connections that eliminate many local minimizers [35, 36].

7.2.5 *DenseNet*

DenseNet (Dense Convolutional Network) [37] exploits the extreme form of skip connection as shown in Fig. 7.4. In DenseNet, at each layer there exists skip connections from all preceding layers to obtain additional inputs.

Since each layer receives inputs from all preceding layers, the representation power of the network increases significantly, which makes the network compact, thereby reducing the number of channels. With dense connections, the authors demonstrated that fewer parameters and higher accuracy are achieved compared to ResNet [37].

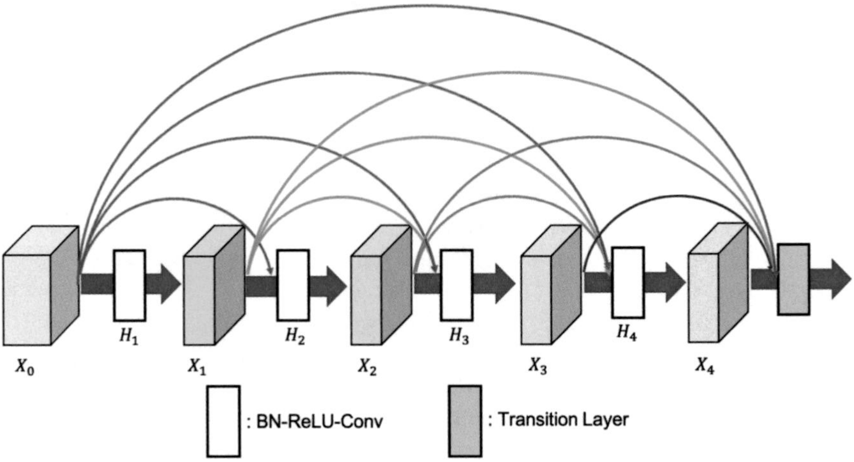

Fig. 7.4 Architecture of DenseNet

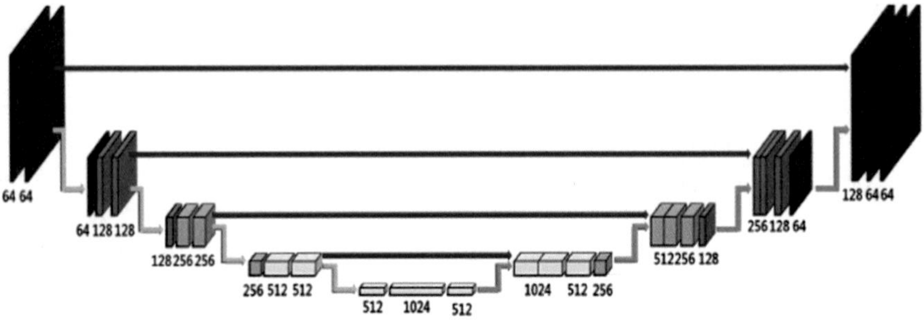

Fig. 7.5 Architecture of U-Net

7.2.6 U-Net

Unlike the aforementioned networks that are designed for ImageNet classification task, the U-Net architecture [38] in Fig. 7.5 was originally proposed for biomedical image segmentation, and is widely used for inverse problems [39, 40].

One of the unique aspects of U-Net is its symmetric encoder–decoder architecture. The encoder part consists of 3×3 convolution, batch normalization [41], and the ReLU. In the decoder part, upsampling and 3×3 convolution are used. Also, there are max pooling layers and skip connections through channel concatenation.

The multi-scale architecture of U-Net significantly increases the receptive field, which may be the main reason for the success of U-Net for segmentation, inverse problems, etc., where global information from all over the images is necessary to update the local image information. This issue will be discussed later. Moreover, the skip connection is important to retain the high-frequency content of the input signal.

The symmetric and multi-scale architecture of U-Net inspired many signal processing discoveries [42], providing important insights into understanding the geometry of deep neural networks.

7.3 Basic Building Blocks of CNNs

Although the aforementioned CNN architectures appear complicated, a closer look at them reveals that they are nothing but cascaded combinations of simple building blocks such as convolution, pooling/unpooling, ReLU, etc. These components are even considered as basic or "primitive" tools in signal processing. In fact, the emergence of the superior performance from the combination of the basic tools is one of the mysteries of deep neural networks, which will be discussed extensively later. In the meanwhile, this section provides a detailed explanation of the basic building blocks of CNNs.

7.3.1 *Convolution*

The convolution is an operation that originates from fundamental properties of linear time invariant (LTI) or linear spatially invariant (LSI) systems. Specifically, for a given LSI system, let $\boldsymbol{h}$ denote the impulse response, then the output image $\boldsymbol{y}$ with respect to the input image $\boldsymbol{x}$ can be computed by

$$\boldsymbol{y} = \boldsymbol{h} * \boldsymbol{x}, \tag{7.1}$$

where $*$ denotes the convolution operation. For example, the 3×3 convolution case for 2-D images can be represented element by element as follows:

$$y[m,n] = \sum_{p,q=-1}^{1} h[p,q]x[m-p,n-q], \tag{7.2}$$

where $y[m,n]$, $h[m,n]$ and $x[m,n]$ denote the (m,n)-element of the matrices $\boldsymbol{Y}$, $\boldsymbol{H}$ and $\boldsymbol{X}$, respectively. One example of computing this convolution is illustrated in Fig. 7.6, where the filter is already flipped for visualization.

It is important to note that the convolution used in CNNs is richer than the simple convolution in (7.1) and Fig. 7.6. For example, a three channel input signal can generate a single channel output as shown in Fig. 7.7a, which is often referred to as multi-input single-output (MISO) convolution. In another example shown in Fig. 7.7b, a 5×5 filter kernel is used to generate 6 (resp. 10) output channels from 3 (resp. and 6) input channels. This is often called the multi-input multi-output

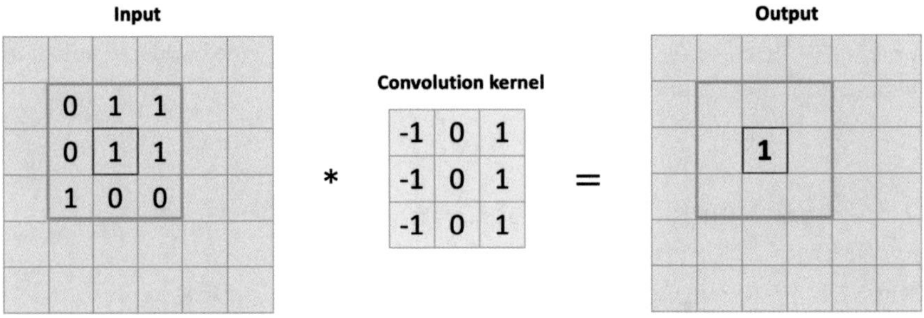

Fig. 7.6 An example of convolution with 3 × 3 filter

(MIMO) convolution. Finally, in Fig. 7.7c, the 1 × 1 filter kernel is used to generate 32 output channels from 64 input channels.

All these seemingly different convolutional operations can be written in a general MIMO convolution form:

$$\boldsymbol{y}_i = \sum_{j=1}^{c_{in}} \boldsymbol{h}_{i,j} * \boldsymbol{x}_j, \quad i = 1, \cdots, c_{out}, \tag{7.3}$$

where c_{in} and c_{out} denote the number of input and output channels, respectively, $\boldsymbol{x}_j$, $\boldsymbol{y}_i$ refer to the j-th input and the i-th output channel image, respectively, and $\boldsymbol{h}_{i,j}$ is the convolution kernel that contributes to the i-th channel output by convolving with the j-th input channel images. For the case of 1 × 1 convolution, the filter kernel becomes

$$\boldsymbol{h}_{i,j} = w_{ij}\delta[0, 0],$$

so that (7.3) becomes the weighted sum of input channel images as follows:

$$\boldsymbol{y}_i = \sum_{j=1}^{c_{in}} w_{ij}\boldsymbol{x}_j, \quad i = 1, \cdots, c_{out}. \tag{7.4}$$

7.3.2 Pooling and Unpooling

A pooling layer is used to progressively reduce the spatial size of the representation to reduce the number of parameters and amount computation in the network. The pooling layer operates on each feature map independently. The most common approaches used in pooling are max pooling and average pooling as shown in Fig. 7.8b. In this case, the pooling layer will always reduce the size of each feature

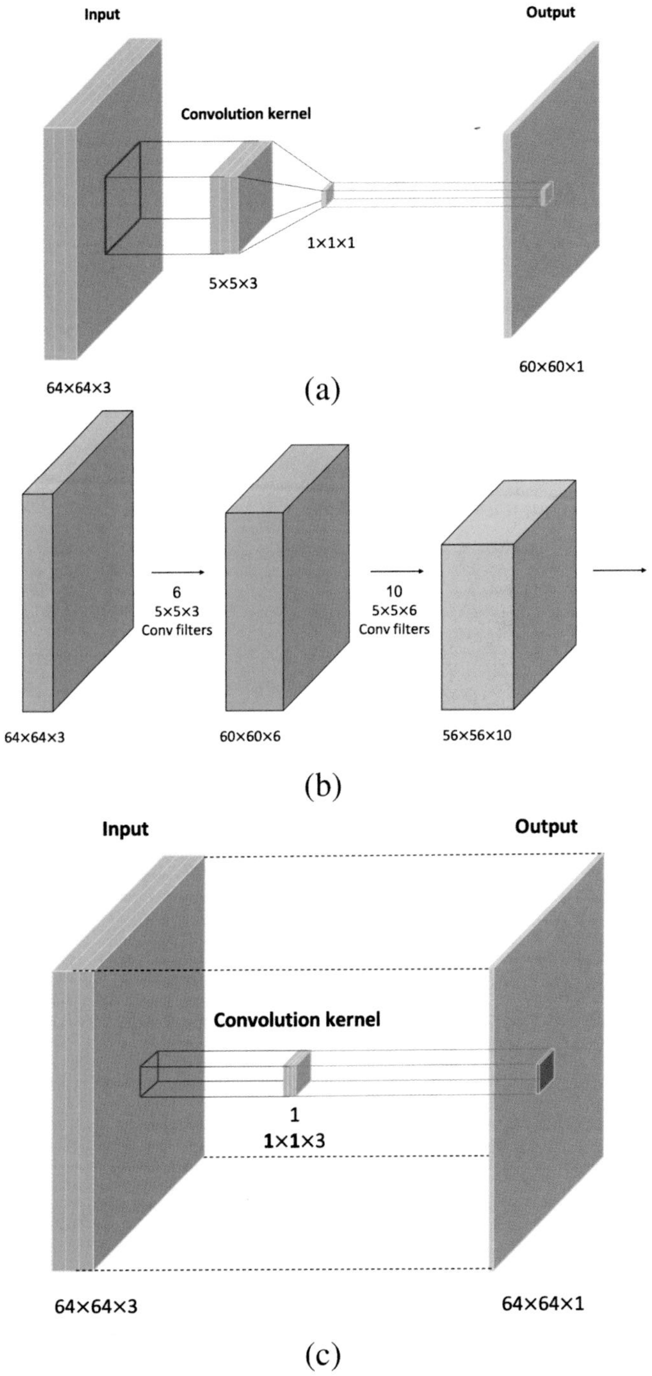

Fig. 7.7 Various convolutions used in CNNs. (**a**) Multi-input single-output (MISO) convolution, (**b**) Multi-input multi-output (MIMO) convolution, (**c**) 1 × 1 convolution

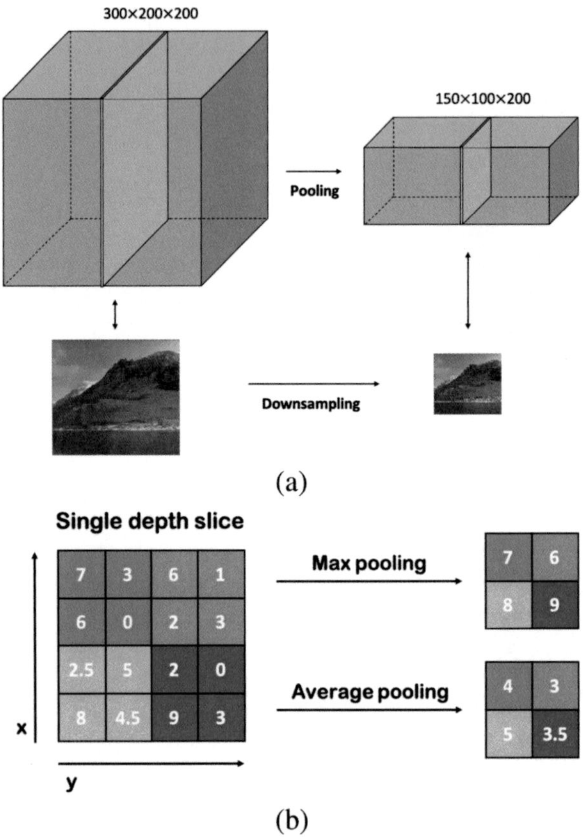

Fig. 7.8 (**a**) Pooling and unpooling operation, (**b**) max and average pooling operation

map by a factor of 2. For example, a max (average) pooling layer in Fig. 7.8b applied to an input image of 16 × 16 produces an output pooled feature map of 8 × 8.

On the other hand, unpooling is an operation for image upsampling. For example, in a narrow meaning of unpooling with respect to max pooling, one can copy the max pooled signal at the original location as shown in Fig. 7.9a. Or one could perform a transpose operation to copy all the pooled signal to the enlarged area as shown in Fig. 7.9b, which is often called the deconvolution. Regardless of the definition, unpooling tries to enlarge the downsampled image.

It was believed that a pooling layer is necessary to impose the spatial invariance in classification tasks [43]. The main ground for this claim is that small movements in the position of the feature in the input image will result in a different feature map after the convolution operation, so that spatially invariant object classification may be difficult. Therefore, downsampling to a lower resolution version of an input signal without the fine detail may be useful for the classification task by imposing invariance to translation.

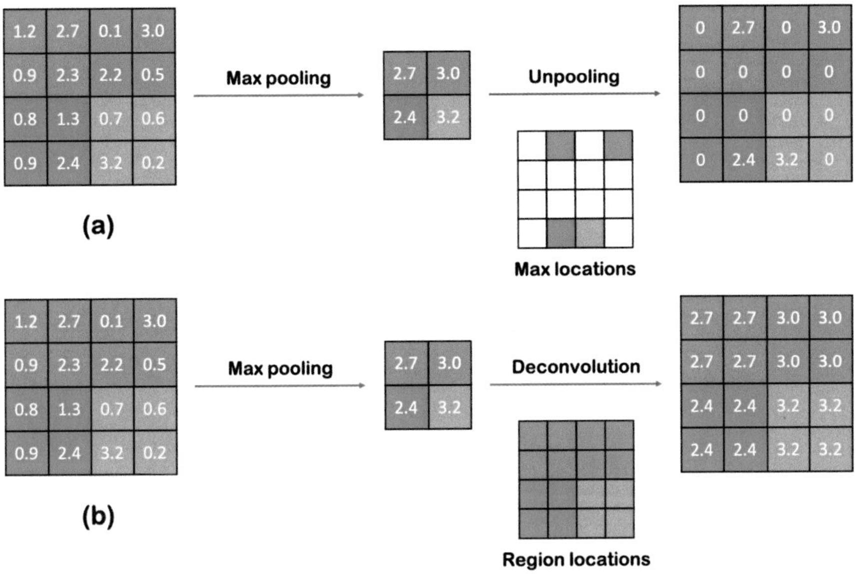

Fig. 7.9 Two ways of unpooling. (**a**) Copying to the original location (unpooling), (**b**) copying to all neighborhood (deconvolution)

However, these classical views have been challenged even by the deep learning godfather, Geoffrey Hinton. In "Ask Me Anything" column on Reddit he said, "the pooling operation used in convolutional neural networks is a big mistake and the fact that it works so well is a disaster. If the pools do not overlap, pooling loses valuable information about where things are. We need this information to detect precise relationships between the parts of an object…".

Regardless of Geoffrey Hinton's controversial comment, the undeniable advantage of the pooling layer results from the increased size of the receptor field. For example, in Fig. 7.10a,b we compare the effective receptive field sizes, which determine the areas of input image affecting a specific point at the output image of a single resolution network and U-Net, respectively. We can clearly see that the receptive field size increases linearly without pooling, but can be expanded exponentially with the help of a pooling layer. In many computer vision tasks, a large receptive field size is useful to achieve better performance. So the pooling and unpooling are very effective in these applications.

Before we move on to the next topic, a remaining question is whether there exists a pooling operation which does not lose any information but increases the receptive field size exponentially. If there is, then it does address Geoffrey Hinton's concern. Fortunately, the short answer is yes, since there exists an important advance in this field from the geometric understanding of deep neural networks [40, 42]. We will cover this issue later when we investigate the mathematical principle.

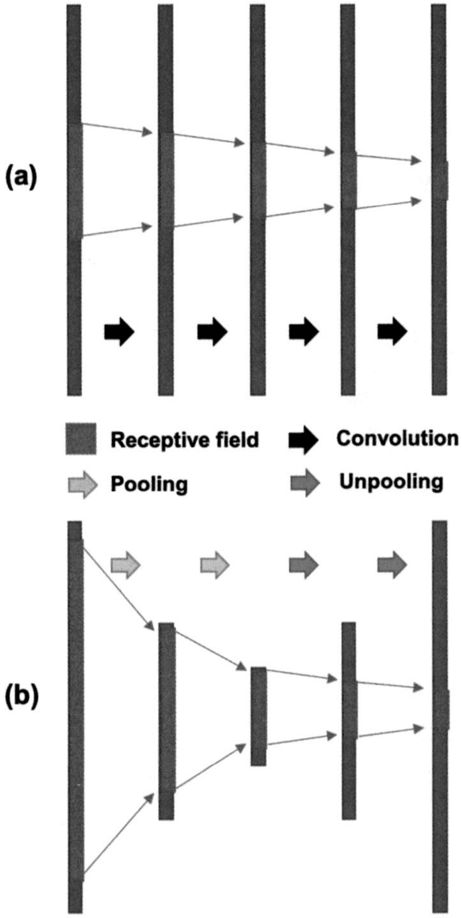

Fig. 7.10 Receptive fields of networks (**a**) without pooling layers, (**b**) with pooling layers

7.3.3 *Skip Connection*

Another important building block, which has been pioneered by ResNet [33] and also by U-Net [38], is the skip connection. For example, as shown in Fig. 7.11, the feature map output from the internal block is given by

$$\boldsymbol{y} = \mathcal{F}(\boldsymbol{x}) + \boldsymbol{x},$$

where $\mathcal{F}(\boldsymbol{x})$ is the output of the standard layers in the CNN with respect to the input $\boldsymbol{x}$, and the additional term $\boldsymbol{x}$ at the output comes directly from the input.

Thanks to the skipped branch, ResNet [33] can easily approximate the identity mapping, which is difficult to do using the standard CNN blocks. Later we will show that additional advantages of the skip connection come from removing local minimizers, which makes the training much more stable [35, 36].

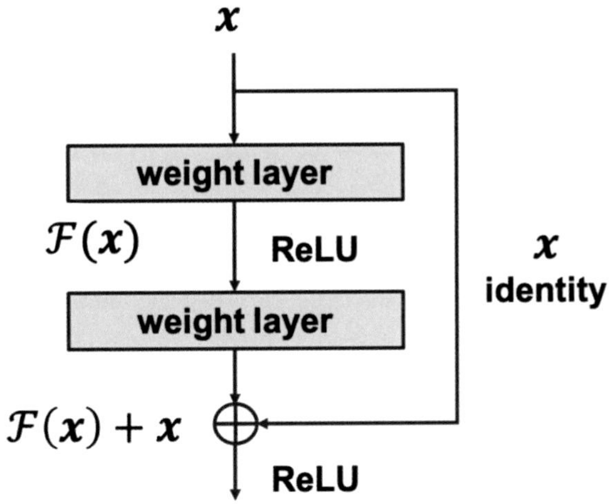

Fig. 7.11 Skip connection in ResNet

7.4 Training CNNs

7.4.1 Loss Functions

When a CNN architecture is chosen, the filter kernel should be estimated. This is usually done during a training phase by minimizing a loss function. Specifically, given input data $\boldsymbol{x}$ and its label $\boldsymbol{y} \in \mathbb{R}^m$, an average loss is defined by

$$c(\boldsymbol{\Theta}) := E[\ell\left(\boldsymbol{y}, \boldsymbol{f}_{\boldsymbol{\Theta}}(\boldsymbol{x})\right)], \tag{7.5}$$

where $E[\cdot]$ denotes the mean, $\ell(\cdot)$ is a loss function, and $\boldsymbol{f}_{\boldsymbol{\Theta}}(\boldsymbol{x})$ is a CNN with input $\boldsymbol{x}$, which is parameterized by the filter kernel parameter set $\boldsymbol{\Theta}$. In (7.5), the mean is usually taken empirically from training data.

For the multi-class classification problem using CNNs, one of the most widely used losses is the softmax loss [44]. This is a multi-class extension of the binary logistic regression classifier we studied before. A softmax classifier produces normalized class probabilities, and also has a probabilistic interpretation. Specifically, we perform the softmax transform:

$$\widehat{\boldsymbol{p}}(\boldsymbol{\Theta}) = \frac{\boldsymbol{e}^{\boldsymbol{f}_{\Theta}(\boldsymbol{x})}}{\boldsymbol{1}^\top \boldsymbol{e}^{\boldsymbol{f}_{\Theta}(\boldsymbol{x})}}, \tag{7.6}$$

where $\boldsymbol{e}^{\boldsymbol{f}_\Theta(\boldsymbol{x})}$ denotes the element-by-element application of the exponential. Then, using the softmax loss, the average loss is computed by

$$c(\boldsymbol{\Theta}) = -E\left[\sum_{i=1}^{m} y_i \log \widehat{p}_i(\boldsymbol{\Theta})\right], \tag{7.7}$$

where y_i and $\widehat{p}_i$ denote the i-th elements of $\boldsymbol{y}$ and $\widehat{\boldsymbol{p}}$, respectively. If the class label $\boldsymbol{y} \in \mathbb{R}^m$ is normalized to have probabilitistic meaning, i.e. $\mathbf{1}^\top \boldsymbol{y} = 1$, then (7.7) is indeed the cross entropy between the target class distribution and the estimated class distribution.

For the case of regression problems using CNNs, which are quite often used for image processing tasks such as denoising, the loss function is usually defined by the norm, i.e.

$$c(\boldsymbol{\Theta}) = E\|\boldsymbol{y} - \boldsymbol{f}_\Theta(\boldsymbol{x})\|_p^p \tag{7.8}$$

where $p = 1$ for the l_1 loss and $p = 2$ for the l_2 loss.

7.4.2 *Data Split*

In training CNNs, available data sets should be first split into three categories: training, validation, and test data sets, as shown in Fig. 7.12. The training data is also split into *mini-batches* so that each mini-batch can be used for stochastic gradient computation. The training data set is then used to estimate the CNN filter kernels, and the validation set is used to monitor whether there exists any overfitting issue in the training.

For example, Fig. 7.13a shows the example of overfitting that can be monitored during the training using the validation data. If this type of overfitting happens, several approaches should be taken to achieve stable training behavior as shown in Fig. 7.13b. Such a strategy will be discussed in the following section.

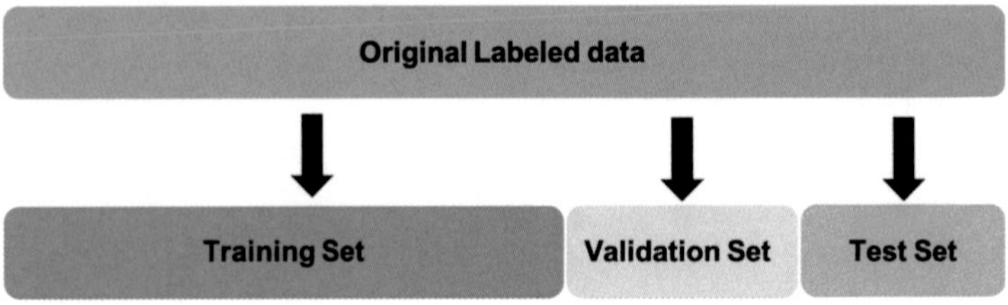

Fig. 7.12 Available data split into training, validation, and test data sets

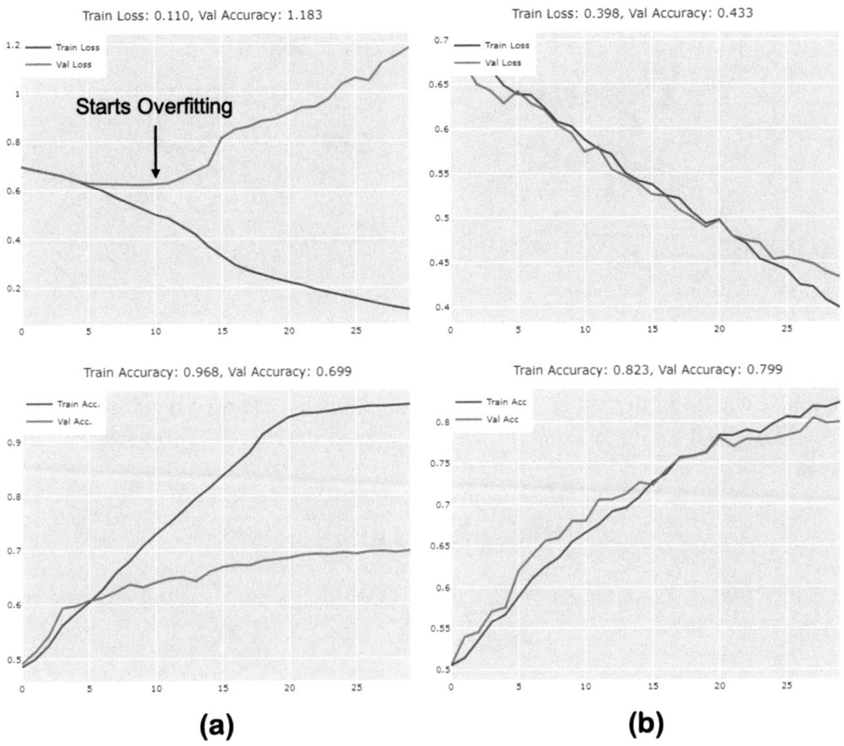

Fig. 7.13 Neural network training dynamics: (**a**) overfitting problems, (**b**) no overfitting

7.4.3 Regularization

When we observe the overfitting behaviors similar to Fig. 7.13a, the easiest solution is to increase the training data set. However, in many real-world applications, the training data are scarce. In this case, there are several ways to regularize the neural network training.

7.4.3.1 Data Augmentation

Using data augmentation we generate artificial training instances. These are new training instances created, for example, by applying geometric transformations such as mirroring, flipping, rotation, on the original image so that it doesn't change the label information.

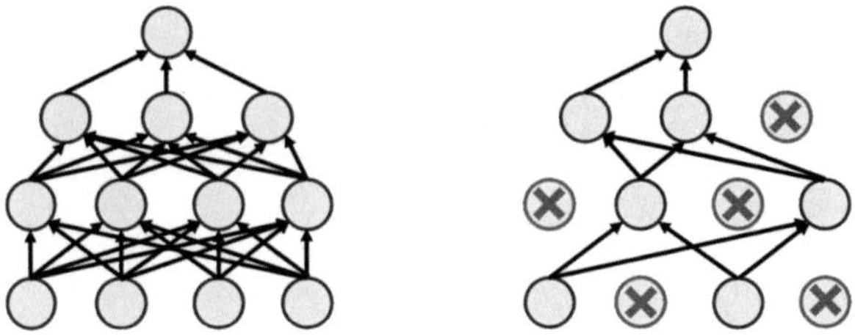

Fig. 7.14 Example of dropout

7.4.3.2 Parameter Regularization

Another way to mitigate the overfitting problem is by adding a regularization term for the original loss. For example, we can convert the loss in (7.5) to the following form:

$$c_{reg}(\boldsymbol{\Theta}) := E[\ell\left(\boldsymbol{y}, \boldsymbol{f}_{\boldsymbol{\Theta}}(\boldsymbol{x})\right)] + R(\boldsymbol{\Theta}), \tag{7.9}$$

where $R(\boldsymbol{\Theta})$ is a regularization function. Recall that similar techniques were used in the kernel machines.

7.4.3.3 Dropout

Another unique regularization used for deep learning is the dropout [45]. The idea of a dropout is relatively simple. During the training time, at each iteration, a neuron is temporarily "dropped" or disabled with probability p. This means all the inputs and outputs to some neurons will be disabled at the current iteration. The dropped-out neurons are resampled with probability p at every training step, so a dropped-out neuron at one step can be active at the next one. See Fig. 7.14. The reason that the dropout prevents overfitting is that during the random dropping, the input signal for each layer varies, resulting in additional data augmentation effects.

7.5 Visualizing CNNs

As already mentioned, hierarchical features arise in the brain during visual information processing. A similar phenomenon can be observed in the convolution neural network, once it is properly trained. In particular, VGGNet provides very intuitive information that is well correlated with the visual information processing in the brain.

For example, Fig. 7.15 illustrates the input signal that maximizes the filter response at specific channels and layers of VGGNet [31]. Remember that the filters

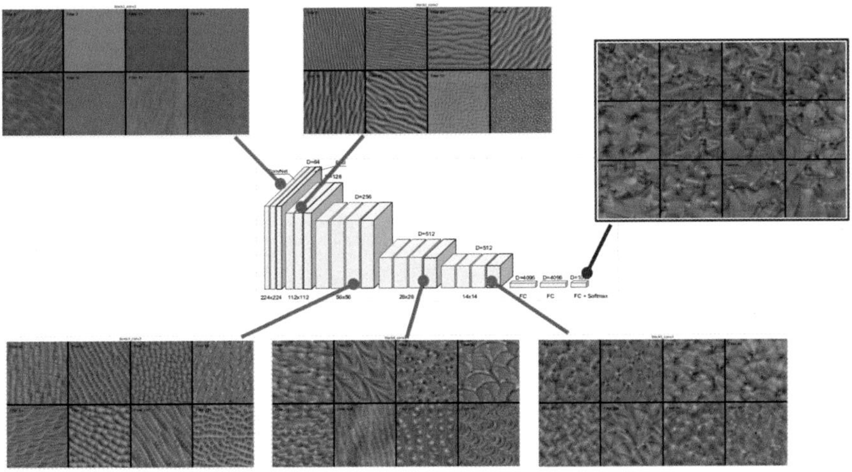

Fig. 7.15 Input images that maximize filter responses at specific channels and layers of VGGNet

are of size 3×3, so rather than visualizing the filters, an input image where this filter activates the most is displayed for specific channel and layer filters. In fact, this is similar to the Hubel and Wiesel experiments where they analyzed the input image that maximizes the neuronal activation.

Figure 7.15 shows that at the earlier layers the input signal maximizing filter response is composed of directional edges similar to the Hubel and Wiesel experiment. As we go deeper into the network, the filters build on each other and learn to code more complex patterns. Interestingly, the input images that maximize the filter response get more complicated as the depth of the layer increases. In one of the filter sets, we can see several objects in different orientations, as the particular position in the picture is not important as long as it is displayed somewhere where the filter is activated. Because of this, the filter tries to identify the object in multiple positions by encoding it in multiple places in the filter.

Finally, the blue box in Fig. 7.15 shows the input images that maximize the response on the last softmax level in the specific classes. In fact, this corresponds to the visualization of the input images that maximize the class categories. In a certain category, an object is displayed several times in the images. The emergence of the hierarchical feature from simple edges to the high-level concept is similar to visual information processing in the brain.

Finally, Fig. 7.16 visualizes the feature maps on the different levels of VGGNets in relation to a cat picture. Since the output of a convolution layer is a 3D volume, we will only visualize some of the images. As can be seen from Fig. 7.16, a feature map develops from edge-like features of the cat to information with the lower-resolution, which describes the location of the cat. In the later levels, the feature map works with a probability map in which the cat is located.

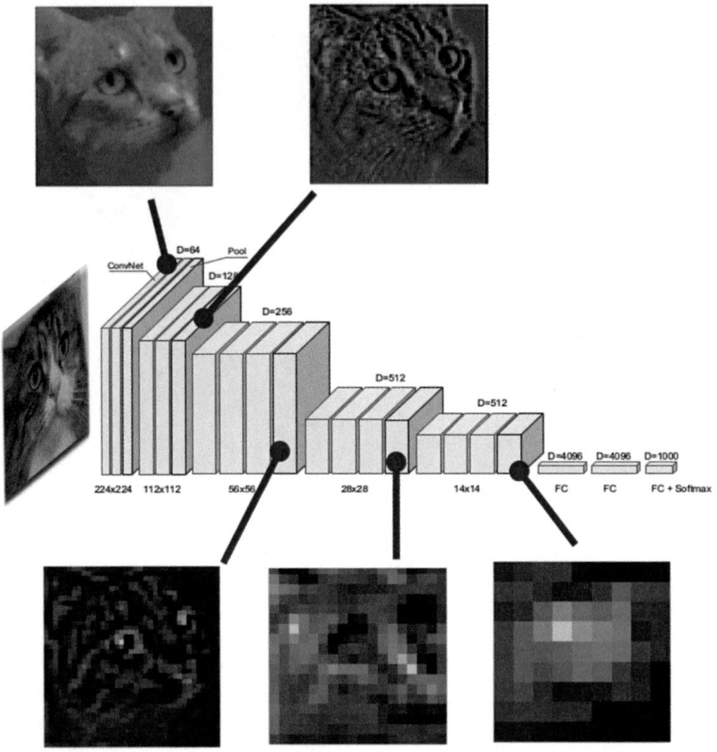

Fig. 7.16 Visualization of feature maps at several channels and layers of VGGNets when the input image is a cat

7.6 Applications of CNNs

CNN is the most widely used neural network architecture in the age of modern AI. Similar to the visual information processing in the brain, the CNN filters are trained in such a way that hierarchical features can be captured effectively. This can be one of the reasons for CNN's success with many image classification problems, low-level image processing problems, and so on.

In addition to commercial applications in unmanned vehicles, smartphones, commercial electronics, etc., another important application is in the field of medical imaging. CNN has been successfully used for disease diagnosis, image segmentation and registration, image reconstruction, etc.

For example, Fig. 7.17 shows a segmentation network architecture for cancer segmentation. Here, the label is the binary mask for cancer, and the backbone CNN is based on the U-Net architecture, where there exists a softmax layer at the end for pixel-wise classification. Then, the network is trained to classify the background

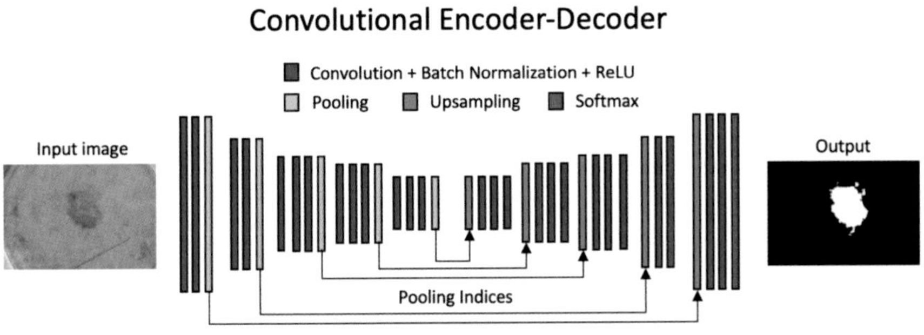

Fig. 7.17 Cancer segmentation using U-Net

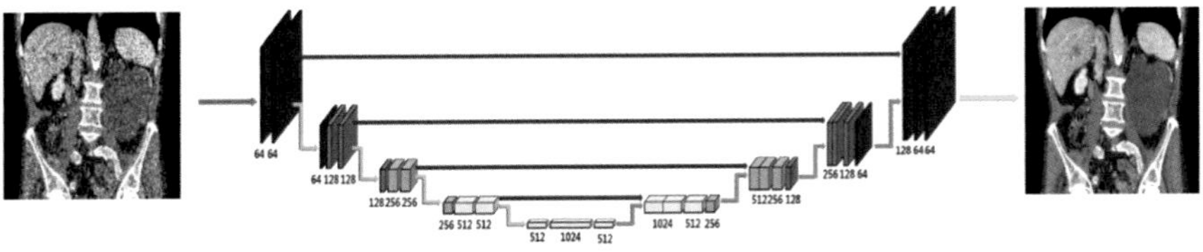

Fig. 7.18 CNN-based low-dose CT denoising

and the cancer regions. Very similar architecture can be also used for noise removal in low-dose CT images, as shown in Fig. 7.18. Instead of using the softmax layer, the network is trained with a regression loss of l_1 or l_2 using the high-quality, low-noise images as a reference. In fact, one of the amazing and also mysterious parts of deep learning is that a similar architecture works for different problems simply by changing the training data.

Because of this simplicity in designing and training CNNs, there are many exciting new startups targeting novel medical applications of AI. As the importance of global health care increases with the COVID-19 pandemic, medical imaging and general health care are undoubtedly among the most important areas of AI. Therefore, for the application of AI to health, opportunities are so numerous that we need many young, bright researchers who can invest their time and effort in AI research to improve human health care.

7.7 Exercises

1. Consider the VGGNet in Fig. 7.2. In its original implementation, the convolution kernel was 3×3.
   a. What is the total number of convolution filter sets in VGGNet?
   b. Then, what is the total number of trainable parameters in VGGNet including convolution filters and fully connected layers? (Hint: for the fully connected

layers, the number of parameters should be input dimension $\times$ output dimension).

2. Let your neural network code for Modified National Institute of Standards and Technology database (MNIST) classification be denoted by $\boldsymbol{f}_{\boldsymbol{\Theta}}(\boldsymbol{x})$, where $\boldsymbol{\Theta}$ represents trainable parameters and $\boldsymbol{x}$ is the input image. The last layer of your neural network should be the softmax layer given by

$$\widehat{\boldsymbol{p}}(\boldsymbol{\Theta}) = \frac{\boldsymbol{e}^{\boldsymbol{f}_{\Theta}(\boldsymbol{x})}}{\mathbf{1}^\top \boldsymbol{e}^{\boldsymbol{f}_{\Theta}(\boldsymbol{x})}}, \tag{7.10}$$

where $\boldsymbol{e}^{\boldsymbol{f}_{\Theta}(\boldsymbol{x})}$ denotes the element-by-element application of the exponential.

 a. What is the meaning of the softmax layer?
 b. Suppose you define the loss function for the MNIST classifier by

$$c(\boldsymbol{\Theta}) = -E\left[\sum_{i=1}^{10} y_i \log \widehat{p}_i(\boldsymbol{\Theta})\right], \tag{7.11}$$

 where $\widehat{p}_i$ denotes the i-th element of $\widehat{\boldsymbol{p}}$. Then, what is $\{y_i\}_{i=1}^{10}$? Provide answers when the label has the values 1 and 5.

3. For the given U-Net architecture in Fig. 7.5, compute the effective receptive field size. Now, suppose that there exist no pooling layers. What is the effective receptive field size?
4. Let $\boldsymbol{u} = [u[0], \cdots, u[n-1]]^\top \in \mathbb{R}^n$ and $\boldsymbol{v} = [v[0], \cdots, v[n-1]]^\top \in \mathbb{R}^n$. We define a circular convolution between the two vectors:

$$(u \circledast v)[n] = \sum_{i=0}^{n-1} u[n-i]v[n],$$

where the periodic boundary condition is assumed. Now, for any vector $\boldsymbol{x} \in \mathbb{R}^{n_1}$ and $\boldsymbol{y} \in \mathbb{R}^{n_2}$ with $n_1, n_2 \leq m$, define their circular convolution in $\mathbb{R}^n$:

$$\boldsymbol{x} \circledast \boldsymbol{y} = \boldsymbol{x}^0 \circledast \boldsymbol{y}^0,$$

where $\boldsymbol{x}^0 = [\boldsymbol{x}, \mathbf{0}^{n-n_1}]^\top$ and $\boldsymbol{y}^0 = [\boldsymbol{y}, \mathbf{0}^{n-n_2}]^\top$. Finally, for any $\boldsymbol{v} \in \mathbb{R}^{n_1}$ with $n_1 \leq n$, define the flip $\overline{v}[n] = v^0[-n]$.

 a. For an input signal $\boldsymbol{x} \in \mathbb{R}^n$ and a filter $\overline{\boldsymbol{\psi}} \in \mathbb{R}^n$, show that

$$\boldsymbol{y} = \boldsymbol{x} \circledast \overline{\boldsymbol{\psi}} = \mathbb{H}_r^n(\boldsymbol{x})\boldsymbol{\psi}\ , \tag{7.12}$$

where $\mathbb{H}_r^n(\boldsymbol{x}) \in \mathbb{R}^{n\times r}$ is a wrap-around Hankel matrix:

$$\mathbb{H}_r^n(\boldsymbol{x}) = \begin{bmatrix} x[0] & x[1] & \cdots & x[r-1] \\ x[1] & x[2] & \cdots & x[r] \\ \vdots & \vdots & \ddots & \vdots \\ x[n-1] & x[n] & \cdots & x[r-2] \end{bmatrix}. \tag{7.13}$$

b. For an input signal $\boldsymbol{x} \in \mathbb{R}^n$ and a filter $\boldsymbol{\psi} \in \mathbb{R}^r$ with $r \leq n$, show the following commutative relationship for the circular convolution in $\mathbb{R}^n$:

$$\boldsymbol{x} \circledast \overline{\boldsymbol{\psi}} = \mathbb{H}_r^n(\boldsymbol{x})\boldsymbol{\psi} = \mathbb{H}_n^n(\boldsymbol{\psi})\boldsymbol{x} = \boldsymbol{\psi} \circledast \overline{\boldsymbol{x}}. \tag{7.14}$$

c. For a given $\boldsymbol{f}, \boldsymbol{u} \in \mathbb{R}^n$ and $\boldsymbol{v} \in \mathbb{R}^r$ with $r \leq n$, show that

$$\boldsymbol{u}^\top \boldsymbol{F}\boldsymbol{v} = \boldsymbol{u}^\top (\boldsymbol{f} \circledast \overline{\boldsymbol{v}}) = \boldsymbol{f}^\top (\boldsymbol{u} \circledast \boldsymbol{v}) = \langle \boldsymbol{f}, \boldsymbol{u} \circledast \boldsymbol{v}\rangle, \tag{7.15}$$

where $\boldsymbol{F} = \mathbb{H}_r^n(\boldsymbol{f})$.

d. Let the multi-input single-output (MISO) circular convolution for the p-channel input $\boldsymbol{Z} = [\boldsymbol{z}_1, \cdots, \boldsymbol{z}_p] \in \mathbb{R}^{n\times p}$ and the output $\boldsymbol{y} \in \mathbb{R}^n$ be defined by

$$\boldsymbol{y} = \sum_{j=1}^{p} \boldsymbol{z}_j \circledast \overline{\boldsymbol{\psi}}^j, \tag{7.16}$$

where $\boldsymbol{\psi}_i \in \mathbb{R}^r$ denotes a r-dimensional vector and $\overline{\boldsymbol{\psi}}_i \in \mathbb{R}^n$ refers to its flip. Then, show that (7.16) can be represented in a matrix form:

$$\boldsymbol{y} = \boldsymbol{Z} \circledast \boldsymbol{\Psi} = \mathbb{H}_{r|p}^n(\boldsymbol{Z})\,\boldsymbol{\Psi}, \tag{7.17}$$

where

$$\boldsymbol{\Psi} = \begin{bmatrix} \boldsymbol{\psi}^1 \\ \vdots \\ \boldsymbol{\psi}^p \end{bmatrix}$$

and

$$\mathbb{H}_{r|p}^n(\boldsymbol{Z}) := \left[\mathbb{H}_r^n(\boldsymbol{z}_1)\ \mathbb{H}_r^n(\boldsymbol{z}_2)\ \cdots\ \mathbb{H}_r^n(\boldsymbol{z}_p)\right]. \tag{7.18}$$

e. Let the multi-input multi-output (MIMO) circular convolution for the p-channel input $\boldsymbol{Z} = [\boldsymbol{z}_1, \cdots, \boldsymbol{z}_p] \in \mathbb{R}^{n\times p}$ and q-channel output $\boldsymbol{Y} = [\boldsymbol{y}_1, \cdots, \boldsymbol{y}_q] \in \mathbb{R}^{n\times q}$ be defined by

$$\boldsymbol{y}_i = \sum_{j=1}^{p} \boldsymbol{z}_j \circledast \overline{\boldsymbol{\psi}}_{i,j}, \quad i = 1, \cdots, q, \tag{7.19}$$

where p and q are the number of input and output channels, respectively; $\boldsymbol{\psi}_{i,j} \in \mathbb{R}^r$ denotes a r-dimensional vector and $\overline{\boldsymbol{\psi}}_{i,j} \in \mathbb{R}^n$ refers to its flip. Then, show that (7.19) can be represented in a matrix form by

$$\boldsymbol{Y} = \sum_{j=1}^{p} \mathbb{H}_r^n(\boldsymbol{z}_j)\boldsymbol{\Psi}_j = \mathbb{H}_{r|p}^n(\mathbf{Z})\,\boldsymbol{\Psi},$$

where

$$\boldsymbol{\Psi} = \begin{bmatrix} \boldsymbol{\Psi}_1 \\ \vdots \\ \boldsymbol{\Psi}_p \end{bmatrix} \quad \text{where} \quad \boldsymbol{\Psi}_j = \begin{bmatrix} \boldsymbol{\psi}_{1,j} \cdots \boldsymbol{\psi}_{q,j} \end{bmatrix}.$$

f. In convolutional neural networks (CNNs), a 1×1 convolution often follows the convolution layer. For 1-D signals, this operation can be written as

$$\boldsymbol{y}_i = \sum_{j=1}^{p} w_j \left(\boldsymbol{z}_j \circledast \overline{\boldsymbol{\psi}}_{i,j}\right), \quad i = 1, \cdots, q, \tag{7.20}$$

where w_j denotes the j-th index of 1×1 convolution filter weighting. Show that this can be represented in a matrix form by

$$\boldsymbol{Y} = \sum_{j=1}^{p} w_j \mathbb{H}_r^n(\boldsymbol{z}_j)\boldsymbol{\Psi}_j = \mathbb{H}_{r|p}^n(\mathbf{Z})\,\Psi^w, \tag{7.21}$$

where

$$\boldsymbol{\Psi}^w = \begin{bmatrix} w_1\boldsymbol{\Psi}_1 \\ \vdots \\ w_p\boldsymbol{\Psi}_p \end{bmatrix}. \tag{7.22}$$

Chapter 8
Graph Neural Networks

8.1 Introduction

Many important real-world data sets are available in the form of graphs or networks: social networks, world-wide web (WWW), protein-interaction networks, brain networks, molecule networks, etc. See some examples in Fig. 8.1. In fact, the complex interaction in real systems can be described by different forms of graphs, so that graphs can be a ubiquitous tool for representing complex systems.

A graph consists of nodes and edges as shown in Fig. 8.2. Although it looks simple, the main technical problem is that the number of nodes and edges in many interesting real-world problems is very large, and cannot be traced by simple inspection. Accordingly, people are interested in different forms of machine learning approaches to extract useful information from diagrams.

With a machine learning tool, for example, a *node classification* can be carried out in which different labels are assigned to each node in a complex diagram. This could be used to classify the function of proteins in the interaction network (see Fig. 8.3a). *Link analysis* is another important problem in graph machine learning, which is about finding missing links between nodes. As shown in Fig. 8.3b, link analysis can be used for repurposing drugs for new types of pathogens or diseases. Yet another important goal of graph analysis is *community detection*. For example, one could identify a subnetwork that consists of disease proteins (see Fig. 8.3c).

Despite the wide range of possible applications, approaches to neural networks in graphs are not as mature as other studies of neural networks for images, voices, etc. This is because the processing and learning of graph data require new perspectives on neural networks.

For example, as shown in Fig. 8.4, the basic assumption of convolutional neural networks (CNNs) is that images have pixel values on regular grids, but graphs have irregular node and edge structure so that the applications of basic modules such as convolution, pooling, etc., are not easy. Another serious problem is that, although CNN training data consists of images or their patches of the same size, the

J. C. Ye, *Geometry of Deep Learning*, Mathematics in Industry 37,
https://doi.org/10.1007/978-981-16-6046-7_8

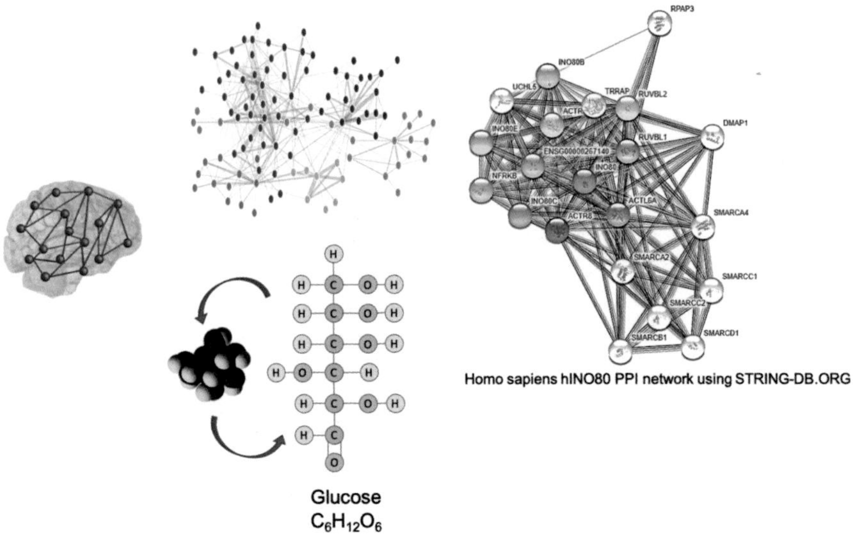

Fig. 8.1 Examples of graphs in real life

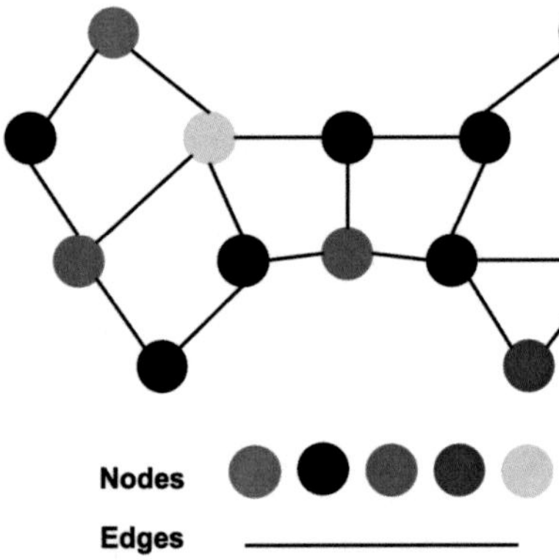

Fig. 8.2 Nodes and edges in a graph

training data of the graph neural network usually consists of graphs with different numbers of nodes, network topology, and so on. For example, in graphical neural network approaches for examining the toxicity of drug candidates, the chemicals in the training data set can have a different number of molecules. This leads to the fundamental question in the graph machine learning task: What do we learn from the training data?

In fact, the main advantage of neural network approaches over other machine learning approaches like compressed sensing [46] and low-rank matrix factorization [47], etc. is that the neural network approaches are *inductive*, which means that the

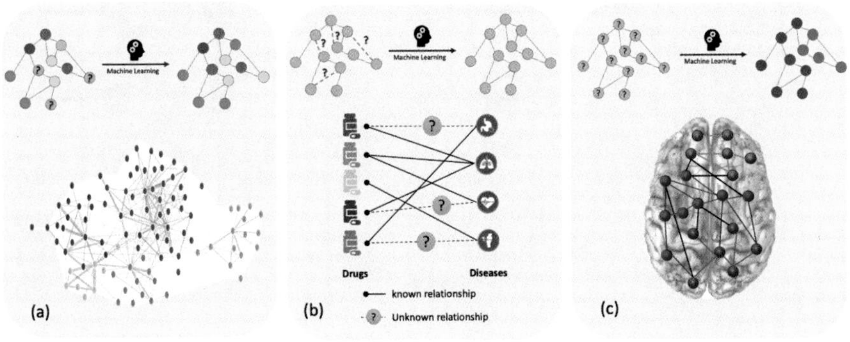

Fig. 8.3 Several application goals of machine learning on graphs: (**a**) node classification, (**b**) link analysis, (**c**) community detection

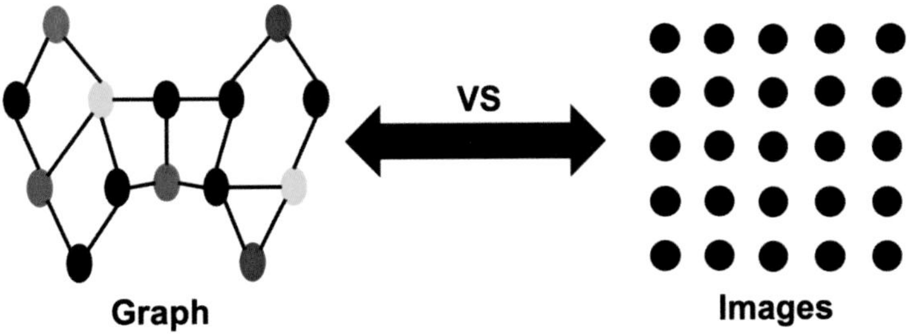

Fig. 8.4 Difference between image domain CNN and graph neural network

trained neural network is not just applied to the data on which the network resides and was originally trained, but also to other unseen data during training.

However, given that each graph in training data is different in its structure (for example, with different node and edge numbers and even topology), what kind of inductive information can we get from the graph neural network training? Although the universal approximation theorem [48] guarantees that neural networks can approximate any nonlinear function, it is not even clear *which* nonlinear function a graph neural network tries to approximate.

Hence the main aim of this chapter is to answer these puzzling questions. In fact, we will focus on how machine learning researchers came up with brilliant ideas to enable inductive learning independent of different graph structures in the training phase.

8.2 Mathematical Preliminaries

Before we discuss graph neural networks, we review basic mathematical tools from graph theory.

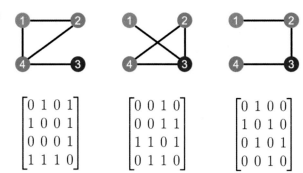

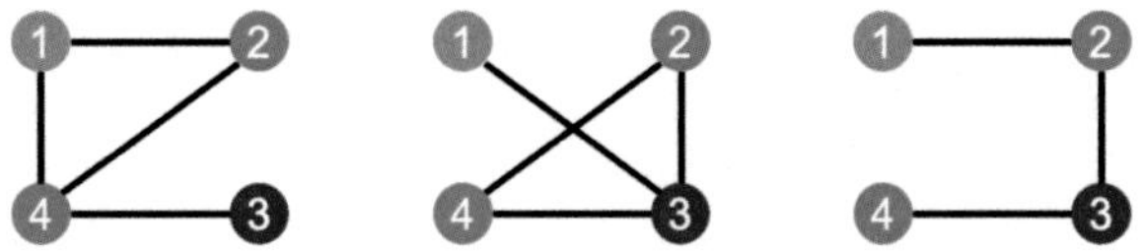

$$\begin{bmatrix} 0&1&0&1\\ 1&0&0&1\\ 0&0&0&1\\ 1&1&1&0 \end{bmatrix} \quad \begin{bmatrix} 0&0&1&0\\ 0&0&1&1\\ 1&1&0&1\\ 0&1&1&0 \end{bmatrix} \quad \begin{bmatrix} 0&1&0&0\\ 1&0&1&0\\ 0&1&0&1\\ 0&0&1&0 \end{bmatrix}$$

Fig. 8.5 Examples of graphs and their adjacency matrices

8.2.1 Definition

We denote a graph $G = (V, E)$ with a set of vertices $V(G) = \{1, \cdots, N\}$ with $N := |V|$ and edges $E(G) = \{e_{ij}\}$, where an edge e_{ij} connects vertices i and j if they are adjacent or neighbors. The set of neighborhoods of a vertex v is denoted by $\mathcal{N}(v)$. For weighted graphs, the edge e_{ij} has a real value. If G is an unweighted graph, then E is a sparse matrix with elements of either 0 or 1.

For a simple unweighted graph with vertex set V, the adjacency matrix is a square $|V| \times |V|$ matrix $\boldsymbol{A}$ such that its element a_{uv} is one when there is an edge from vertex u to vertex v, and zero when there is no edge. See Fig. 8.5 for some examples of adjacency matrices for undirected graphs. Note that the dimension of the adjacency matrix varies depending on the number of nodes in the graph.

8.2.2 Graph Isomorphism

A graph can exist in different forms having the same number of vertices, edges, and also the same edge connectivity. Such graphs are called isomorphic graphs. Formally, two graphs G and H are said to be isomorphic if (1) their numbers of components (vertices and edges) are equal, and (2) their edge connections are identical. Some examples of isomorphic graphs are shown in Fig. 8.6.

Graph isomorphism is widely used in many areas where identifying similarities between graphs is important. In these areas, the graph isomorphism problem is often referred to as the graph matching problem. Some practical uses of graph isomorphism include identifying identical chemical compounds in different configurations, checking equivalent circuits in electronic design, etc.

Unfortunately, testing graph isomorphism is not a trivial task. Even if the number of nodes is the same, two isomorphic graphs, for example, can have different adjacency matrices, since the order of the nodes in the isomorphic graph can be arbitrary, but the structure of their adjacency matrices is critically determined by the order of the nodes. In fact, the graph isomorphism problem is one of the few standard problems whose complexity remains unsolved.

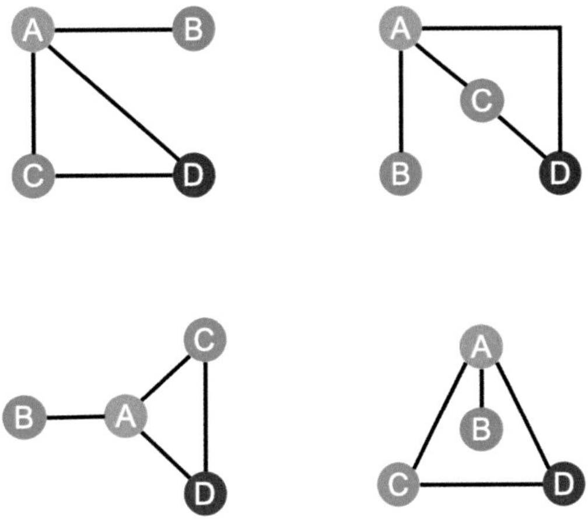

Fig. 8.6 Examples of isomorphic graphs. All four graphs are isomorphic to each other

8.2.3 *Graph Coloring*

A node coloring is a function $V(G) \mapsto \Sigma$ with arbitrary codomain Σ. Then a node colored or colored graph (G, l) is a graph G endowed with a node coloring $l : V(G) \mapsto \Sigma$. We say that $l(v)$ is a color of $v \in V(G)$.

Figure 8.7 shows an example of graph coloring in a molecular system [49]. In the initial phase, each node is colored with feature vectors that consist of various chemical properties. In this case, the codomain is $\Sigma \subset \mathbb{R}^5$. Using machine learning approaches, the node colors can be updated sequentially by taking into account the color information of neighboring nodes to extract useful global properties of the molecule.

8.3 Related Works

Since each diagram in the training data has a different configuration, the main concern of machine learning of graphs is to assign latent vectors in the common latent space to graphs, subgraphs, or nodes so that standard CNN, perceptron, etc. can be applied to the latent space for inference or regression. This procedure is often called *graph embedding*, as shown in Fig. 8.8. One of the most important research topics in graph neural networks is to find an *inductive rule* for the graph embedding that can be applied to graphs with a different number of nodes, topologies, etc.

Unfortunately, one of the difficulties associated with the graphs is that they are unstructured. In fact, there is a lot of unstructured data that we encounter in everyday life, and one of the most important classes of unstructured data is natural language. Therefore, many of the graphics machine learning techniques are borrowed from natural language processing (NLP). So this section explains the key idea of natural language processing.

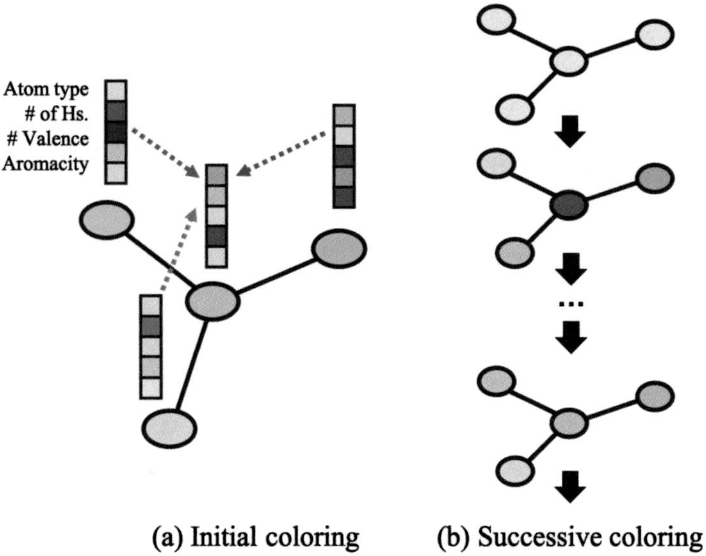

Fig. 8.7 Node coloring example in a molecular system. (**a**) Initial coloring with feature vectors, (**b**) its successive update using a machine learning approach

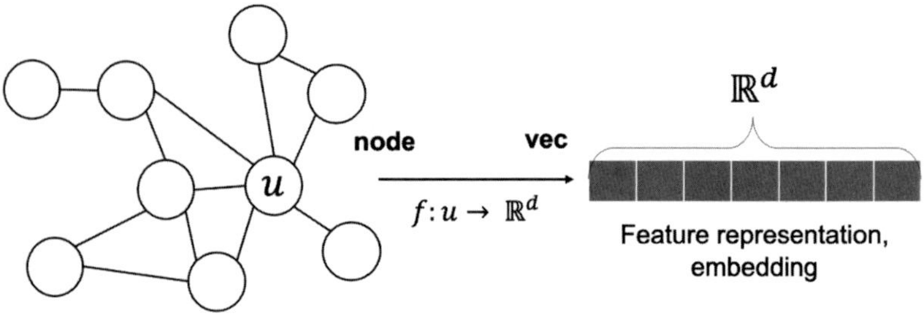

Fig. 8.8 Concept of graph embedding to a latent vector

8.3.1 Word Embedding

Word embedding is one of the most popular representations for natural language processing. Basically, it is a vector representation of a particular word that can capture the context of a word in a document, its semantic and syntactic similarity, its relationship to other words, and so on.

For example, consider a vocabulary "king". From its semantic meaning, one could come to the following conclusion:

$$\text{King} - \text{Man} + \text{Woman} = \text{Queen}. \tag{8.1}$$

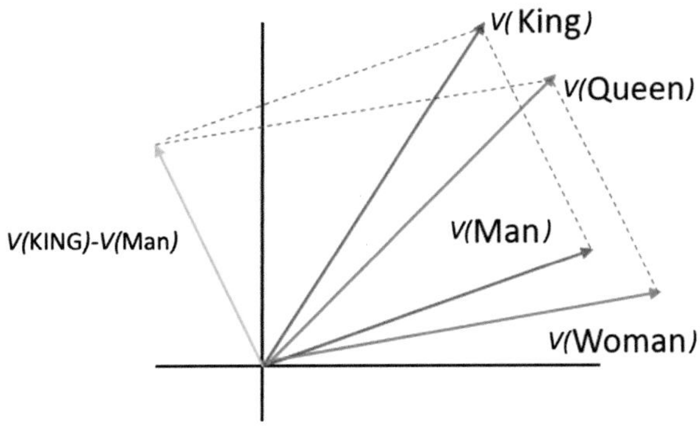

Fig. 8.9 Example of vector operation via word embedding

However, there is no mathematical operation in natural language to formally derive (8.1). Hence, the idea of word embedding is to perform this operation through vector operations in latent space. Specifically, let $\mathcal{V}(\cdot)$ denote a mapping of a vocabulary to a vector in $\mathbb{R}^d$. Then, the goal of the word embedding is to find the mappings $\mathcal{V}$ so that

$$\mathcal{V}(\text{King}) - \mathcal{V}(\text{Man}) + \mathcal{V}(\text{Woman}) = \mathcal{V}(\text{Queen}). \tag{8.2}$$

This concept is illustrated in Fig. 8.9. There are several ways to embed a word. The main problem here is to represent each word in large text as a vector so that similar words are close together in latent space.

Among the various ways of performing word embedding, the so-called *word2vec* is one of the most frequently used methods [50, 51]. Word2vec is composed of a two-layer neural network. The network is trained in two complementary ways: continuous bag of words (CBOW) and skip-gram. The key idea of these approaches is that there are significant causal relationships and redundancies between words in natural languages, the information of which can be used to embed words in vector space. In the following, we describe them in detail.

8.3.1.1 CBOW

CBOW begins with the assumption that a missing word can be found from its surrounding words in the sentence. For example, consider a sentence: *The big dog is chasing the small rabbit.* The idea of CBOW is that a target word in the sentence (which is usually the center word), for example, "dog" as shown in Fig. 8.10, can be estimated from the nearby words within the context window (for example, using "big" and "is" for the case of context window size $c = 1$). In general, for a given context window size c, the i-th word $\boldsymbol{x}_i$ is assumed to be estimated using the adjacent words within a window, i.e. $\{\boldsymbol{x}_j \mid j \in \mathcal{I}_c(i)\}$, as shown in Fig. 8.10, where

$$\mathcal{I}_c(i) := \{i - c, \cdots, i - 1, i + 1, \cdots, i + c\}. \tag{8.3}$$

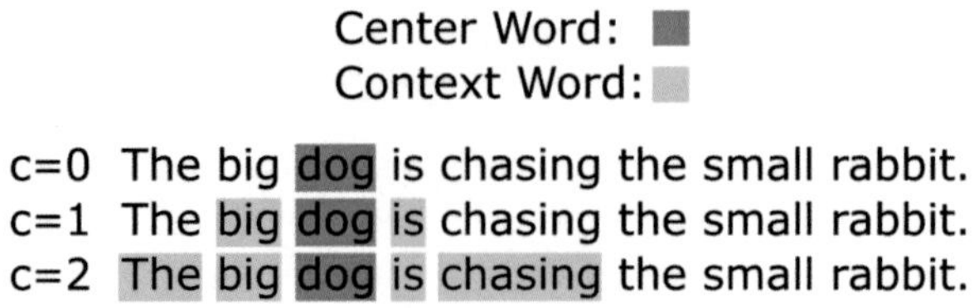

Fig. 8.10 Example of context and center words in CBOW

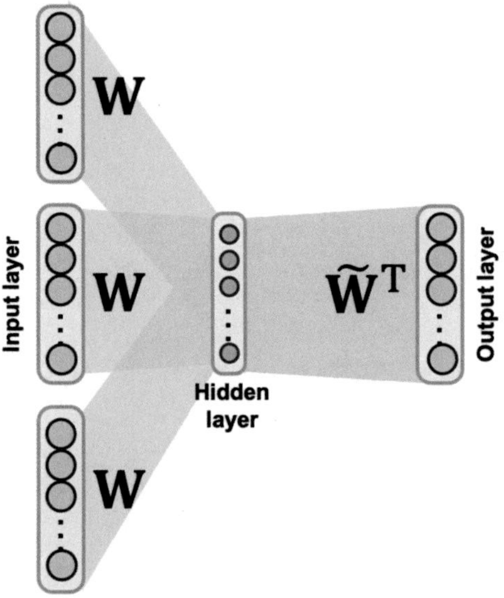

Fig. 8.11 Encoder–decoder structure of CBOW

Now, here comes the fun part. In CBOW, rather than directly estimating the word $\boldsymbol{x}_i$, it employs an encoder-decoder structure as depicted in Fig. 8.11. Specifically, an encoder, represented by the shared weight $\boldsymbol{W}$, converts input $\boldsymbol{x}_n$ into a corresponding latent space vector, and then the decoder with the weight $\widetilde{\boldsymbol{W}}$ converts the latent vector into the estimate of the target word $\hat{\boldsymbol{x}}_i$.

Furthermore, one of the most important assumptions of CBOW is that the latent vector of the missing word is represented as the average value of the latent vectors of the adjacent words, i.e.

$$\boldsymbol{h}_i = \frac{1}{2c-1} \sum_{k \in \mathcal{I}_c(i)} \boldsymbol{W}\boldsymbol{x}_k. \tag{8.4}$$

Unique words

I
play
games
every
day

One-hot encoding

	I	play	games	every	day
I	1	0	0	0	0
play	0	1	0	0	0
games	0	0	1	0	0
every	0	0	0	1	0
day	0	0	0	0	1

Fig. 8.12 Example of one-hot vector encoding for vocabularies

Specifically, using the $2c - 1$ input vectors and the shared encoder weight, we generate $2c - 1$ latent vectors, after which their average value is generated. Then, the center word is estimated by decoding from the averaged latent vector with the weight $\widetilde{\boldsymbol{W}}$:

$$\hat{\boldsymbol{x}}_i = \widetilde{\boldsymbol{W}}^\top \boldsymbol{h}_i. \tag{8.5}$$

Note that other than the softmax unit in the network output, which will be explained later, there are no non-linearities in the hidden layer of CBOW.

To start off, one should first build the corpus vocabulary, where we could map each vocabulary to a unique numeric identifier $\boldsymbol{x}_i$. For example, if the corpus size is M, then $\boldsymbol{x}_i$ is an M-dimensional vector with one-hot vector encoding as shown in Fig. 8.12. Once the neural network in CBOW is trained, the word embedding can be simply done using the encoder part of the network.

The very strict assumption that the center word may be similar to the average of the surrounding vocabularies in the latent space works amazingly well, and CBOW is one of the most popular classical word embedding techniques [50, 51].

8.3.1.2 Skip-Gram

Skip-gram can be seen as a complementary idea of CBOW. The main idea behind the skip-gram model is this: once the neural network is trained, the latent vector generated by the focus word can predict every word in the window with high probability. For example, Fig. 8.13 shows the example of how we extract the focus word and the target word within different window sizes. Here the green word is the focus word from which the target words in the window are estimated.

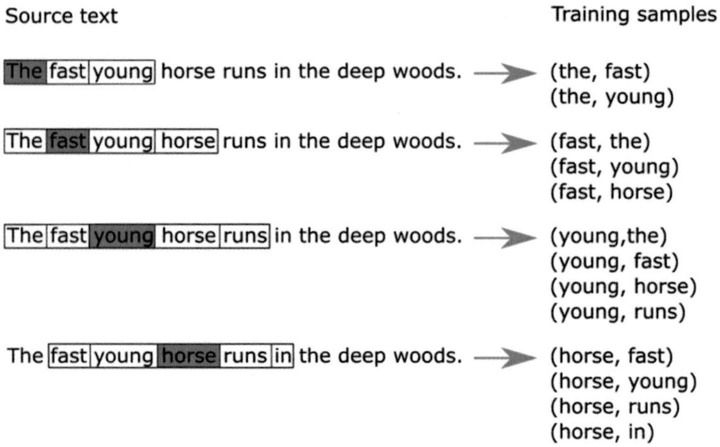

Fig. 8.13 The focus and target vocabularies in skip-gram training

Similar to CBOW, the neural network training is carried out in the form of latent vectors. In particular, the focus word encoded with a one-hot vector is converted to a latent vector using an encoder with the weight $\boldsymbol{W}$, and then the latent vector is decoded via a parallel decoder network with the shared weight $\widetilde{\boldsymbol{W}}^{\top}$, as shown in Fig. 8.14. So the basic assumption of skip-gram can be written by

$$\boldsymbol{x}_j \simeq \widetilde{\boldsymbol{W}}^{\top} \boldsymbol{h}_i, \quad \forall j \in \mathcal{I}_c(i), \tag{8.6}$$

where the latent vector $\boldsymbol{h}_i$ is given by

$$\boldsymbol{h}_i = \boldsymbol{W} \boldsymbol{x}_i. \tag{8.7}$$

Again, there are no non-linearities in the hidden layer of skip-gram other than the softmax unit in the network output.

8.3.2 *Loss Function*

The loss function for the neural network training in word2vec deserves further discussion. Similar to the classification problem, the loss function is based on the cross entropy between the target word and the generated word from the decoder.

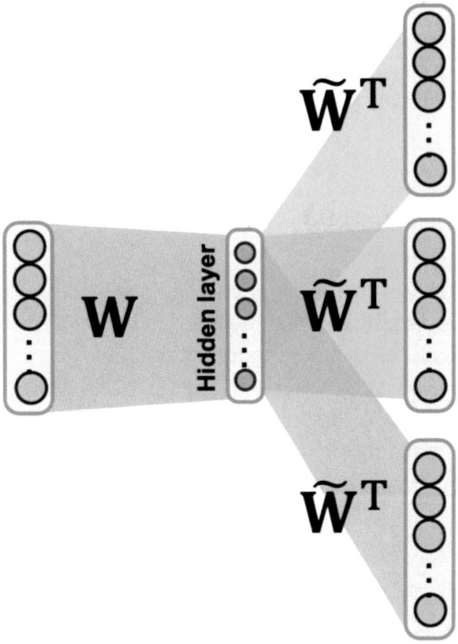

Fig. 8.14 Encoder–decoder structure of skip-gram

In the case of CBOW in particular, it should be remembered that the target vector $\boldsymbol{x}_i$ is also a one-hot encoded vector. Let t_k denote the nonzero index of the vocabulary vector $\boldsymbol{x}_k$. Then, the loss function of CBOW can be written as a softmax function:

$$\begin{aligned}\ell_{CBOW}(\boldsymbol{W}, \widetilde{\boldsymbol{W}}) &= -\log\left(\frac{e^{\widetilde{\boldsymbol{w}}_{t_i}^{\top}\boldsymbol{h}_i}}{\sum_{k=1}^{M} e^{\widetilde{\boldsymbol{w}}_{t_k}^{\top}\boldsymbol{h}_i}}\right) \\ &= -\widetilde{\boldsymbol{w}}_{t_i}^{\top}\boldsymbol{h}_i + \log\left(\sum_{k=1}^{M} e^{\widetilde{\boldsymbol{w}}_{t_k}^{\top}\boldsymbol{h}_i}\right), \end{aligned} \tag{8.8}$$

where the latent vector $\boldsymbol{h}_i$ is given by the average latent vector in (8.4). On the other hand, the loss function for the skip-gram is given by

$$\begin{aligned}\ell_{skipgram}(\boldsymbol{W}, \widetilde{\boldsymbol{W}}) &= -\log\left(\prod_{j\in\mathcal{I}_c(i)}^{C} \frac{e^{\widetilde{\boldsymbol{w}}_{t_j}^{\top}\boldsymbol{h}_i}}{\sum_{k=1}^{M} e^{\widetilde{\boldsymbol{w}}_{t_k}^{\top}\boldsymbol{h}_i}}\right) \\ &= -\sum_{j\in\mathcal{I}_c(i)} \widetilde{\boldsymbol{w}}_{t_j}^{\top}\boldsymbol{h}_i + C\log\left(\sum_{k=1}^{M} e^{\widetilde{\boldsymbol{w}}_{t_k}^{\top}\boldsymbol{h}_i}\right), \end{aligned} \tag{8.9}$$

where the latent vector $\boldsymbol{h}_i$ is given by (8.7).

In both approaches, the computationally intensive step is the calculation of the denominator terms, since we have to calculate them for each corpus of size M. One of the main research efforts is to approximate this term without sacrificing the accuracy [50, 51].

8.4 Graph Embedding

Similar to word embedding, graph embedding is used to convert nodes, subgraphs, and their features into vectors in latent space so that similar nodes, subgraphs, and features are close together in latent space.

As summarized in Fig. 8.15, currently there exist three types of approaches for graph embedding: matrix factorization, random walks, and neural network approaches [52]. In the following, we first briefly review the first two approaches, then we discuss neural network approaches in detail.

8.4.1 Matrix Factorization Approaches

The main assumption of matrix factorization approaches for graph embedding is that an adjacency matrix can be decomposed into low rank matrices. More specifically, for a given adjacency matrix $\boldsymbol{A} \in \mathbb{R}^{N \times N}$, its low rank matrix decomposition is to find $\boldsymbol{U}, \boldsymbol{V} \in \mathbb{R}^{N \times d}$ such that

$$\boldsymbol{A} \simeq \boldsymbol{U}\boldsymbol{V}^{\top}, \tag{8.10}$$

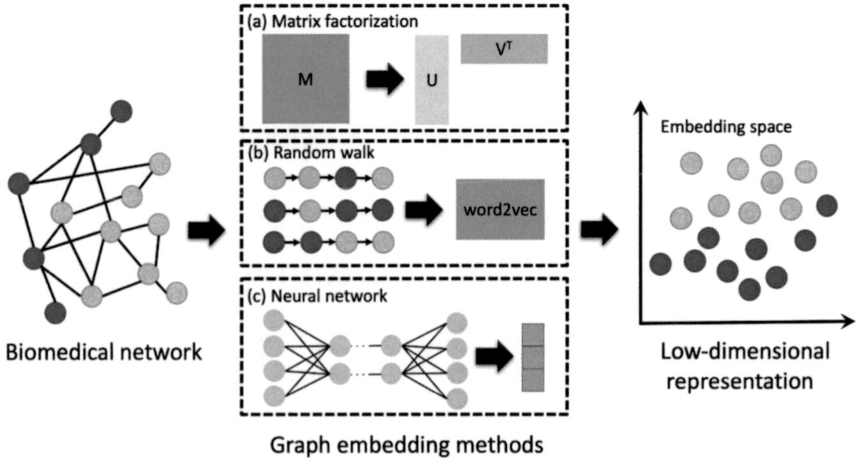

Fig. 8.15 Various approaches for graph embedding

where d is the latent space dimension. Then, the i-th node embedding in the latent space $\mathbb{R}^d$ is given by

$$\boldsymbol{h}_i = \boldsymbol{V}^\top \boldsymbol{x}_i \in \mathbb{R}^d,$$

where $\boldsymbol{x}_i \in \mathbb{R}^N$ is again the one-hot vector encoded i-th node vector.

Aside from the computational complexity of matrix decomposition, there are several limitations in matrix factorization approaches as a graph embedding. First, to use a matrix factorization approach, the number of the nodes should be the same. Second, the approach is not inductive, but rather *transductive*. This means that the learned embedding transform only works for the graph with the same adjacency matrix and if the connectivity changes, the embedding does not hold anymore.

8.4.2 Random Walks Approaches

Random walks approaches for graph embedding are very closely related to the word embedding, in particular, word2vec [50, 51]. Here, we review two powerful random walk approaches: DeepWalks [53] and node2vec [54].

8.4.2.1 DeepWalks

The main intuition of DeepWalks [53] is that random walks are comparable to sentences in the word2vec approach so that word2vec can be used for embedding each node of a graph. More specifically, as depicted in Fig. 8.16, the method basically consists of three steps:

- Sampling: A graph is sampled with random walks. A few random walks with specific length are performed from each node.
- Training skip-gram: The skip-gram network is trained by accepting a node from the random walk as a one-hot vector as an input and target.
- Node embedding: From the encoder part of the trained skip-gram, each node in a graph is embedded into a vector in the latent space.

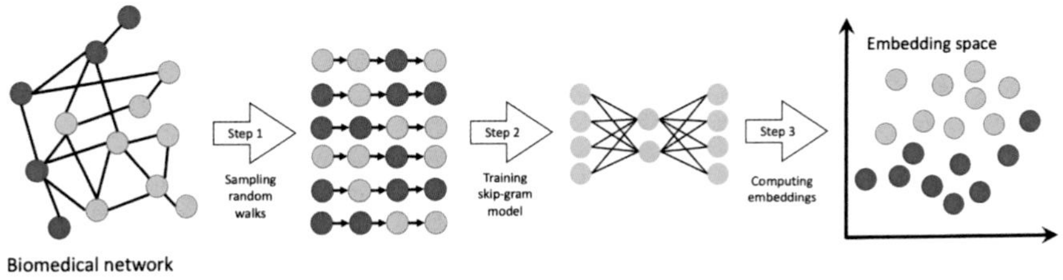

Fig. 8.16 Graph node embedding using DeepWalks

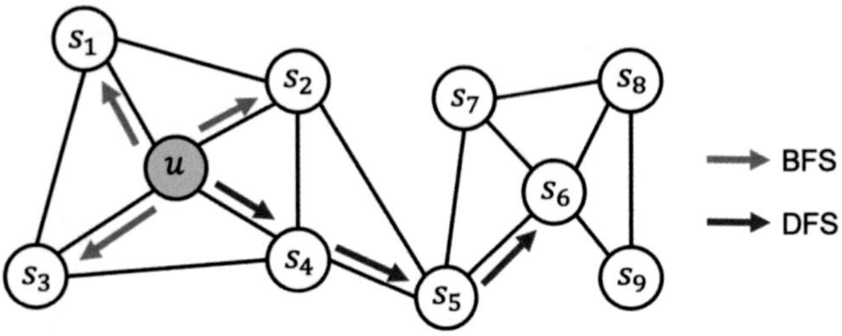

Fig. 8.17 BFS and DFS random walks in node2vec

8.4.2.2 Node2vec

Node2vec is a modification of DeepWalks with subtle but significant differences. Node2vec is parameterized by two parameters p and q. The parameter p prioritizes a breadth-first-search (BFS) procedure, while the parameter q prioritizes a depth-first-search (DFS) procedure. The decision of where to walk next is therefore influenced by probabilities $1/p$ or $1/q$. As shown in Fig. 8.17, BFS is ideal for learning local neighbors, while DFS is better for learning global variables. Node2vec can switch to and from the two priorities depending on the task. Other procedures, such as the use of skip-gram, are exactly the same as DeepWalks.

8.4.3 Neural Network Approaches

Recently, there has been significant progress and growing interest in graph neural networks (GNNs), which comprise graph operations performed by deep neural networks. For example, spectral graph convolution approaches [55], graph convolution network (GCN) [56], graph isomorphism network (GIN) [57], graphSAGE [58], to just name a few.

Although these approaches have been derived from different assumptions and approximations, common GNNs typically integrate the features on each layer in order to embed each node features into a predefined feature vector of the next layer. The integration process is implemented by selecting suitable functions for aggregating features of the neighborhood nodes. Since a level in the GNN aggregates its 1-hop neighbors, each node feature is embedded with features in its k-hop neighbor of the graph after k aggregation layers. These features are then extracted by applying a readout function to obtain a nodal embedding.

Specifically, let $\boldsymbol{x}_v^{(t)}$ denote the t-th iteration feature vector at the v-th node. Then, this graph operation is generally composed of the $AGGREGATE$, and $COMBINE$ functions:

$$\boldsymbol{a}_v^{(t)} = AGGREGATE\left(\left\{\!\!\left\{\boldsymbol{x}_u^{(t-1)} : u \in \mathcal{N}(v)\right\}\!\!\right\}\right),$$

$$\boldsymbol{x}_v^{(t)} = COMBINE\left(\boldsymbol{x}_v^{(t-1)}, \boldsymbol{a}_v^{(t)}\right),$$

where the $AGGREGATE$ function collects features of the neighborhood nodes to extract the aggregated feature vector $\boldsymbol{a}_v^{(t)}$, and $COMBINE$ function then combines the previous node feature $\boldsymbol{x}_v^{(t-1)}$ with aggregated node features $\boldsymbol{a}_v^{(t)}$ to output the node feature $\boldsymbol{x}_v^{(t)}$.

One of the most important considerations in designing GNNs as a graph embedding method is that the $AGGREGATE$ function is a function of $\{\{\cdot\}\}$ that denotes the *multiset*. Multiset is a set (a collection of elements where the order is not important) where elements may appear multiple times. Therefore, the $AGGREGATE$ function should be operated with various sets of nodes and should be independent of the order of the elements in the sets.

The importance of the condition is well illustrated in Fig. 8.18. For example, at $t = 1$, each node has distinct set of neighborhood nodes, so the neural network should be applicable for all these node configurations with the shared weight. Similar situations can happen at $t = 2$, since the nodes A and B have three and two connecting nodes, respectively. One simple example of an $AGGREGATE$

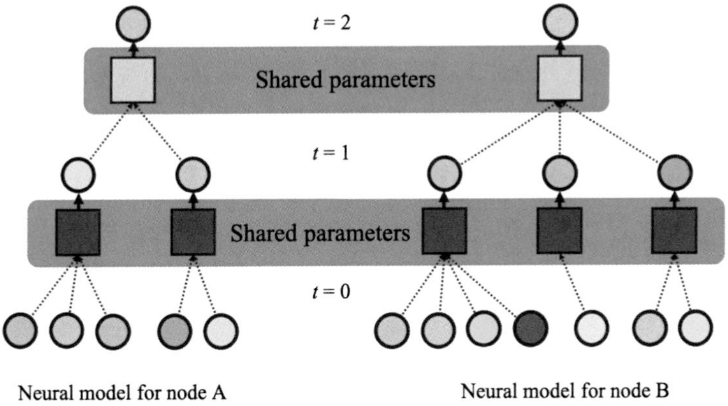

Fig. 8.18 Example of aggregation function operation in a GNN

function that satisfies this requirement is a sum operation:

$$\begin{aligned} \boldsymbol{a}_v^{(t)} &= AGGREGATE\left(\left\{\!\!\left\{\boldsymbol{x}_u^{(t-1)} : u \in \mathcal{N}(v)\right\}\!\!\right\}\right) \\ &= \sum_{u \in \mathcal{N}(v)} \boldsymbol{x}_u^{(t-1)}. \end{aligned} \tag{8.11}$$

Although this sum operation is one of the most popular approaches in GNNs, we can consider a more general form of the operation with desirable properties. This is the main topic in the following section.

8.5 WL Test, Graph Neural Networks

Compared to the matrix factorization and random walks approaches, the success of graph embedding using neural networks appears mysterious. This is because in order to be a valid embedding, the semantically similar input should be closely located in the latent space, but it is not clear whether the graph neural network produces such behaviors.

For the case of matrix factorization, the embedding transform is obtained from the assumption that the latent vector should live in the low-dimensional subspace. For the case of random walks, the underlying intuition for the embedding is similar to that of word2vec. Therefore, these approaches are guaranteed to retain semantic information in the latent space. Then, how do we know that the neural-network-based graph embedding also conveys the semantic information?

This understanding is particularly important because a GNN algorithm is usually designed as an empirical algorithm and not based on the top-down principle in order to achieve the desired embedding properties. Recently, a number of authors [57, 59–62] has shown that the GNN is indeed a neural network implementation of Weisfeiler–Lehman (WL) graph isormorphism test [63]. This implies that if the embedding vectors of a GNN are distinct from each other, then the corresponding graphs are not isormophic. Therefore, GNNs may retain useful semantic information during the embedding. In this section, we review this exciting discovery in more detail.

8.5.1 Weisfeiler–Lehman Isomorphism Test

As discussed before, determining whether two graphs are isomorphic is a challenging problem. It is not even known whether there is a polynomial time algorithm for determining whether graphs are isomorphic.

In this sense, the Weisfeiler–Lehman (WL) algorithm [63] is a mechanism to efficiently assign fairly unique attributes. The core idea of the Weisfeiler–Lehman isomorphism test is to find a signature for each node in each graph based on the neighborhood around the node. These signatures can then be used to find the correspondence between nodes in the two graphs. Specifically, if the signatures of two graphs are not equivalent, then the graphs are definitively not isomorphic.

We now describe the WL algorithm formally. For a given colored graph G, the WL computes a node coloring $\boldsymbol{c}_v^{(t)} : V(G) \mapsto \Sigma$, depending on the coloring from the previous iteration. To iterate the algorithm, we assign each node a tuple that contains the old compressed label (or color) of the node and a multiset of the compressed labels (colors) of the neighbors of the node:

$$\boldsymbol{m}_v^{(t)} = \left\{\boldsymbol{c}_v^{(t)}, \left\{\!\!\left\{\boldsymbol{c}_u^{(t)} \mid u \in \mathcal{N}(v)\right\}\!\!\right\}\right\}, \tag{8.12}$$

where $\{\!\{\cdot\}\!\}$ denotes the multiset, which is a set (a collection of elements where order is not important) in which elements may appear more than once. Then, $HASH(\cdot)$ bijectively assigns the above pair to a unique compressed label that was not used in previous iterations:

$$\boldsymbol{c}_v^{(t+1)} = HASH\left(\boldsymbol{m}_v^{(t)}\right). \tag{8.13}$$

If the number of colors does not change between two iterations, then the algorithm ends. This procedure is illustrated in Fig. 8.19.

To test two graphs G and H for isomorphism, we run the above algorithm in "parallel" on both graphs. If the two graphs have different numbers of nodes, which are colored in the WL algorithm, it is concluded that the graphs are not isomorphic. In the algorithm described above, the "compressed labels" serve as

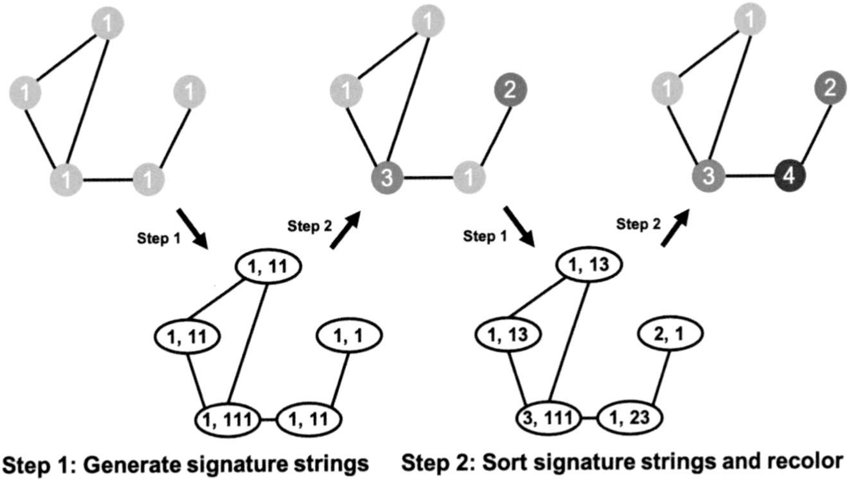

Fig. 8.19 WL algorithm for graph isormorphism test

signatures. However, it is possible that two non-isomorphic graphs have the same signatures, so this test alone cannot provide conclusive evidence that two graphs are isomorphic. However, it has been shown that the WL test can be successful in the graph isomorphism test with a high degree of probability. This is the main reason the WL test is so important [63].

8.5.2 Graph Neural Network as WL Test

Recall that a GNN computes a sequence $\{\boldsymbol{x}_v^{(t)}\}_{v\in V}$ for $t \geq 0$ of vector embeddings of a graph $G = (V, E)$. In the most general form, the embedding is recursively computed as

$$\boldsymbol{a}_v^{(t)} = AGGREGATE\left(\left\{\!\!\left\{\boldsymbol{x}_u^{(t-1)} : u \in \mathcal{N}(v)\right\}\!\!\right\}\right), \tag{8.14}$$

where $\{\!\{\cdot\}\!\}$ is the multi-set and the aggregation function is symmetric in its arguments, and the updated feature vector is given by

$$\boldsymbol{x}_v^{(t)} = COMBINE\left(\boldsymbol{x}_v^{(t-1)}, \boldsymbol{a}_v^{(t)}\right). \tag{8.15}$$

From (8.14) and (8.15) in comparison with (8.12) and (8.13), if we identify $\boldsymbol{x}_v^{(t)}$ as the coloring at the t-th iteration, i.e. $\boldsymbol{c}_v^{(t)}$, then we can see that there are remarkable similarities between GNN updates and the WL algorithm in terms of their arguments, which are made up of multiset neighborhoods and the previous node. In fact, these are not incidental findings; there is a fundamental equivalence between them.

For example, in graph convolutional neural networks (GCNs) [56] and graph-SAGE [58], the $AGGREGATE$ function is given by an average operation, whereas it is just a simple sum in the graph isormorphism network (GIN) [57]. One could use the element-by-element max operation as the $AGGREGATE$ function, or even a long short-term memory (LSTM) can be used [58]. Similarly, a simple sum followed by a multilayer percentron (MLP) can be used as the $COMBINE$ function, or the weighted sum or concatenation followed by an MLP could be used [58, 59]. In general, the GNN operation can be represented by

$$\boldsymbol{x}_v^{(t+1)} = \sigma\left(\boldsymbol{W}_1^{(t)}\boldsymbol{x}_v^{(t)} + \sum_{u\in\mathcal{N}(v)} \boldsymbol{W}_2^{(t)}\boldsymbol{x}_u^{(t)}\right), \tag{8.16}$$

for some matrices $\boldsymbol{W}_1^{(t)}, \boldsymbol{W}_2^{(t)}$ and the nonlinearity $\sigma(\cdot)$ [59]. One of the important discoveries in [59] is that for a given coloring $\{\boldsymbol{x}_v^{(t-1)}\}_{v\in V}$, there always exist matrices $\boldsymbol{W}_1^{(t)}$ and $\boldsymbol{W}_2^{(t)}$ which makes the update (8.16) equivalent to the WL algorithm in (8.12) and (8.13). Therefore, the GNN is indeed a neural network implementation of the WL algorithm for the graph isomorphism test, and the way GNNs produce node embedding is to map the graph to a signature that can be used to test the graph matching.

8.6 Summary and Outlook

So far we have discussed the graphical neural network approach as a modern method of performing graph embedding. The most important finding is that the GNN is actually a neural network implementation of the WL test. Therefore, GNN fulfills the important properties of embedding: if the two feature vectors in latent space are different, the underlying graph is different.

The embedding of the graph with GNNs is by no means complete. In order to get a really meaningful graph embedding, the vector operation in latent space should have the same semantic meaning as in the original diagram, similar to that of word embedding. However, it is still not clear whether the current GNN-based embedding of graphs can lead to such versatile properties.

Hence, the field of graphic neural networks is still a wide open area of research and the next level of breakthroughs will require many good ideas from young and enthusiastic researchers.

8.7 Exercises

1. Show that every connected graph with n vertices has at least $n - 1$ edges.
2. For the case of CBOW, recall that the target vector $\boldsymbol{x}_i$ is also a one-hot encoded vector. Let t_k denote the nonzero index of the vocabulary vector $\boldsymbol{x}_k$. Then, show that the loss function of CBOW can be written as a softmax function:

$$\ell_{CBOW}(\boldsymbol{W}, \widetilde{\boldsymbol{W}}) = -\log\left(\frac{e^{\widetilde{\boldsymbol{w}}_{t_i}^\top \boldsymbol{h}_i}}{\sum_{k=1}^{M} e^{\widetilde{\boldsymbol{w}}_{t_k}^\top \boldsymbol{h}_i}}\right)$$
$$= -\widetilde{\boldsymbol{w}}_{t_i}^\top \boldsymbol{h}_i + \log\left(\sum_{k=1}^{M} e^{\widetilde{\boldsymbol{w}}_{t_k}^\top \boldsymbol{h}_i}\right), \quad (8.17)$$

   where the latent vector $\boldsymbol{h}_i$ is given by the average latent vector.
3. Classify, up to isomorphism, all connected graphs (simple or not simple) with 5 vertices and 5 edges. You may find that every simple, connected graph with 5 vertices and 5 edges is isomorphic to exactly one of the five cases.
4. Let G be a graph with 4 connected components and 20 edges. What is the maximum possible number of vertices in G?

5. The GIN was proposed as a special case of spatial GNNs suitable for graph classification tasks. The network implements the aggregate and combine functions as the sum of the node features:

$$\boldsymbol{x}_v^{(k)} = MLP^{(k)}\Big((1+\epsilon^{(k)})\cdot \boldsymbol{x}_v^{(k-1)} + \sum\nolimits_{u\in\mathcal{N}(v)} \boldsymbol{x}_u^{(k-1)}\Big), \tag{8.18}$$

where $\epsilon^{(k)} = 0.1$, and MLP is a multilayer perceptron with ReLU nonlinearity.

a. Draw the corresponding graph, whose adjacency matrix is given by

$$\boldsymbol{A} = \begin{bmatrix} 0 & 1 & 1 & 0 \\ 1 & 0 & 1 & 1 \\ 1 & 1 & 0 & 0 \\ 0 & 1 & 0 & 0 \end{bmatrix}.$$

b. Suppose that the input node feature is a one-hot feature matrix:

$$\boldsymbol{X}^{(0)} = \begin{bmatrix} 1 & 0 & 0 & 0 \\ 0 & 1 & 0 & 0 \\ 0 & 0 & 1 & 0 \\ 0 & 0 & 0 & 1 \end{bmatrix}$$

and the MLP weight matrix $\boldsymbol{W}^{(1)} = \boldsymbol{W}^{(2)}$ is given by

$$\boldsymbol{W}^{(1)} = \begin{bmatrix} 0.1 & -0.2 & -0.3 & 0.4 \\ -0.1 & 0.2 & -0.3 & 0.4 \\ 0.4 & 0.3 & 0.2 & -0.1 \\ -0.4 & 0.3 & 0.2 & -0.1 \end{bmatrix}.$$

Then, obtain the next layer feature matrices $\boldsymbol{X}^{(1)}$ and $\boldsymbol{X}^{(2)}$ assuming that there exists no bias at each MLP.

Chapter 9
Normalization and Attention

9.1 Introduction

In this chapter, we will discuss very exciting and rapidly evolving technical fields of deep learning: *normalization* and *attention*.

Normalization originated from the batch normalization technique [41] that accelerates the convergence of stochastic gradient methods by reducing the covariate shift. The idea has been extended further to various forms of normalization, such as layer norm [64], instance norm [65], group norm [66], etc. In addition to the original use of normalization for better convergence of stochastic gradients, adaptive instance normalization (AdaIN) [67] is another example where the normalization technique can be used as a simple but powerful tool for style transfer and generative models.

On the other hand, attention has been drawn to computer vision applications based on intuition that we "attend to" a particular part when processing a large amount of information [68–72]. Attention has played the key role in the recent breakthroughs in natural language processing (NLP), such as Transformer [73], Google's Bidirectional Encoder Representations from Transformers (BERT) [74], OpenAI's Generative Pre-trained Transformer (GPT)-2 [75] and GPT-3 [76], etc.

For beginners, the normalization and attention mechanisms look very heuristic without any clue for systematic understanding, which is even more confusing due to their similarities. In addition, understanding AdaIN, Transformer, BERT, and GPT is like reading recipes the researchers developed with their own secret sauces. However, an in-depth study reveals a very nice mathematical structure behind their intuition.

In this chapter, we first review classical and current state-of-the art normalization and attention techniques, and then discuss their specific realization in various deep learning architectures, such as style transfer [77–83], multi-domain image transfer [84–87], generative adversarial network (GAN) [71, 88, 89], Transformer

J. C. Ye, *Geometry of Deep Learning*, Mathematics in Industry 37,
https://doi.org/10.1007/978-981-16-6046-7_9

[73], BERT [74], and GPT [75, 76]. Then, we conclude by providing a unified mathematical view to understand both normalization and attention.

9.1.1 Notation

In deep neural networks, a feature map is defined as a filter output at each layer. For example, feature maps from VGGNet are shown in Fig. 9.1, where the input image is a cat. Since there exist multiple channels at each layer, the feature map is indeed a 3D volume. Moreover, during the training, multiple 3D feature maps are obtained from a mini-batch.

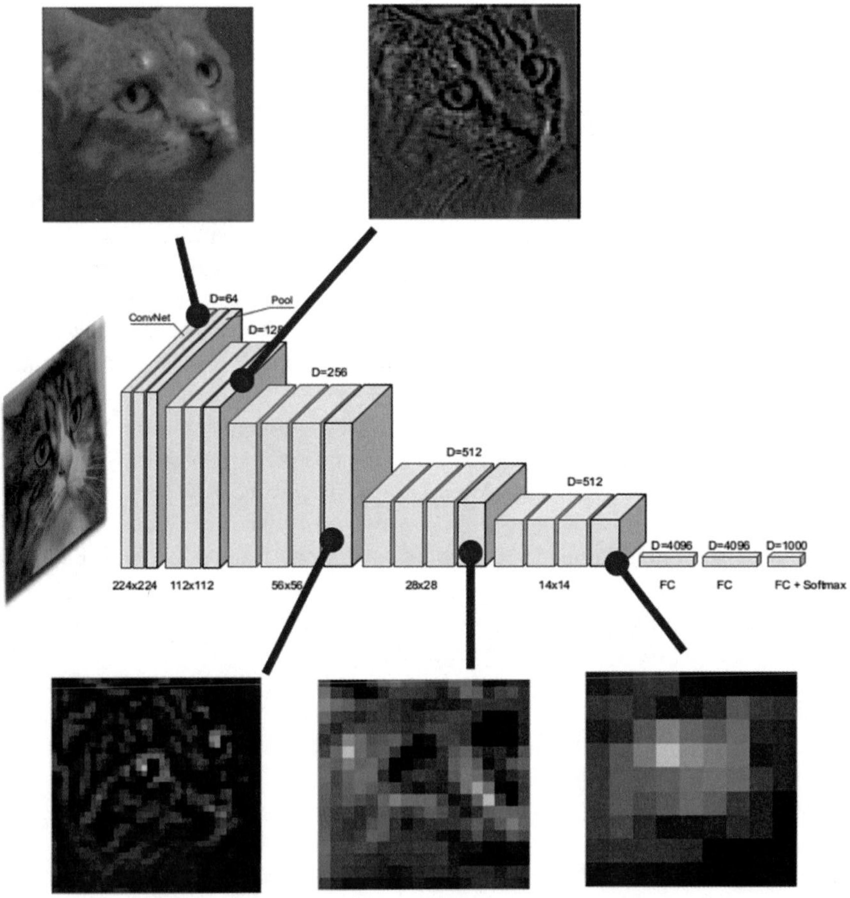

Fig. 9.1 Examples of feature maps on one channel of each layer of VGGNet

To make the notation simple for mathematical analysis, in this chapter a feature map for each channel is vectorized. Moreover, we often ignore the layer-dependent indices in the features. Specifically, the feature map on a layer is represented by

$$X = \begin{bmatrix} \boldsymbol{x}_1 \cdots \boldsymbol{x}_C \end{bmatrix} \in \mathbb{R}^{HW \times C}, \tag{9.1}$$

where $\boldsymbol{x}_c \in \mathbb{R}^{HW \times 1}$ refers to the c-th column vector of $\boldsymbol{X}$, which refers to the vectorized feature map of size of $H \times W$ at the c-th channel. We often use $N := HW$ to denote the number of pixels. Equation (9.1) is often represented with row vectors to explicitly show the row dependency:

$$X = \begin{bmatrix} \boldsymbol{x}^1 \\ \vdots \\ \boldsymbol{x}^{HW} \end{bmatrix} \in \mathbb{R}^{HW \times C}, \tag{9.2}$$

where $\boldsymbol{x}^i \in \mathbb{R}^{1 \times C}$ refers to the i-th row vector, representing the channel dimensional feature at the i-th pixel location.

9.2 Normalization

The basic idea of normalization is to normalize the input/feature layer by recentering and rescaling, although specific details differ depending on algorithms. Perhaps the most influential paper that has opened up the research field of normalization is on *batch normalization* [41], reflected in the total number of 25k citations as of Feb., 2021. Thus, we first review the batch normalization techniques, and discuss how this evolves into different forms of the normalization techniques.

9.2.1 Batch Normalization

Batch normalization was originally proposed to reduce the internal covariate shift and improve the speed, performance, and stability of artificial neural networks. During the training phase of the networks, the distribution of the input on the current layer changes accordingly if the distribution of the feature on the previous layers changes, so that the current layer has to be constantly adapted to new distributions. This problem is particularly severe for deep networks because small changes in shallower hidden layers are amplified as they propagate through the network, causing a significant shift in deeper hidden layers. The method of batch normalization is therefore proposed to reduce these undesirable shifts by recentering and scaling.

Specifically, the batch normalization is carried out by the following transform:

$$\boldsymbol{y}_c = \frac{\gamma_c}{\bar{\sigma}_c}(\boldsymbol{x}_c - \bar{\mu}_c\mathbf{1}) + \beta_c\mathbf{1}, \tag{9.3}$$

for all $c = 1, \cdots, C$, where $\mathbf{1} \in \mathbb{R}^{HW}$ denotes the vector of ones, γ_c and β_c are trainable parameters for the c-th channel, and $\bar{\mu}_c$ and $\bar{\sigma}_c$ are the mini-batch statistics defined by

$$\bar{\mu}_c = \frac{1}{HW}\mathbb{E}[\mathbf{1}^\top \boldsymbol{x}_c], \tag{9.4}$$

$$\bar{\sigma}_c = \sqrt{\frac{1}{HW}\mathbb{E}[\|\boldsymbol{x}_c - \bar{\mu}_c\mathbf{1}\|^2]}, \tag{9.5}$$

where the expectation $\mathbb{E}[\cdot]$ is taken over the mini-batch. In matrix form, (9.3) can be represented by

$$\boldsymbol{Y} = \boldsymbol{X}\boldsymbol{T} + \boldsymbol{B}, \tag{9.6}$$

where

$$\boldsymbol{T} = \begin{bmatrix} \frac{\gamma_1}{\bar{\sigma}_1} & \cdots & 0 \\ \vdots & \ddots & \vdots \\ 0 & \cdots & \frac{\gamma_C}{\bar{\sigma}_C} \end{bmatrix}, \quad \in \mathbb{R}^{C\times C} \tag{9.7}$$

$$\boldsymbol{B} = \overbrace{[\mathbf{1} \cdots \mathbf{1}]}^{C} \begin{bmatrix} \beta_1 - \frac{\gamma_1\bar{\mu}_1}{\bar{\sigma}_1} & \cdots & 0 \\ \vdots & \ddots & \vdots \\ 0 & \cdots & \beta_C - \frac{\gamma_C\bar{\mu}_C}{\bar{\sigma}_C} \end{bmatrix}.$$

In addition to reducing the internal covariate shift, it is believed that batch normalization has many other advantages. With this additional operation, the network can use a higher learning rate without gradients vanishing or exploding. In addition, the batch normalization appears to have a regularization effect so that the network improves its generalization properties and therefore there is no need to use dropout to reduce overfitting. It has also been observed that with the batch normalization, the network becomes more robust towards different initialization schemes and learning rates.

For example, Fig. 9.2 shows that the batch norm (BN) layer is used within the structure of DenseNet [37] to improve the learning rate of the ImageNet classification task. Similarly, a powerful CNN image denoiser was proposed in [90] by just cascading BN layer, ReLU, and filter layers as shown in Fig. 9.3.

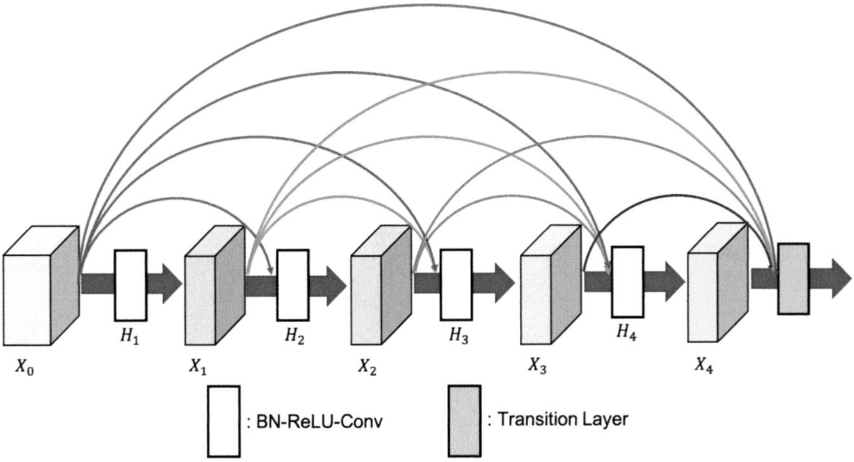

Fig. 9.2 Batch norm layer in DenseNet

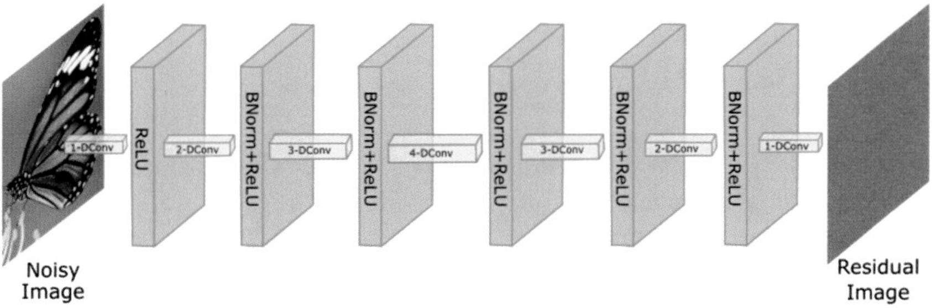

Fig. 9.3 The use of batch norm in CNN denoiser

9.2.2 *Layer and Instance Normalization*

Batch normalization is a powerful tool, but not without its limitations. The main limitation of batch normalization is that it depends on the mini-batch when calculating (9.4) and (9.5). Then, how can we mitigate the problem of batch normalization?

To understand this question, let us look into the volume of the feature maps that are stacked along the mini-batch in Fig. 9.4. The left column of Fig. 9.4 shows the normalization operation in batch norm, whereby the shadow area is used to calculate the mean and standard deviation for centering and rescaling. Here, B denotes the size of the mini-batch.

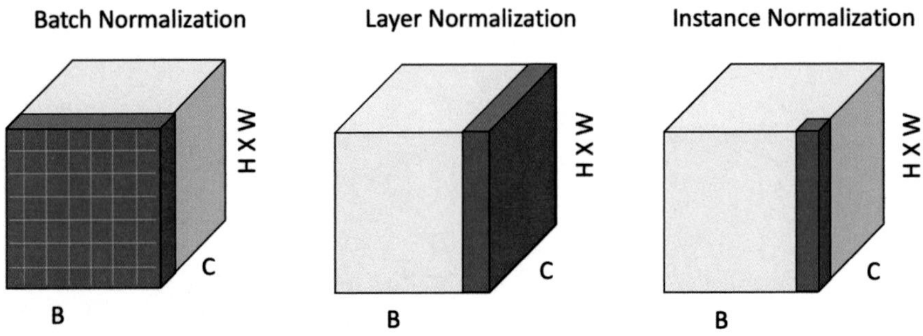

Fig. 9.4 Various forms of feature normalization methods. B: batch size, C: number of channels, and H, W: height and width of the feature maps

In fact, the picture of batch norm shows that there are several normalization options. For example, the layer normalization [64] computes the mean and standard deviation along the channel and image direction without considering the mini-batch. More specifically, we have

$$\boldsymbol{y}_c = \frac{\gamma}{\sigma}(\boldsymbol{x}_c - \mu\mathbf{1}) + \beta\mathbf{1}, \tag{9.8}$$

for all $c = 1, \cdots, C$. Here, γ and β are channel-independent trainable parameters, while μ and σ are computed by

$$\mu = \frac{1}{HWC}\sum_{c=1}^{C}\mathbf{1}^\top \boldsymbol{x}_c, \tag{9.9}$$

$$\sigma = \sqrt{\frac{1}{HWC}\sum_{c=1}^{C}\|\boldsymbol{x}_c - \mu\mathbf{1}\|^2}. \tag{9.10}$$

In the layer normalization, each sample within the mini-batch has a different normalization operation, allowing arbitrary mini-batch sizes to be used. The experimental results show that layer normalization performs well for recurrent neural networks [64].

On the other hand, the instance normalization normalizes the feature data for each sample and channel as shown on the right-hand side of Fig. 9.4. More specifically, we have

$$\boldsymbol{y}_c = \frac{\gamma_c}{\sigma_c}(\boldsymbol{x}_c - \mu_c\mathbf{1}) + \beta_c\mathbf{1}, \tag{9.11}$$

for all $c = 1, \cdots, C$, where

$$\mu_c = \frac{1}{HW}\mathbf{1}^\top \boldsymbol{x}_c, \tag{9.12}$$

$$\sigma_c = \sqrt{\frac{1}{HW}\|\boldsymbol{x}_c - \mu_c\mathbf{1}\|^2}, \tag{9.13}$$

whereas γ_c and β_c are trainable parameters for the channel c. In matrix form, (9.11) can be represented by

$$\boldsymbol{Y} = \boldsymbol{X}\boldsymbol{T} + \boldsymbol{B}, \tag{9.14}$$

where $\boldsymbol{T}$ and $\boldsymbol{B}$ are similar to (9.7) but calculated for each sample.

9.2.3 Adaptive Instance Normalization (AdaIN)

With AdaIN [67], a new chapter of normalization method has opened, which goes beyond the classical normalization methods that were designed to improve the performance and reduce the dependency on learning rate. The most important finding of AdaIN is that the instance normalization transformation in (9.11) provides an important hint for the style transfer.

Before we discuss the details of AdaIN, we first explain the concept of image style transfer. Figure 9.5 shows an example of image style transfer using AdaIN [67]. Here, the top row shows the content images associated with the content feature $\boldsymbol{X} = [\boldsymbol{x}_1, \cdots, \boldsymbol{x}_C]$, while the left-most column corresponds to style images that are associated with the style feature $\boldsymbol{S} = [\boldsymbol{s}_1, \cdots, \boldsymbol{s}_C]$. The aim of the image style transfer is then to convert the content images into a stylized image that is guided by a certain style image. How does AdaIN manage the style transfer in this context?

The main idea is to use the instance normalization in (9.11), but instead of using γ_c and β_c that are calculated by its own feature, these values are calculated as the standard deviation and the mean value of the style image, i.e.

$$\beta_c^s = \frac{1}{HW}\mathbf{1}^\top \boldsymbol{s}_c, \tag{9.15}$$

$$\gamma_c^s = \sqrt{\frac{1}{HW}\|\boldsymbol{s}_c - \beta_c^2\mathbf{1}\|^2}, \tag{9.16}$$

where $\boldsymbol{s}_c$ is the c-th channel feature map from the style image. In matrix form, AdaIN can be represented by

$$\boldsymbol{Y} = \boldsymbol{X}\boldsymbol{T}_x\boldsymbol{T}_s + \boldsymbol{B}_{x,s}, \tag{9.17}$$

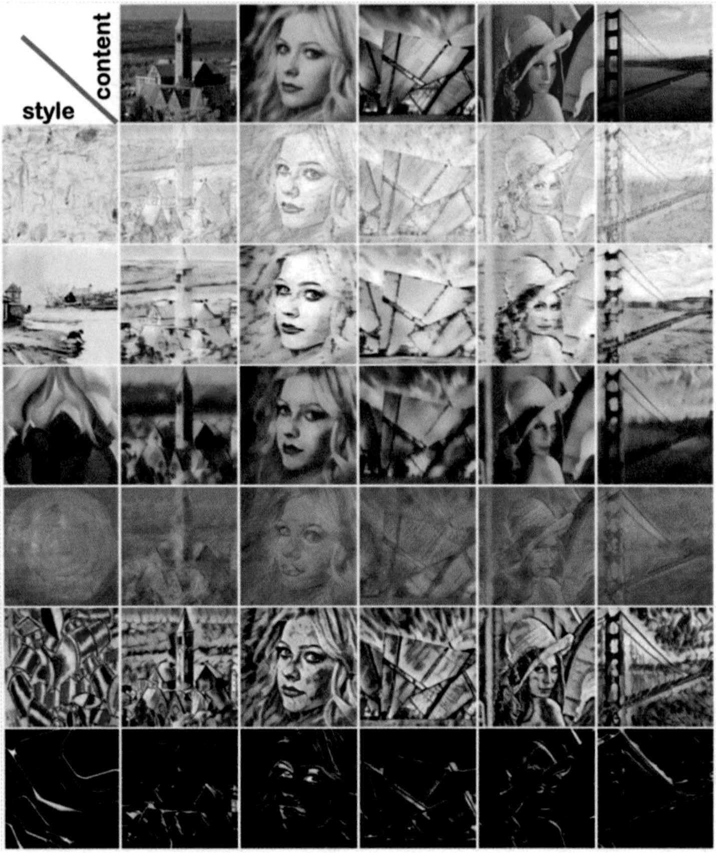

Fig. 9.5 Examples of image style transfer using AdaIN [67]

where $\boldsymbol{T}_x$ and $\boldsymbol{T}_s$ are diagonal matrices computed from $\boldsymbol{X}$ and $\boldsymbol{S}$, respectively:

$$\boldsymbol{T}_x = \begin{bmatrix} \frac{1}{\sigma_1} & \cdots & 0 \\ \vdots & \ddots & \vdots \\ 0 & \cdots & \frac{1}{\sigma_C} \end{bmatrix} \in \mathbb{R}^{C \times C} \tag{9.18}$$

$$\boldsymbol{T}_s = \begin{bmatrix} \gamma_1^s & \cdots & 0 \\ \vdots & \ddots & \vdots \\ 0 & \cdots & \gamma_C^s \end{bmatrix} \in \mathbb{R}^{C \times C}, \tag{9.19}$$

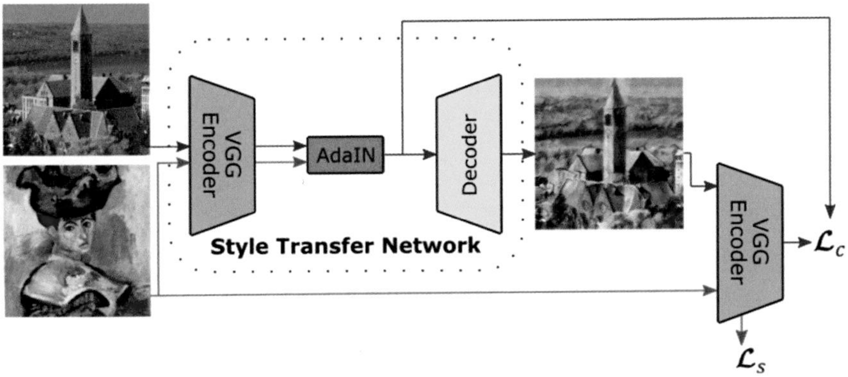

Fig. 9.6 The network architecture of AdaIN style transfer

whereas $\boldsymbol{B}_{x,s}$ is the bias term computed using both $\boldsymbol{X}$ and $\boldsymbol{S}$:

$$\boldsymbol{B}_{x,s} = \overbrace{[\mathbf{1} \cdots \mathbf{1}]}^{C} \begin{bmatrix} \beta_1^s - \frac{\gamma_1^s}{\sigma_1}\mu_1 & \cdots & 0 \\ \vdots & \ddots & \vdots \\ 0 & \cdots & \beta_C^s - \frac{\gamma_C^s}{\sigma_C}\mu_C \end{bmatrix} \tag{9.20}$$

The generation of the style feature map can be done with the same encoder, as shown in Fig. 9.6, whereby both content and style images are given as inputs for the VGG encoder for feature vector extraction, from which the AdaIN layer changes the style using the AdaIN operation described above.

9.2.4 Whitening and Coloring Transform (WCT)

The whitening and coloring transform (WCT) is another powerful method of image style transfer [79], which is composed of a whitening transform followed by a coloring transform. Mathematically, this can be written by

$$\boldsymbol{Y} = \boldsymbol{X}\boldsymbol{T}_x\boldsymbol{T}_s + \boldsymbol{B}_{x,s}, \tag{9.21}$$

where $\boldsymbol{B}_{x,s}$ is the same as (9.20), and the whitening transform $\boldsymbol{T}_x$ and the coloring transform $\boldsymbol{T}_x$ are computed by $\boldsymbol{X}$ and $\boldsymbol{S}$, respectively:

$$\boldsymbol{T}_x = \boldsymbol{U}_x\boldsymbol{\Sigma}_x^{-\frac{1}{2}}\boldsymbol{U}_x^\top, \quad \boldsymbol{T}_s = \boldsymbol{U}_s\boldsymbol{\Sigma}_s^{\frac{1}{2}}\boldsymbol{U}_s^\top, \tag{9.22}$$

where $\boldsymbol{U}_x, \boldsymbol{\Sigma}_x$ and $\boldsymbol{U}_s, \boldsymbol{\Sigma}_s$ are from the eigen-decomposition of the covariance matrices of $\boldsymbol{X}$ and $\boldsymbol{S}$:

$$\boldsymbol{X}^\top \boldsymbol{X} = \boldsymbol{U}_x \boldsymbol{\Sigma}_x \boldsymbol{U}_x^\top, \quad \boldsymbol{S}^\top \boldsymbol{S} = \boldsymbol{U}_s \boldsymbol{\Sigma}_s \boldsymbol{U}_s^\top \tag{9.23}$$

Therefore, we can easily see that AdaIN is a special case of WCT, when the covariance matrix is diagonal.

9.3 Attention

In cognitive neuroscience, attention is defined as the behavioral and cognitive process in which one selectively focuses on one aspect of information and ignores other perceptible information. In this section we describe a biological analogy of attention at the neuronal level and discuss its mathematical formulation.

9.3.1 *Metabotropic Receptors: Biological Analogy*

It is known that there are two types of neurotransmitter receptors: ionotropic and metabotropic receptors [91]. Ionotropic receptors are transmembrane molecules that can "open" or "close" a channel so that different types of ions can migrate in and out of the cell, as shown in Fig. 9.7a. On the other hand, the activation of the metabotropic receptors only indirectly influences the opening and closing of ion channels. In particular, a receptor activates the G-protein as soon as a ligand binds to the metabotropic receptor. Once activated, the G-protein itself goes on and activates another molecule called a "secondary messenger". The secondary messenger moves until it binds to ion channels, located at different points on the membrane, and opens them (see Fig. 9.7b). It is important to remember that metabotropic receptors do not have ion channels and the binding of a ligand may or may not lead to the opening of ion channels at different locations on the membrane.

Mathematically, this process can be modeled as follows. Let x_n be the number of neurotransmitters that bind to the n-th synapse. G-proteins generated at the n-th synapse are proportional to the sensitivity of the metabotropic receptor, which is denoted by k_n. Then, the G-proteins generate the secondary messengers that bind to the ion channel at the m-th synapse with the sensitivity of q_m. Since the secondary messengers are generated from metabotropic receptors at various synapses, the total amount of ion influx from the m-th synapse is determined by the sum given by

$$y_m = \sum_{n=1}^{N} q_m k_n x_n, \quad m = 1, \cdots, N, \tag{9.24}$$

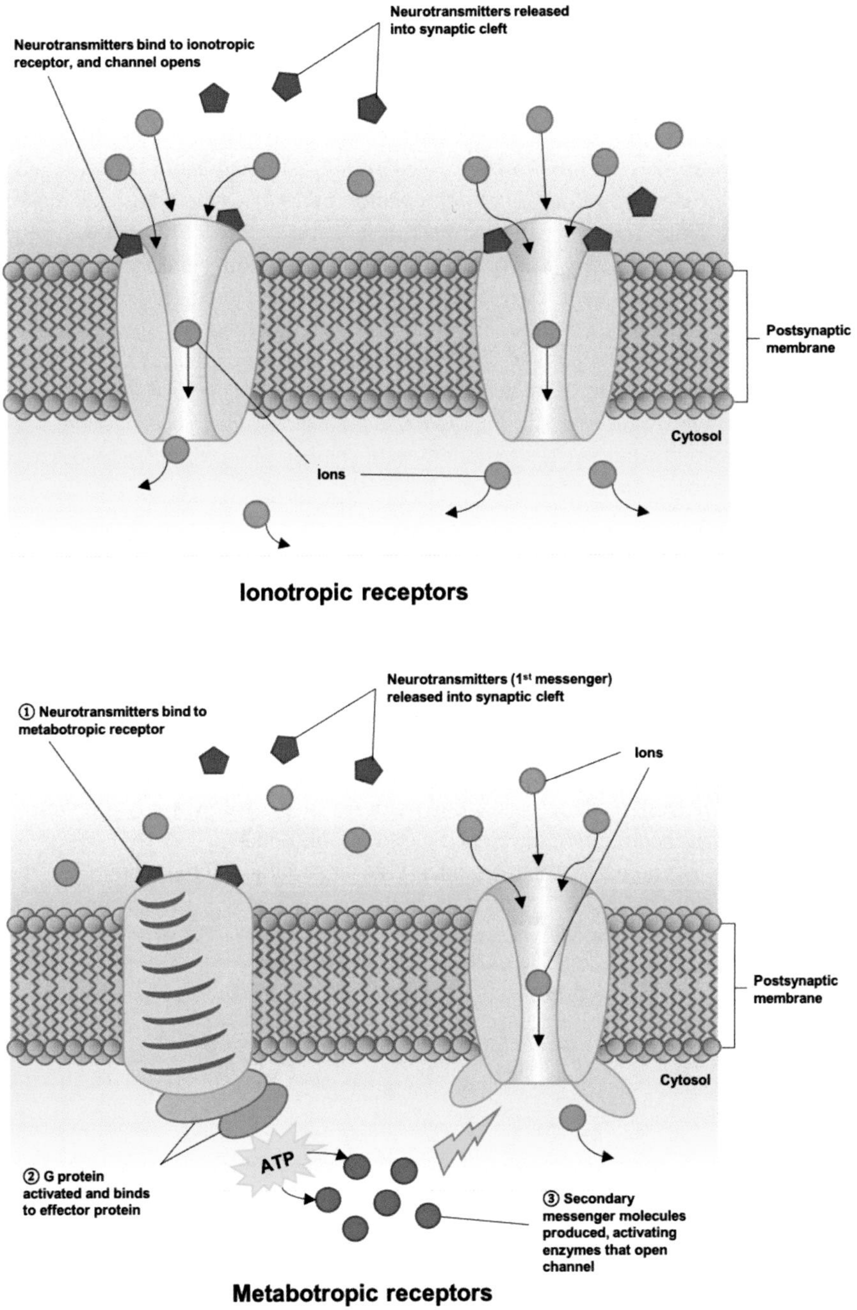

Fig. 9.7 Two different types of neurotransmitter receptors and their mechanisms

which can be represented by a vector form

$$\boldsymbol{y} = \boldsymbol{T}\boldsymbol{x}, \quad \text{where } \boldsymbol{T} := \boldsymbol{q}\boldsymbol{k}^\top. \tag{9.25}$$

Note that the matrix $\boldsymbol{T}$ in (9.25) is a transform matrix from $\boldsymbol{x}$ to $\boldsymbol{y}$. Indeed, the transform matrix $\boldsymbol{T}$ is a rank-1 matrix. Accordingly, the output $\boldsymbol{y}$ is constrained to live in the linear subspace of the column vector, i.e. $\mathcal{R}(\boldsymbol{q})$, where $\mathcal{R}(\cdot)$ denotes the range space. This implies that the activation patterns in the neuron follow the ion channel sensitivity patterns, $\boldsymbol{q}$, while their magnitude is modulated by $\boldsymbol{k}$.

This could explain another role for the metabotropic receptors. In particular, metabotropic receptors act more for their prolonged activation than for a short-term activation as in the case of ionotropic receptors, since the activation pattern is determined by the ion channel distributions to which the secondary messengers bind rather than by the specific location at which the original neurotransmitter is released. Thus, the synergistic combination of the $\boldsymbol{q}$ and $\boldsymbol{k}$ determines the general behavior of neuronal activation.

9.3.2 Mathematical Modeling of Spatial Attention

In (9.25), the vectors $\boldsymbol{q}$ and $\boldsymbol{k}$ are often referred to as *query* and *key*. It is remarkable that even with the same key $\boldsymbol{k}$, a totally different activation pattern can be obtained by changing the query vector $\boldsymbol{q}$. In fact, this is the core idea of the attention mechanism. By decoupling the query and key, we can dynamically adapt the neuronal activation patterns for our purpose. In the following, we review the general form of the attention developed based on this concept.

In artificial neural networks, the model (9.24) is generalized for vector quantities. Specifically, the row vector output at the m-th pixel $\boldsymbol{y}^m \in \mathbb{R}^C$ is determined by the vector version of query $\boldsymbol{q}^m \in \mathbb{R}^d$, keys $\boldsymbol{k}^n \in \mathbb{R}^d$, and values $\boldsymbol{x}^n \in \mathbb{R}^C$:

$$\boldsymbol{y}^m = \sum_{n=1}^{N} a_{mn}\boldsymbol{x}^n, \tag{9.26}$$

where $m = 1, \cdots, N$ and

$$a_{mn} := \frac{\exp\left(\text{score}(\boldsymbol{q}^m, \boldsymbol{k}^n)\right)}{\sum_{n'=1}^{N} \exp\left(\text{score}(\boldsymbol{q}^m, \boldsymbol{k}^{n'})\right)}. \tag{9.27}$$

Here, score$(\cdot,\cdot)$ determines the similarity between the two vectors. In matrix form, (9.26) can be represented by

$$\boldsymbol{Y} = \boldsymbol{A}\boldsymbol{X}, \tag{9.28}$$

where

$$X = \begin{bmatrix} x^1 \\ \vdots \\ x^N \end{bmatrix}, \quad Y = \begin{bmatrix} y^1 \\ \vdots \\ y^N \end{bmatrix}, \tag{9.29}$$

and

$$A = \begin{bmatrix} a_{11} & \cdots & a_{1N} \\ \vdots & \ddots & \vdots \\ a_{N1} & \cdots & a_{NN} \end{bmatrix}. \tag{9.30}$$

Various forms of the score functions are used for attention:

- Dot product: $\text{score}(\boldsymbol{q}^m, \boldsymbol{k}^n) := \langle \boldsymbol{q}^m, \boldsymbol{k}^n \rangle$.
- Scaled dot product: $\text{score}(\boldsymbol{q}^m, \boldsymbol{k}^n) := \langle \boldsymbol{q}^m, \boldsymbol{k}^n \rangle / \sqrt{d}$.
- Cosine similarity: $\text{score}(\boldsymbol{q}^m, \boldsymbol{k}^n) := \frac{\langle \boldsymbol{q}^m, \boldsymbol{k}^n \rangle}{\|\boldsymbol{q}^m\| \|\boldsymbol{k}^n\|}$.

For example, in dot production attention, the query and key vectors are usually generated using linear embeddings. More specifically,

$$\boldsymbol{q}^n = \boldsymbol{x}^n \boldsymbol{W}_Q, \quad \boldsymbol{k}^n = \boldsymbol{x}^n \boldsymbol{W}_K, \quad n = 1, \cdots, N, \tag{9.31}$$

where $\boldsymbol{W}_Q, \boldsymbol{W}_K \in \mathbb{R}^{C \times d}$ are shared across all indices. Matrix form representation of the query and key are then given by

$$\boldsymbol{Q} = \boldsymbol{X}\boldsymbol{W}_Q, \quad \boldsymbol{K} = \boldsymbol{X}\boldsymbol{W}_K, \tag{9.32}$$

where $\boldsymbol{Q}, \boldsymbol{K} \in \mathbb{R}^{N \times d}$ are given by

$$\boldsymbol{Q} = \begin{bmatrix} \boldsymbol{q}^1 \\ \vdots \\ \boldsymbol{q}^N \end{bmatrix}, \quad \boldsymbol{K} = \begin{bmatrix} \boldsymbol{k}^1 \\ \vdots \\ \boldsymbol{k}^N \end{bmatrix}, \tag{9.33}$$

We are often interested in the embedding of $\boldsymbol{x}^n$ to a smaller-dimensional vector $\boldsymbol{v}^n \in \mathbb{R}^{d_v}$, which leads to the matrix representation of values:

$$\boldsymbol{v}^n = \boldsymbol{x}^n \boldsymbol{W}_V \quad \in \mathbb{R}^{d_v}, \tag{9.34}$$

where $\boldsymbol{W}_V \in \mathbb{R}^{C \times d_v}$ is the linear embedding matrix for the values. Then, attention is computed by

$$\boldsymbol{y}^m = \sum_{n=1}^{N} a_{mn} \boldsymbol{v}^n, \tag{9.35}$$

where

$$a_{mn} := \frac{\exp\left(\langle \boldsymbol{x}^m \boldsymbol{W}_Q, \boldsymbol{x}^n \boldsymbol{W}_K \rangle\right)}{\sum_{n'=1}^{N} \exp\left(\langle \boldsymbol{x}^m \boldsymbol{W}_Q, \boldsymbol{x}^{n'} \boldsymbol{W}_K \rangle\right)}, \tag{9.36}$$

or in matrix form, we have

$$\boldsymbol{Y} = \boldsymbol{A}\boldsymbol{X}\boldsymbol{W}_V, \tag{9.37}$$

where $\boldsymbol{X}$, $\boldsymbol{Y}$ and $\boldsymbol{A}$ are defined by (9.29) and (9.30), respectively.

9.3.3 Channel Attention

So far, we have discussed the mathematical formulation of spatial attention. One downside of spatial attention is that we need a matrix multiplication of $N \times N$ size of attention map $\boldsymbol{A}$, which can be computationally intensive. To address the problem, channel attention techniques have been developed. One of the most well-known methods for channel attention is the so-called squeeze and excitation network (SENet), which won the 2017 ImageNet challenge [68].

The SENet is composed of two steps: the squeeze and the excitation (see Fig. 9.8). In the squeeze step, a $1 \times C$-dimensional vector z is generated by average pooling as follows:

$$z = \frac{1}{N}\mathbf{1}^\top \boldsymbol{X}. \tag{9.38}$$

At the excitation step, a $1 \times C$ weight vector $\boldsymbol{w}$ is generated from z using a neural network $\boldsymbol{F}_{\boldsymbol{\Theta}}$ which is parameterized by $\boldsymbol{\Theta}$:

$$\boldsymbol{w} = \boldsymbol{F}_{\boldsymbol{\Theta}}(z). \tag{9.39}$$

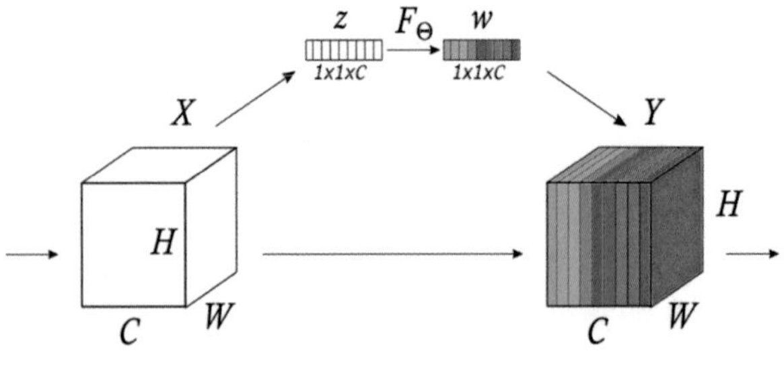

Fig. 9.8 Architecture of SENet

Then, the final attended map is given by

$$Y = XW, \quad \text{where } W := [\mathrm{diag}(w)], \tag{9.40}$$

where $\mathrm{diag}(w)$ is a diagonal matrix whose diagonal component is obtained by the vector w. One can easily see that associated computational complexity is minimal. Still, the SENet provides efficient channel attention mechanism, which significantly improves the performance of the neural network [68].

9.4 Applications

In this section, we provide a review of the exciting applications of normalization and attention in modern deep learning.

9.4.1 StyleGAN

One of the most exciting developments in CVPR 2019 was the introduction of a novel generative adversarial network (GAN) called StyleGAN from Nvidia [89]. As shown in Fig. 9.9, StyleGAN can generate high-resolution images that were realistic enough to shock the world.

Although generative models, specifically GANs, will be discussed later in Chap. 13, we are introducing StyleGAN here, as the main breakthrough of style-GAN comes from AdaIN. The right-hand neural network of Fig. 9.10 generates the

Fig. 9.9 Examples of fake faces generated by StyleGAN

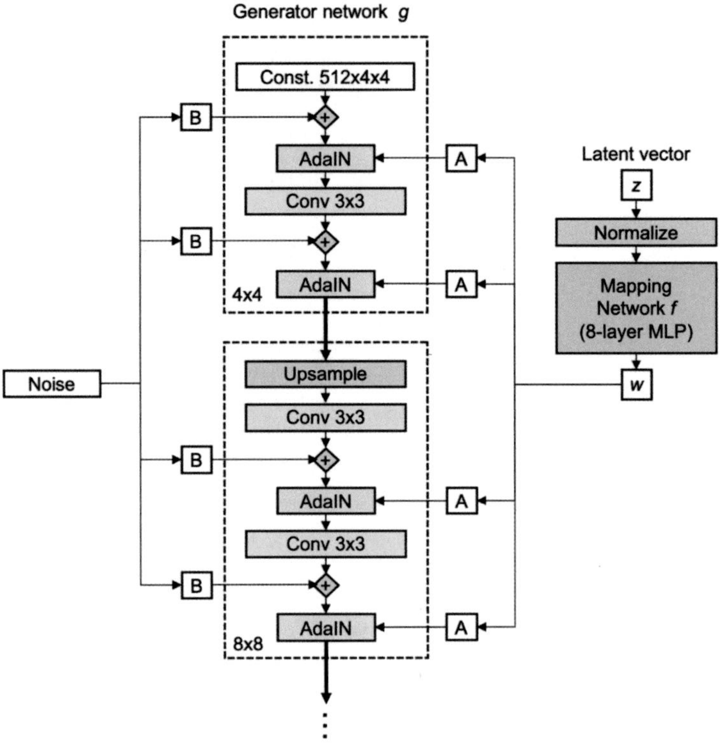

Fig. 9.10 Architecture of StyleGAN

latent codes used as the style image feature vector, while the left-hand network generates the content feature vectors from random noise. The AdaIN layer then combines the style features and the content features in order to generate more realistic features for each resolution. In fact, this architecture is fundamentally different from the standard GAN architecture that we will review later, with the fake image only being generated by a content generator (for example, the one on the left). Through the synergistic combination with another style generator, StyleGAN successfully produces very realistic images.

9.4.2 Self-Attention GAN

One important advantage of the attention mechanism is the separate control of query and key vectors. In the case of self-attention, both the query and the key are obtained from the same data set. In this case, the attention tries to extract the global information from the same input signal in order to find out which part of the signal needs to be focused.

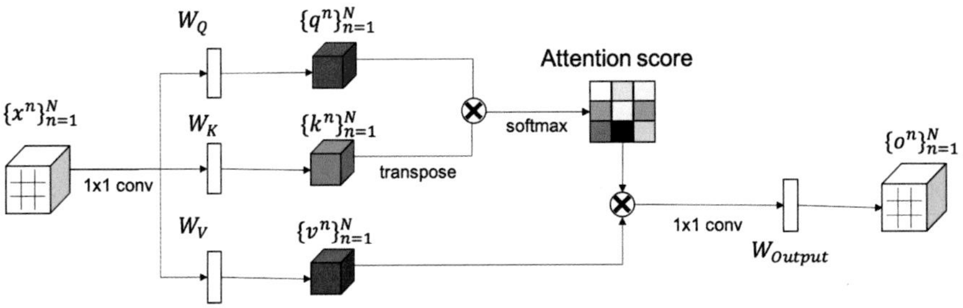

Fig. 9.11 Architecture of self-attention GAN. Both key and query are generated by the input features

In a self-attention GAN (SAGAN) [71], self-attention layers are added into the GAN so that both the generator and the discriminator can better capture model relationships between spatial regions (see Fig. 9.11). It should be remembered that in convolutional neural networks, the size of the receiving field is limited by the size of the filter. With this in mind, self-attention is a great way to learn the relationship between a pixel and all other positions, even regions that are far apart so that global dependencies can be easily grasped. Hence, a GAN endowed with self-attention is expected to handle details better.

More specifically, let $\boldsymbol{X} \in \mathbb{R}^{N\times C}$ be the feature map with N pixels and C channels, and $\boldsymbol{x}^m \in \mathbb{R}^C$ denote the m-th row vector of $\boldsymbol{X}$, which represents the feature vector at the m-th pixel location. The query, key, and the value images are then generated as follows:

$$\boldsymbol{q}^m = \boldsymbol{x}^m \boldsymbol{W}_Q, \quad \boldsymbol{k}^m = \boldsymbol{x}^m \boldsymbol{W}_K, \quad \boldsymbol{v}^m = \boldsymbol{x}^m \boldsymbol{W}_V \tag{9.41}$$

for all pixel indices $m = 1, \cdots, N$. Note that $\boldsymbol{W}_Q, \boldsymbol{W}_K, \boldsymbol{W}_V \in \mathbb{R}^{C\times C}$ matrices can be implemented using 1×1 convolution (see Fig. 9.11). Then, similar to (9.37), the attended image is represented by

$$\boldsymbol{Y} = \boldsymbol{A}\boldsymbol{V} = \boldsymbol{A}\boldsymbol{X}\boldsymbol{W}_V, \tag{9.42}$$

where

$$\boldsymbol{V} = \begin{bmatrix} \boldsymbol{v}^1 \\ \vdots \\ \boldsymbol{v}^N \end{bmatrix}, \tag{9.43}$$

and the (m, n)-th element of $\boldsymbol{A}$ matrix is given by

$$a_{mn} := \frac{\exp\left(\langle \boldsymbol{q}^m, \boldsymbol{k}^n \rangle\right)}{\sum_{n'=1}^N \exp\left(\langle \boldsymbol{q}^m, \boldsymbol{k}^{n'} \rangle\right)}. \tag{9.44}$$

Then, the final self attended feature map is calculated by

$$O = YW_O, \tag{9.45}$$

which can also be implemented using 1×1 convolution.

As shown in (9.42) and (9.45), the new feature vector $\boldsymbol{o}^m$ is generated at the m-th pixel location by the linear combination of the value vectors $\{\boldsymbol{v}^n\}_{n=1}^N$ across the whole image by weighting the elements of the attention map $\boldsymbol{A}$. Therefore, the receptive field of the self-attention map is an overall image, which makes the image generation more effective. A disadvantage, however, is that we need a matrix multiplication of $N \times N$ size of attention map $\boldsymbol{A}$, which can be computationally expensive.

9.4.3 Attentional GAN: Text to Image Generation

In Attentional GAN (AttnGAN) [72], the authors proposed an attention-driven architecture for text-to-image generation (see Fig. 9.12). In addition to the detailed structure for a fine-grained translation, the key idea of AttnGAN is to use the cross-domain attention. In particular, the query vector is generated from image areas, while the key vector is generated from word features. By combining the query and key, AttnGAN can automatically select the word level condition to generate different parts of the image [72].

9.4.4 Graph Attention Network

In the graph attention network (GAT) [69], the main focus is on a node which a neural network should visit more in order to achieve better embedding in the middle node (Fig. 9.13). To incorporate the graph connectivity, the authors suggested specific constraints on the query, key, and value vectors as follows:

$$\boldsymbol{q}^v = \boldsymbol{x}^v \boldsymbol{W}, \quad \boldsymbol{k}^u = \boldsymbol{v}^u = \boldsymbol{x}^u \boldsymbol{W}, \quad u \in \mathcal{N}(v). \tag{9.46}$$

From this, the attentional coefficients between the nodes are calculated by

$$e_{vu} = \text{score}(\boldsymbol{q}^v, \boldsymbol{k}^u),$$

where score$(\cdot)$ denotes the specific attention mechanism. To make the coefficient easily accessible across different nodes, the coefficients are normalized by

$$\alpha_{vu} = \frac{\exp(e_{vu})}{\sum_{u' \in \mathcal{N}(v)} \exp(e_{vu'})}. \tag{9.47}$$

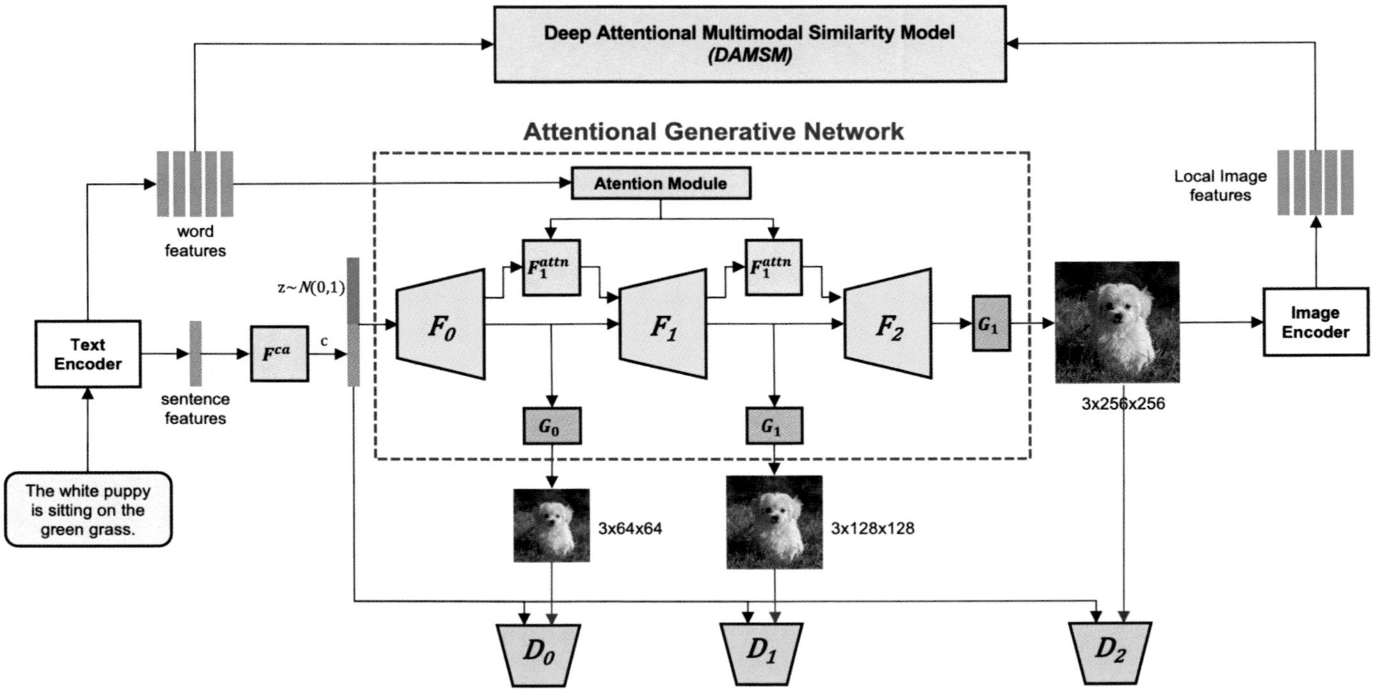

Fig. 9.12 Attentional GAN architecture. Here, the query is generated by image regions, whereas the key is generated by sentence embedding

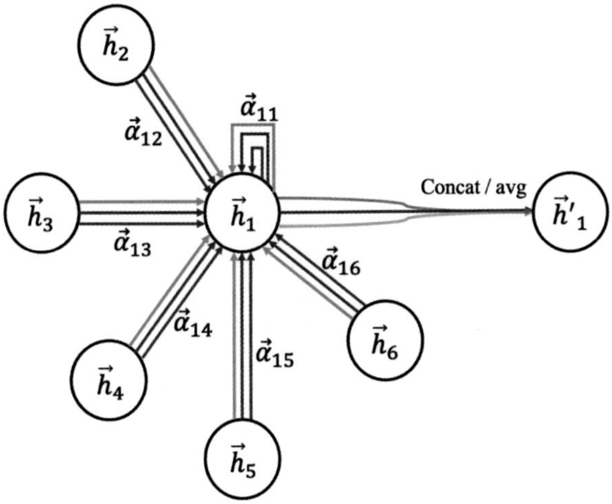

Fig. 9.13 Graph attention network

Then, the graph neural network is represented by the normalized connective coefficient:

$$\boldsymbol{x}^v = \sigma\left(\sum_{u\in\mathcal{N}(v)} \alpha_{vu}\boldsymbol{x}^u \boldsymbol{W}\right). \tag{9.48}$$

9.4.5 Transformer

Transformer is a deep machine learning model that was introduced in 2017 and was originally used for natural language processing (NLP) [73]. In NLP, the recurrent neural networks (RNN) such as Long Short-Term Memory (LSTM) [92] had traditionally been used. In RNN, the data is processed in a sequential order using the memory unit inside. Although Transformers are designed to process ordered data sequences such as speech, unlike the RNN, Transformer processes the entire sequence in parallel to reduce path lengths, making it easier to learn long-distance dependencies in sequences. Since its inception, Transformer has become the building block of most state-of-the-art architectures in NLP, resulting in the development of famous state-of-the-art Bidirectional Encoder Representations from Transformers (BERT) [74], Generative Pre-trained Transformer 3 (GPT-3) [76], etc.

As shown in Fig. 9.14, Transformer-based language translation consists of an encoder and decoder architecture. The main idea of Transformer is the attention mechanism discussed earlier. In particular, the essence of the query, key, and value vectors in the attention mechanism is fully utilized so that the encoder can learn the language embedding and the decoder performs the language translation.

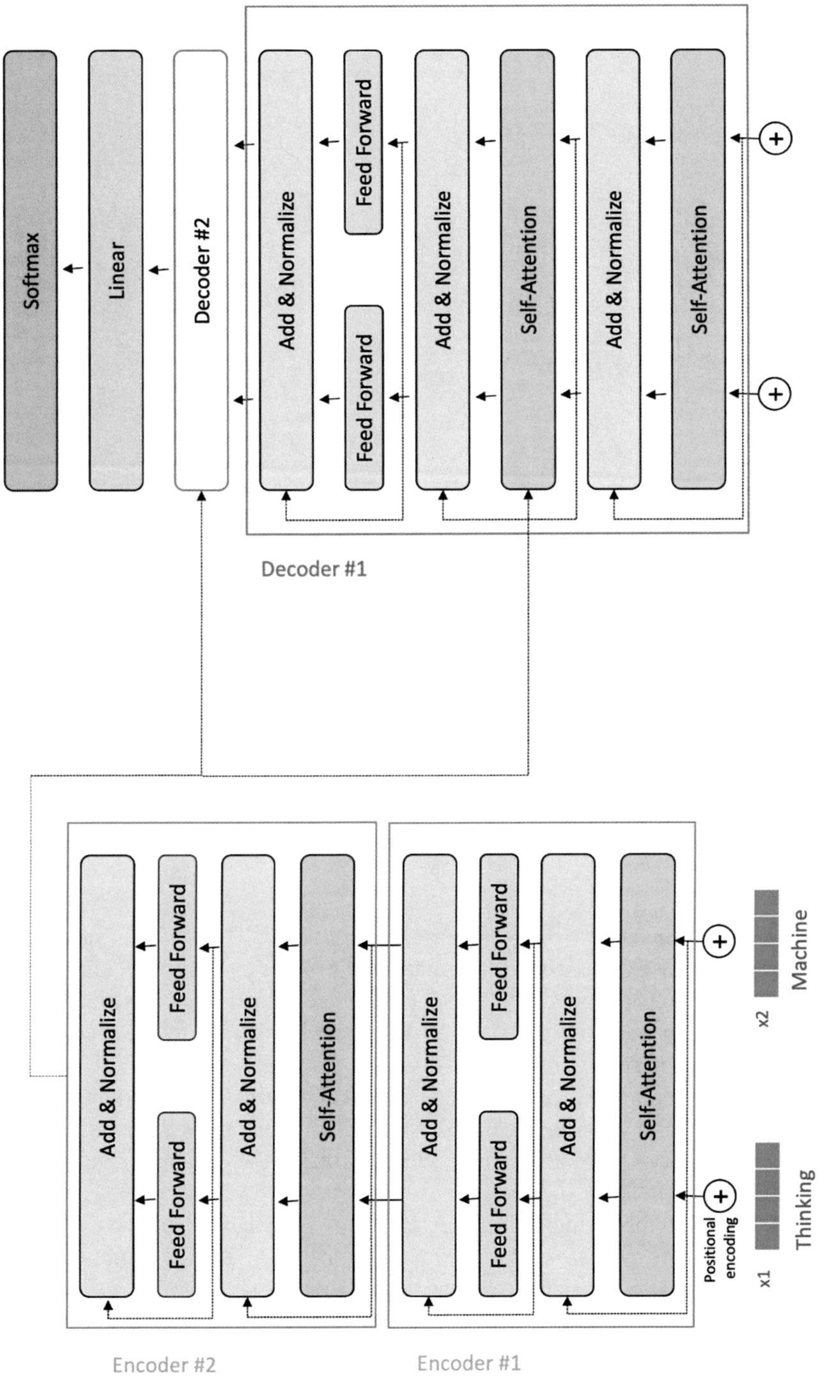

Fig. 9.14 Encoder–decoder structure for language translation

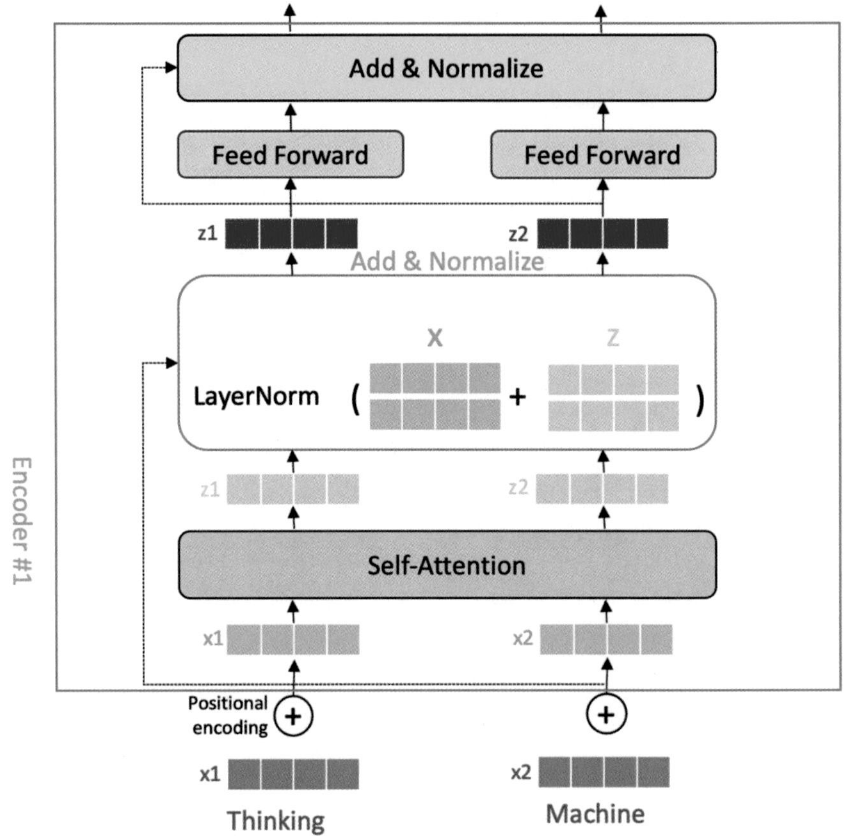

Fig. 9.15 Network architecture of encoder of Transformer

In particular, sentences from, for example, English are used on the encoder to learn how to embed each word in a sentence. In order to learn the long-range dependency between the words within the sentence, a self-attention mechanism is used on the encoder. Of course, self-attention is not enough to perform a complicated speech embedding task. Therefore, there are an additional residual connection, a layer normalization, and a neural feedforward network, followed by additional units of encoder blocks (see Fig. 9.15). Once trained, Transformer's encoder generates the word embedding, which contains the structural role of each word within the sentence.

In the decoder, these embedding vectors from the encoder are now used to generate the key vectors, as shown in Figs. 9.14 and 9.16. This is combined with the query vector that is generated from the target language, like French. This hybrid combination then creates the attention map, which serves as the transformation matrix of the words between the two languages by taking into account their structural roles.

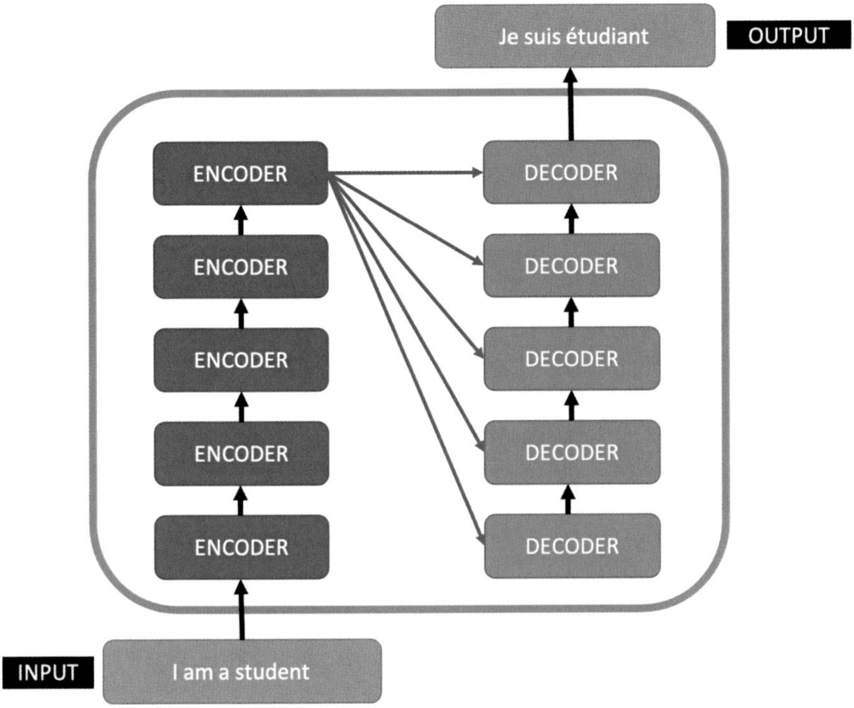

Fig. 9.16 Generation of key vectors for each decoder layer of Transformer

Another important component of Transformer is the positional encoding (see the positional encoding blocks in Figs. 9.14 and 9.15). In contrast to RNN and LSTM, each word in a sentence is processed simultaneously by Transformer in order to capture longer dependencies in a sentence, so the model itself does not have any notion of position for each word. However, the position of a word within a sentence is important as it determines the grammar and the semantics of the sentence. Therefore, there is a need to consider the order of the words, and the positional encoding is used for this. To be a valid positional encoding, a method should output a unique encoding of the position of each word in a sentence and easily generalize to longer sentences.

Among the various possible approaches, the original authors of Transformer used the sine and cosine functions of different frequencies [73]. More specifically, let n be the desired position in an input sentence and $\boldsymbol{p}_n \in \mathbb{R}^d$ be its corresponding encoding, where d is the encoding dimension for which an even number is chosen.

Then, the position encoding vector is given by

$$\boldsymbol{p}_n = \begin{bmatrix} \sin(\omega_1 n) \\ \cos(\omega_1 n) \\ \sin(\omega_2 n) \\ \cos(\omega_2 n) \\ \vdots \\ \sin(\omega_{\frac{d}{2}} n) \\ \cos(\omega_{\frac{d}{2}} n) \end{bmatrix} \in \mathbb{R}^d, \quad \text{where} \quad \omega_k = \frac{1}{10000^{2k/d}}. \tag{9.49}$$

This position encoding vector is then added to the word embedding vector $\boldsymbol{x}_n \in \mathbb{R}^d$ to obtain a position encoded word embedding vector:

$$\boldsymbol{x}_n \leftarrow \boldsymbol{x}_n + \boldsymbol{p}_n, \tag{9.50}$$

which is then fed into the self-attention module in Transform.

Readers may wonder why the positional encoding vector is summed with a word embedding instead of concatenation. Although this was used empirically in the original paper [73], recent theoretical analysis showed that Transformer architecture with additive positional encodings is Turing complete [93] and can be reparametrized to express any convolutional layer [94].

Transformer is an ingenious combination of the full mathematical principle of attention, which uses separate query and key vectors for the specific purpose of language translation. Because of this, Transformer has become the main workhorse for modern NLP.

9.4.6 BERT

One of the latest milestones in NLP is the release of BERT (Bidirectional Encoder Representations from Transformers) [74]. This release of BERT can even be seen as the beginning of a new era in NLP. One of the unique features of BERT is that the resulting structure is as regular as FPGA (Field Programmable Gate Array) chips, so the BERT unit can be used for different purposes and languages by simply changing the training scheme.

The main architecture of BERT is the cascaded connection of bidirectional transformer encoder units, as shown in Fig. 9.17. Due to the use of the encoder-part of the Transformer architecture, the number of input and output features remains the same, while each feature vector dimension may be different. For example, the input feature can be a one-hot coded word, the feature dimension of which is determined by the size of the corpus vocabulary. The output may be the low dimensional embedding that sums up the role of the word in context. The reason for using the bidirectional Transformer encoder is based on the observation that people can understand the sentence even if the order of the words within the sentence is

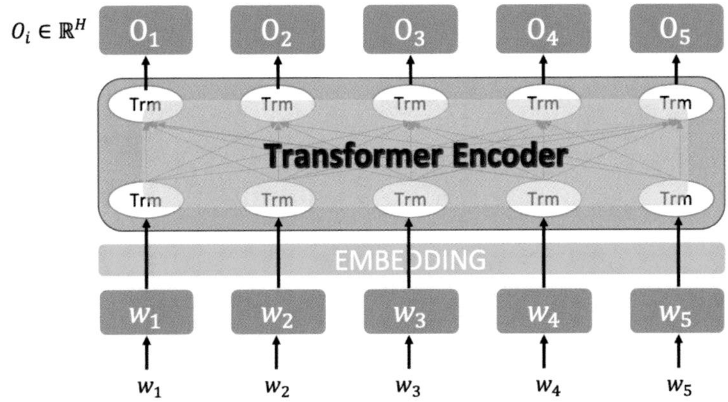

Fig. 9.17 BERT architecture

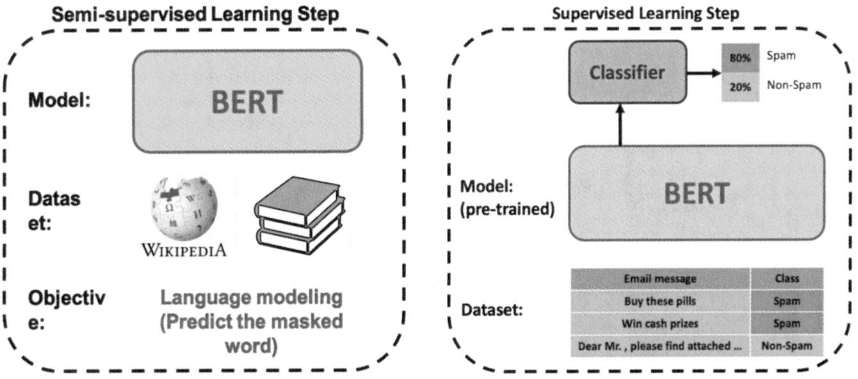

Fig. 9.18 Pre-training and fine-tuning scheme for BERT training

reversed. By considering the reverse order, the role of each word in context is better summarized as an attention map, resulting in more efficient embedding of words.

Yet another beauty of BERT lies in the training. More specifically, as shown in Fig. 9.18, BERT training consists of two steps: pre-training and fine-tuning. In the pre-training step, the goal of the task is to guess the masked word within an input sentence. Figure 9.19 shows a more detailed explanation of this masked word estimation. Approximately 15% of the words in the input sentence from Wikipedia are masked with a specific token (in this case, [MASK]), and the goal of the training is to estimate the masked word from the embedded output in the same place. Since the BERT output is just an embedded feature, we need an additional fully connected neural network (FFNN) and softmax layer to estimate the specific word. With this additional network we can correctly pre-train the BERT unit.

Once BERT pre-training is finished, the BERT unit is fine-tuned using supervised learning tasks. For example, Fig. 9.20 shows a supervised learning task. Here, the

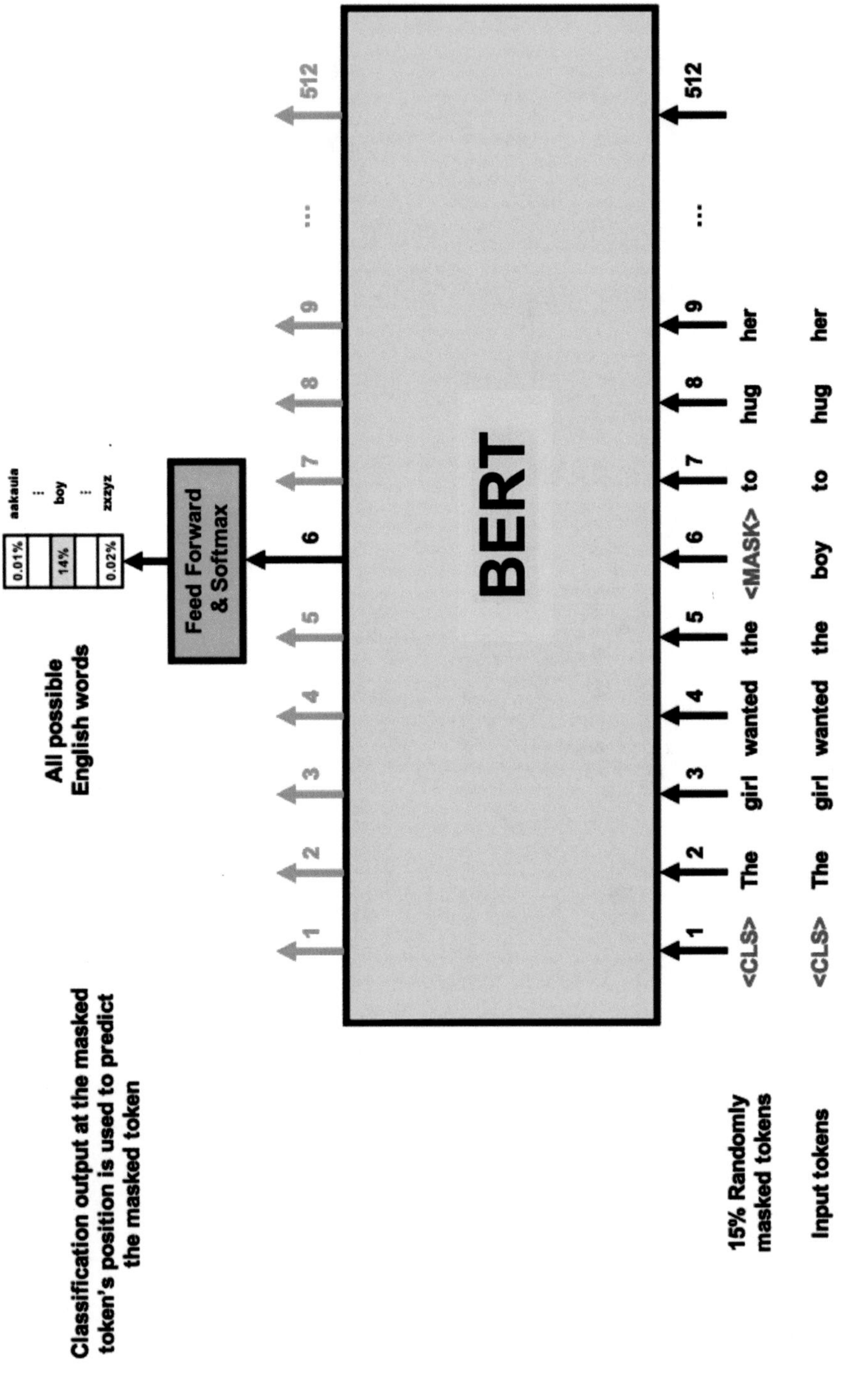

Fig. 9.19 Estimating the masked word for BERT training

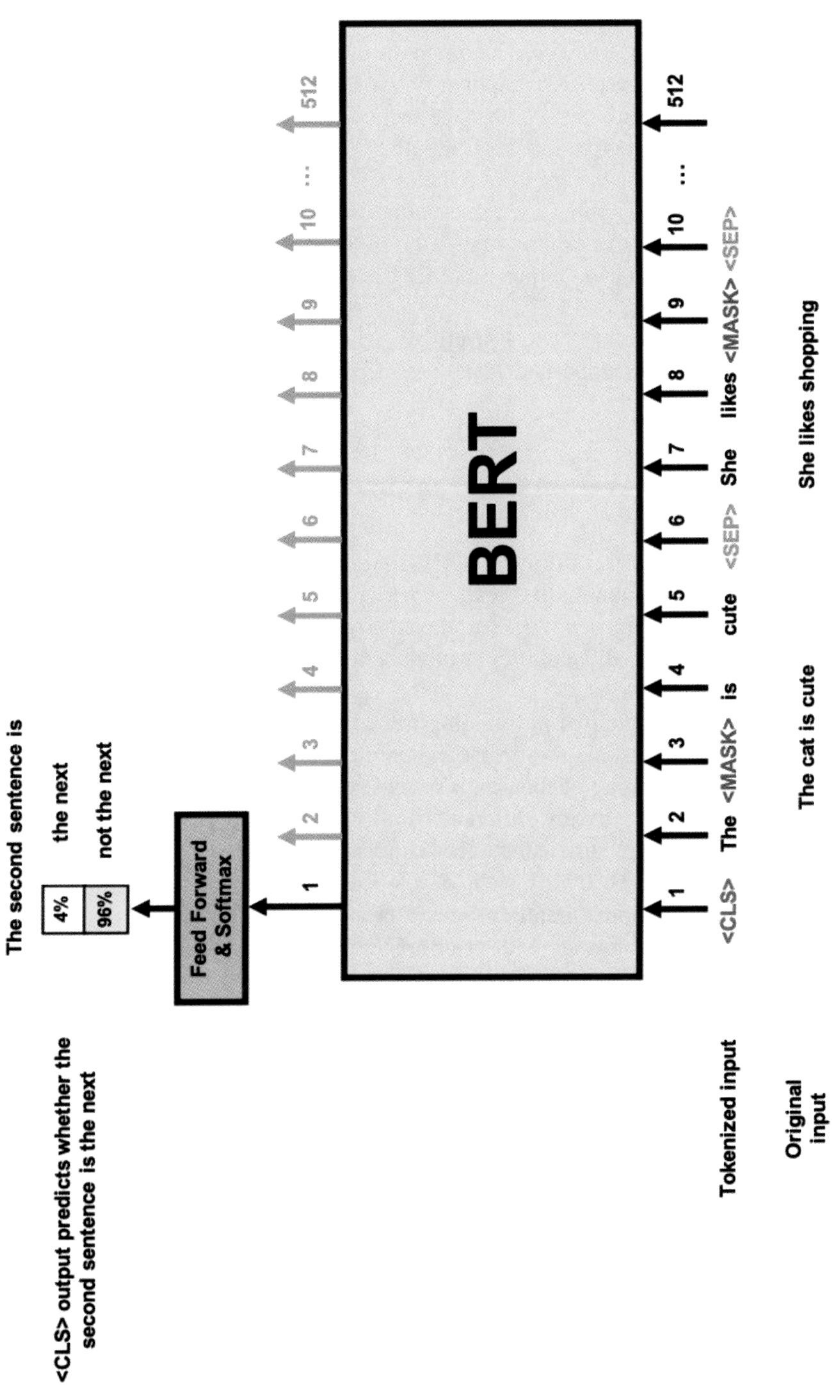

Fig. 9.20 Supervised learning task for BERT fine-tuning for the next sentence estimation

input for BERT consists of two sentences, separately with another token [SEP]. The goal of supervised learning is then to assess whether the second sentence is a correct continuation of the first sentence. The output of this is now embedded in the BERT Output 1, which is then used as an input of fully connected neural network, followed by a softmax layer to estimate whether the second sentence is next. Since the same number is entered and output in BERT, the first word of the input record should be a token that indicates the vacant word [CLS].

Another example of a supervised fine-tuning is the classification of whether the sentence is spam or not, as shown in Fig. 9.21. In this case, only a single sentence is used as the BERT input and Output 1 of BERT is used to classify whether the input sentence is spam or not.

In fact, there are multiple ways of utilizing the BERT unit for supervised fine tuning, which is another important advantage of BERT [74].

9.4.7 Generative Pre-trained Transformer (GPT)

Generative pre-trained transformers (GPTs) are language models developed by OpenAI that produce human-like text. In particular, the third-generation model, GPT-3, is arguably the most powerful and controversial artificial intelligence model for NLP due to its incredible ability to produce text that is indistinguishable from what written by humans [76].

Recall that BERT requires pre-training for a large corpus of text, followed by fine-tuning a specific task. However, the requirement of a task-specific, finely tuned training data set consisting of thousands or tens of thousands of examples is often quite demanding. This is very different from humans, who are usually able to complete a new language task using a few examples.

GPT-2 [75] and GPT-3 [76] were developed based on the observation that scaling the language model greatly improves task-agnostic, few-shot performance, and sometimes even competes with prior art fine-tuning approaches. The goal of GPT training is similar to BERT pre-training, where the next word in a sentence is estimated based on the previous words in a sentence. For this reason, GPT stands for generative *pre-trained* Transformer. For example, the GPT is trained to generate the word "awesome" by using the preceding words "The latest language model GPT-3 is" as input. While this pure pre-training scheme doesn't improve BERT's performance, one of the main reasons for the success of GPT-2, and GPT-3 in particular, is its massive architecture that makes generative pre-training even more powerful than fine-tuning. Compared to the largest BERT architecture with around 340 million parameters, GPT-3 is extremely massive with around 175 billion parameters.

Recall that the generative estimation of the following word can be done by the Transformer decoder in the language translation. Accordingly, GPT-3 consists of a stack of 96 Transformer decoder layers, which differs from the encoder-only architecture in BERT (see Fig. 9.22). Each decoder layer is composed of multiple

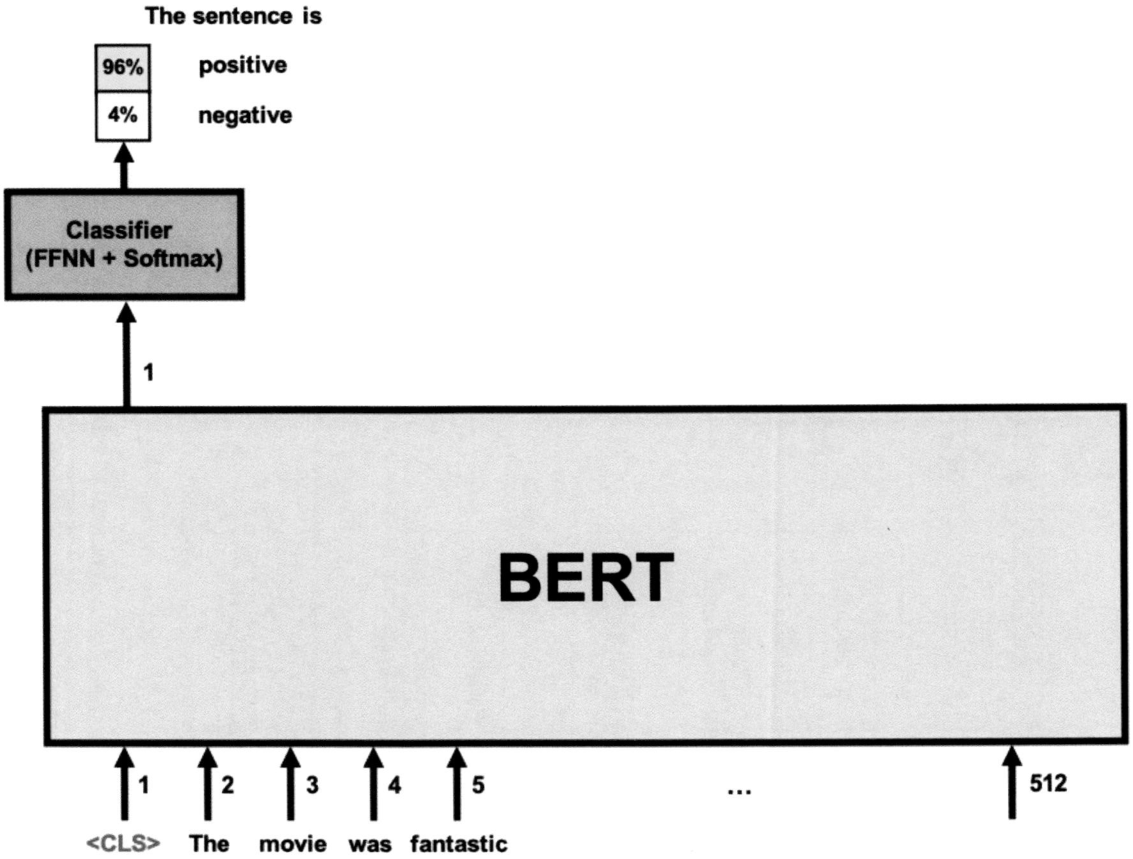

Fig. 9.21 BERT fine-tuning using supervised learning to classify whether the input sentence is spam or not

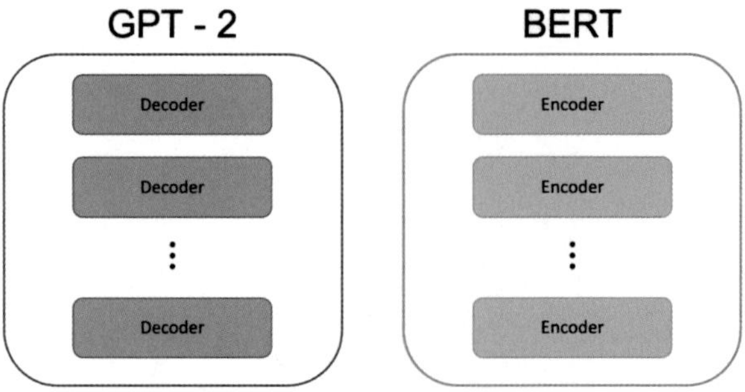

Fig. 9.22 Differences in BERT and GPT architecture

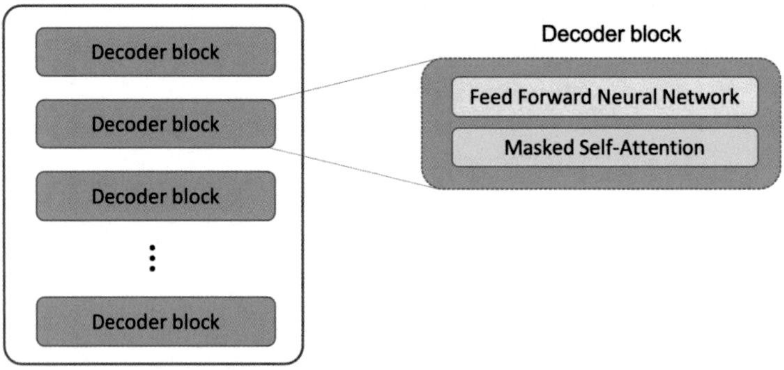

Fig. 9.23 Architecture of GPT decoder block

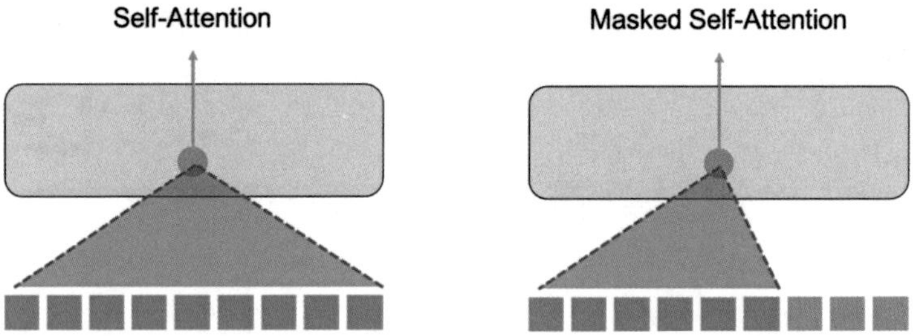

Fig. 9.24 Difference between the self-attention in BERT and masked self-attention in GPT-3

decoder blocks, which consist of masked self-attention blocks with a width of 2048 tokens and a feedforward neural network (see Fig. 9.23). As shown in Fig. 9.24, the masked self-attention calculates the attention matrix using the preceding words in a sentence that can be used to estimate the next word.

To train the 175 billion weights, GPT-3 is trained with 499 billion tokens or words. Sixty percent of the training data set comes from a filtered version of Common Crawl consisting of 410 billion tokens. Other sources are 19 billion tokens from WebText2, 12 billion tokens from Books1, 55 billion tokens from Books2, and

3 billion tokens from Wikipedia [76]. Nonetheless, the performance of GPT-3 can be affected by the quality of the training data. For example, it was reported that GPT-3 generates sexist, racist and other biased and negative language when it was asked to discuss Jews, women, black people, and the Holocaust [95].

9.4.8 Vision Transformer

Inspired by the fact that Transformer architecture has become state of the art for NLP, researchers have explored its applications for computer vision. As mentioned earlier, in computer vision, attention is usually applied in connection with convolutional networks, so that certain components of convolutional networks are replaced with attention while maintaining their overall structure. In [96], the authors have shown that this dependence on CNNs is not necessary and a pure transformer applied directly to sequences of image patches can work very well in image classification tasks.

Their model, called Vision Transformer (ViT), is depicted in Fig. 9.25. To handle 2D images, the input image $\boldsymbol{x}$ is reshaped into a sequence of flattened 2D patches, after each patch is embedded into a D-dimensional vector using a trainable linear projection. Transformer then uses a constant latent vector size D through all of its layers. Position embeddings are added to the patch embeddings to retain positional information. The resulting sequence of embedding vectors serves as input to the encoder. With regard to the [Class] token on the front, a learnable embedding in the sequence of embedded patches at the output of the Transformer encoder serves as the entire image representation. A classification head is attached during both pre-training and fine-tuning to train the network to have the embedded image representation for the best classification results.

The Transformer encoder in ViT consists of alternating layers of multi-headed self-attention and MLP blocks. Layer norm and residual connections are applied before and after every block, respectively. The MLP contains two layers with a GELU non-linearity. Typically, ViT is trained on large data sets, and fine-tuned to (smaller) downstream tasks. For this, we remove the pre-trained prediction head and attach a zero-initialized $D \times K$ feedforward layer, where K is the number of downstream classes.

9.5 Mathematical Analysis of Normalization and Attention

So far we have discussed normalization and attention. Normalization was originally developed for accelerating stochastic gradient methods, and has been extended to style transfer, image generation, etc. On the other hand, due to its ability to learn long-range relationships and its flexibility from manipulating query and key, attention has been successfully extended to various applications, leading to

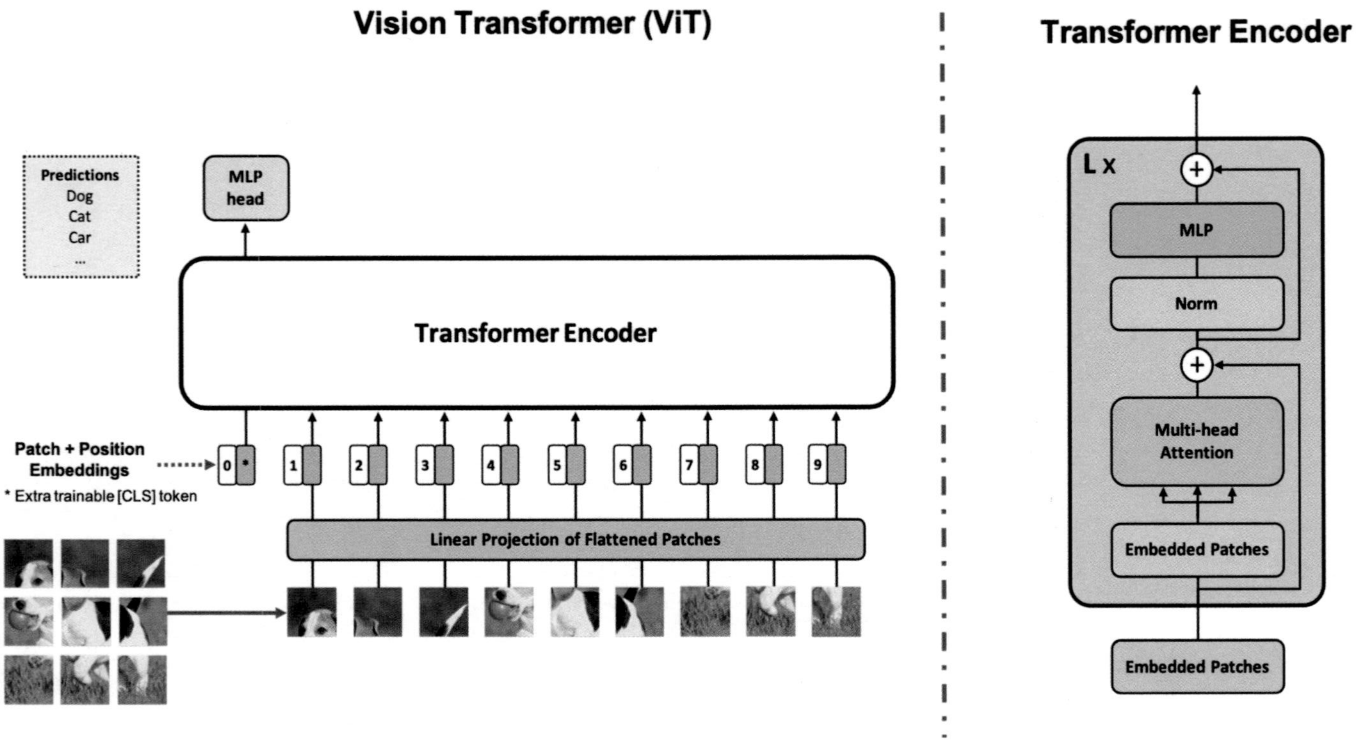

Fig. 9.25 Model overview. We split an image into fixed-size patches, linearly embed each of them, add position embeddings, and feed the resulting sequence of vectors to a standard Transformer encoder. In order to perform classification, we use the standard approach of adding an extra learnable "classification token" to the sequence

breakthroughs in natural language processing approaches such as BERT, GPT-3, etc.

As you may have noticed while reading, normalization and attention may have a very similar mathematical formulation. For example, for a given feature map $\boldsymbol{X} \in \mathbb{R}^{HW \times C}$, the instance normalization, AdaIN, and WCT can be represented as follows:

$$\boldsymbol{Y} = \boldsymbol{X}\boldsymbol{T} + \boldsymbol{B}, \tag{9.51}$$

where the channel-directional transform $\boldsymbol{T}$ and the bias $\boldsymbol{B}$ are learned from the statistics of the feature maps. The only differences between instance normalization, AdaIN, and WCT are their specific ways of estimating $\boldsymbol{T}$ and $\boldsymbol{B}$. For example, all elements of $\boldsymbol{T}$ are estimated from the input features in the case of instance normalization, while they are estimated from the statistics on content and style images in the case of AdaIN and WCT. The main difference between WCT, instance normalization and AdaIN is that $\boldsymbol{T}$ is a densely populated matrix for the case of WCT, while instance norm and AdaIN use a diagonal matrix.

On the other hand, the spatial attention can be represented by

$$\boldsymbol{Y} = \boldsymbol{A}\boldsymbol{X}, \tag{9.52}$$

where $\boldsymbol{A}$ is calculated from its own feature for the case of self-attention, or with the help of other domain features for the case of cross-domain attention. Similarly, the channel attention such as SENet can be computed as

$$\boldsymbol{Y} = \boldsymbol{X}\boldsymbol{T}, \tag{9.53}$$

where the diagonal matrix $\boldsymbol{T}$ is again calculated from $\boldsymbol{X}$.

This implies that normalization and attention, with the exception of the specific differences in the generation of $\boldsymbol{A}$, $\boldsymbol{T}$, $\boldsymbol{W}$, and $\boldsymbol{B}$, can be viewed as a special case of the following transformation:

$$\boldsymbol{Y} = \boldsymbol{A}\boldsymbol{X}\boldsymbol{T} + \boldsymbol{B}. \tag{9.54}$$

Mathematically, $\boldsymbol{A}$ modifies the column space of $\boldsymbol{X}$, whereas $\boldsymbol{T}$ control the row space of $\boldsymbol{X}$. Therefore, the attention map $\boldsymbol{A}$ differs from $\boldsymbol{T}$ and controls different factors and the variations in the feature $\boldsymbol{X}$.

Based on this observation, Kwon et al. [97] proposed the so-called Diagonal GAN. This is based on the following intuition: although $\boldsymbol{A}$ was a dense matrix obtained from $\boldsymbol{X}$ in the original self-attention, the insight from AdaIN can be used to obtain an efficient diagonal attention map $\boldsymbol{A}$ from a novel attention code generator for content control. Specifically, they introduced a novel diagonal attention (DAT) module to manipulate the content feature maps as shown in Fig. 9.26b. One of the important advantages of the method is that thanks to the symmetry in (9.54), both AdaIN and DAT can be applied to each layer, so that the image content and

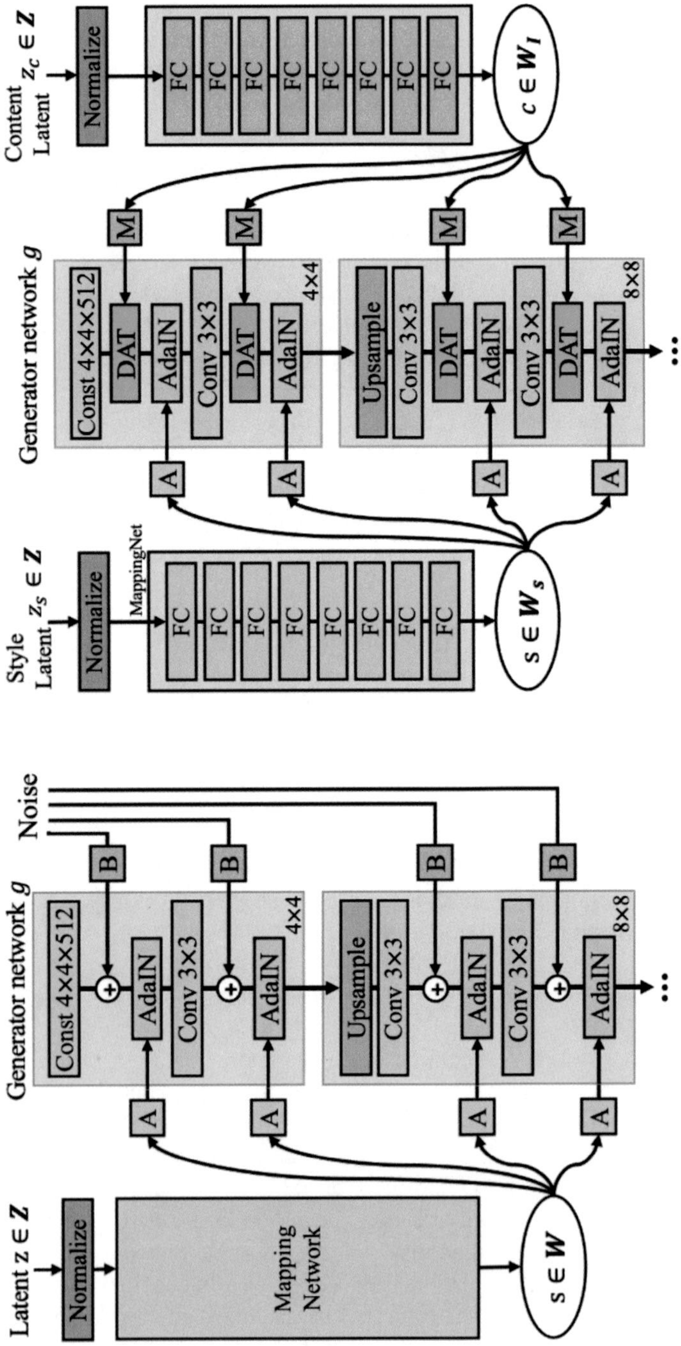

Fig. 9.26 (a) StyleGAN and (b) Diagonal GAN

style can be modulated independently. This leads to an effective disentanglement of the content and style components in generated images. Furthermore, the proposed method has flexibility by selectively controlling the spatial attribute of generated images at arbitrary resolution by changing the hierarchical attention maps.

As shown in Fig. 9.27, the combination of AdaIN and DAT is quite impressive. For given source images in Fig. 9.27a, which are generated from arbitrary style and content code, (b) shows the samples with varying style codes and fixed content code. Note that the hairstyles and identities vary while the face directions and expressions are similar. On the other hand, if we generate samples with varying content codes and fixed style, the face direction and expression for the same person or animal changes. Finally, if the content and style codes are both varied as shown in (c), the face direction, expression, hair styles, and person's identity change accordingly. This clearly shows the disentanglement between style and content.

One may wonder whether additive noise at each layer of styleGAN in Fig. 9.26b may serve a similar role in the content variation. In fact, the addition of the noise for the original styleGAN is from a similar motivation, as indicated by the authors' claim that the right-hand network generates the content feature vectors from random noise. That said, it should be remembered that the additive noise terms are basically additions to the bias term in (9.54), which is fundamentally different from $\boldsymbol{A}$ that modulates the column space of $\boldsymbol{X}$. In fact, the additional bias terms both affect the row and column spaces of $\boldsymbol{X}$, resulting in the entangled modulation between the style and content.

9.6 Exercises

1. Find the conditions when the WCT transform in (9.22) is reduced to AdaIN.
2. Let the feature map with the number of pixels $H \times W = 4$ and the channels $C = 3$ be given by

$$\boldsymbol{X} = \begin{bmatrix} 1 & 2 & 3 \\ -1 & -3 & 0 \\ 5 & -2 & 1 \\ 0 & 0 & -5 \end{bmatrix}. \tag{9.55}$$

   a. Perform the layer normalization of $\boldsymbol{X}$.
   b. Perform the instance normalization of $\boldsymbol{X}$.

CelebA-HQ Generated Samples

AFHQ Generated Samples

Fig. 9.27 (Top) 1024×1024 images generated by our method, trained using CelebA-HQ data set. (Bottom) 512×512 images generated by our method, trained using AFHQ data set. (**a**) A source image generated from arbitrary style and content code. (**b**) Samples with varying style codes and fixed content code. (**c**) Samples generated with varying content codes and fixed style. (**d**) Samples generated with both varying content and style codes

3. Additionally, suppose that the feature map for the style image is given by

$$S = \begin{bmatrix} 0 & 1 & 1 \\ -1 & -1 & 1 \\ 1 & 0 & 0 \\ -1 & 1 & 1 \end{bmatrix}. \tag{9.56}$$

   a. For the given feature map in (9.55), perform the adaptive instance normalization from $\boldsymbol{X}$ to the style of $\boldsymbol{S}$.
   b. For the given feature map in (9.55), perform the WCT style transfer from $\boldsymbol{X}$ to the style of $\boldsymbol{S}$.

4. Using the feature map in (9.55), we are interested in computing the self-attention map. Let $\boldsymbol{W}_Q$ and $\boldsymbol{W}_k$ be the embedding matrices for the query and key, respectively:

$$\boldsymbol{W}_Q = \begin{bmatrix} 2 & 1 \\ 0 & \frac{1}{2} \\ 0 & 0 \end{bmatrix}, \quad \boldsymbol{W}_K = \begin{bmatrix} \frac{1}{3} & 0 \\ 1 & -1 \\ 10 & 5 \end{bmatrix}. \tag{9.57}$$

   a. Using the dot product score function, compute the attention matrix $\boldsymbol{A}$.
   b. What is the attended feature map, i.e. $\boldsymbol{Y} = \boldsymbol{A}\boldsymbol{X}$?
   c. For the case of masked self-attention in GPT-3, compute the attention mask $\boldsymbol{A}$ and attended feature map $\boldsymbol{Y} = \boldsymbol{A}\boldsymbol{X}$.

5. For a given positional encoding in (9.49) for the Transformer with encoding dimension $d = 10$, compute the positional encoding vector $\boldsymbol{p}_n$ for $n = 1, \cdots, 10$.
6. Explain the following sentence in detail: "BERT has encoder only structure, while GPT-3 has decoder only architecture."
7. For a given feature map $\boldsymbol{X} \in \mathbb{R}^{N \times C}$, show that the feature map of styleGAN after the application of AdaIN and noise is represented by

$$\boldsymbol{Y} = \boldsymbol{X}\boldsymbol{T} + \boldsymbol{B}. \tag{9.58}$$

   Specify the structure of the matrices $\boldsymbol{T}$ and $\boldsymbol{B}$.
8. For a given feature map $\boldsymbol{X} \in \mathbb{R}^{N \times C}$, show that the feature map of the Diagonal GAN after the application of AdaIN, DAT, and noise is represented by

$$\boldsymbol{Y} = \boldsymbol{A}\boldsymbol{X}\boldsymbol{T} + \boldsymbol{B}. \tag{9.59}$$

   Specify the structure of the matrices $\boldsymbol{A}$, $\boldsymbol{T}$ and $\boldsymbol{B}$, and their mathematical roles.

Part III
Advanced Topics in Deep Learning

"I am really confused. I keep changing my opinion on a daily basis, and I cannot seem to settle on one solid view of this puzzle. No, I am not talking about world politics or the current U.S. president, but rather something far more critical to humankind, and more specifically to our existence and work as engineers and researchers. I am talking about . . . deep learning."

– Michael Elad

Chapter 10
Geometry of Deep Neural Networks

10.1 Introduction

In this chapter, which is mathematically intensive, we will try to answer perhaps the most important questions of machine learning: what does the deep neural network learn? How does a deep neural network, especially a CNN, accomplish these goals? The full answer to these basic questions is still a long way off. Here are some of the insights we've obtained while traveling towards that destination. In particular, we explain why the classic approaches to machine learning such as single-layer perceptron or kernel machines are not enough to achieve the goal and why a modern CNN turns out to be a promising tool.

Recall that at the early phase of the deep learning revolution, most of the CNN architectures such as AlexNet, VGGNet, ResNet, etc., were mainly developed for the classification tasks such as ImageNet challenges. Then, CNNs started to be widely used for low-level computer vision problems such as image denoising [90, 98], super-resolution [99, 100], segmentation [38], etc., which are considered as regression tasks. In fact, classification and regression are the two most fundamental tasks in machine learning, which can be unified under the umbrella of function approximation. Recall that the representer theorem [15] says that a classifier design or regression problem for a given test data set $\{(\boldsymbol{x}_i, y_i)\}_{i=1}^n$ can be addressed by solving the following optimization problem:

$$\min_{f \in \mathcal{H}_k} \frac{1}{2}\|f\|_{\mathcal{H}}^2 + C\sum_{i=1}^{n} \ell\left(y_i, f(\boldsymbol{x}_i)\right), \tag{10.1}$$

J. C. Ye, *Geometry of Deep Learning*, Mathematics in Industry 37,
https://doi.org/10.1007/978-981-16-6046-7_10

where $\mathcal{H}_k$ denotes the reproducing kernel Hilbert space (RKHS) with the kernel $k(\boldsymbol{x}, \boldsymbol{x}')$, $\|\cdot\|_{\mathcal{H}}$ is the Hilbert space norm, and $\ell(\cdot,\cdot)$ is the loss function. One of the most important results of the representer theorem is that the minimizer f has the following closed-form representation:

$$f(\boldsymbol{x}) = \sum_{i=1}^{n} \alpha_i k(\boldsymbol{x}_i, \boldsymbol{x}), \tag{10.2}$$

where $\{\alpha_i\}_{i=1}^n$ are learned parameters from the training data set. For example, if a hinge function is used as a loss, the solution becomes a kernel SVM, whereas if an l_2 function is used as a loss, it becomes a kernel regression.

In general, the solution $f(\boldsymbol{x})$ in (10.2) is a nonlinear function of the input $\boldsymbol{x}$ based on the kernel $k(\boldsymbol{x}_j, \cdot)$, which is nonlinearly dependent upon $\boldsymbol{x}$. This nonlinearity of the kernel makes the expression in (10.2) more expressive, thereby generating a wide variation of functions within the RKHS $\mathcal{H}_k$.

That said, the expression in (10.2) still has fundamental limitations. First, the RKHS $\mathcal{H}_k$ is specified by choosing the kernel in a top-down manner, and to the best of our knowledge, there is no way to automatically learn from the data. Second, once the kernel machine is trained, the parameters $\{\alpha_i\}_{i=1}^n$ are fixed, and it is not possible to adjust them at the test phase. These drawbacks lead to the fundamental limitations of the *expressivity* of neural networks, which means the capability of approximating any function. Of course, one could increase the expressivity by increasing complexity of the learning machines, for example, by combining multiple kernel machines. However, our goal is to achieve better expressivity for a given complexity constraint, and in this sense the kernel machine has problems.

10.1.1 Desiderata of Machine Learning

Given the limitations of the kernel machine, we can state the following *desiderata*—the desired things that an ultimate learning machine should satisfy:

- **Data-driven model:** The function space that a learning machine can represent should be learned from the data, rather than specified by a top-down mathematical model.
- **Adaptive model:** Even after the machine has learned, the learned model should adapt to the given input data at the test phase.
- **Expressive model:** The expressivity of the model should increase more than the model complexity increases.
- **Inductive model:** The learned information from the training data should be used at the test phase.

In the following, we review two classical approaches—single layer perceptron and frame representation—and explain why these classical models failed to meet the desiderata. Later we will show how the modern deep learning approaches have been developed by overcoming the drawbacks of these classical approaches by exploiting their inherent strengths.

10.2 Case Studies

10.2.1 Single–Layer Perceptron

The single-layer perceptron is a special case of the multilayer perceptron (MLP), which consists of fully connected neurons at the single hidden layer. Specifically, let $\varphi : \mathbb{R} \mapsto \mathbb{R}$ be a nonconstant, bounded, and continuous activation function. Let $\mathcal{X} \subset \mathbb{R}^m$ denote the input space. Then, a single-layer perceptron $f_{\boldsymbol{\Theta}} : \mathcal{X} \mapsto \mathbb{R}$ can be represented by

$$f_{\boldsymbol{\Theta}}(\boldsymbol{x}) = \sum_{i=1}^{d} v_i \varphi\left(\boldsymbol{w}_i^\top \boldsymbol{x} + b_i\right), \quad \boldsymbol{x} \in \mathcal{X}, \tag{10.3}$$

where $\boldsymbol{w}_i \in \mathbb{R}^m$ is a weight vector, $v_i, b_i \in \mathbb{R}$ are real constants, and $\boldsymbol{\Theta} = \{(\boldsymbol{w}_i, v_i, b_i)\}_{i=1}^d$ represents the neural network parameters. Then, the parameters are estimated by solving the following optimization problem using the training data $\{(\boldsymbol{x}_i, y_i)\}_{i=1}^N$:

$$\min_{\boldsymbol{\Theta}} \sum_{i=1}^{n} \ell\left(y_i, f_{\boldsymbol{\Theta}}(\boldsymbol{x}_i)\right) + \lambda R(\boldsymbol{\Theta}), \tag{10.4}$$

where λ is a regularization parameter and $R(\boldsymbol{\Theta})$ is a regularization function with respect to the parameter set $\boldsymbol{\Theta}$.

One of the classical results for the representation power of single-layer perceptrons dates from 1989 [48]. It states that a feed-forward network with a single hidden layer containing a finite number of neurons can approximate continuous functions on compact subsets under mild assumptions on the activation function.

Theorem 10.1 (Universal Approximation Theorem[48]) *Let the space of real-valued continuous functions on a compact set $\mathcal{X}$ be denoted by $C(\mathcal{X})$. Then, given any $\varepsilon > 0$ and any function $g \in C(\mathcal{X})$, there exist an integer d such that the single layer perceptron in* (10.3) *is an approximate realization of the function f; that is,*

$$|f_{\boldsymbol{\Theta}}(\boldsymbol{x}) - g(\boldsymbol{x})| < \varepsilon$$

for all $\boldsymbol{x} \in \mathcal{X}$.

The theorem thus states that simple neural networks can represent a wide variety of interesting functions when given appropriate parameters. In fact, the universal approximation theorem was a blessing for classic machine learning; it promoted the research interests of the neural network as a powerful functional approximation, but also turned out to be a curse for the development of machine learning by preventing understanding of the role of deep neural networks.

More specifically, the theorem only guarantees the existence of d, the number of neurons, but it does not specify how many neurons are required for a given approximation error. Only recently have people realized that the depth matters, i.e. there exists a function that a deep neural network can approximate but a shallow neural network with the same number of parameters cannot [101–105]. In fact, these modern theoretical studies have provided a theoretical foundation for the revival of modern deep learning research.

When compared with the kernel machine (10.2), the pros and cons of the single-layer perception in (10.3) can be easily understood. Specifically, $\varphi\left(\boldsymbol{w}_i^\top \boldsymbol{x} + b_i\right)$ in (10.3) works similarly as a kernel function $k(\boldsymbol{x}_i, \boldsymbol{x})$, and v_i in (10.3) is similar to the weight parameter α_i in (10.2). However, the nonlinear mapping in the perceptron, i.e. $\varphi\left(\boldsymbol{w}_i^\top \boldsymbol{x} + b_i\right)$, does not necessarily satisfy the positive semidefiniteness of the kernel, thereby increasing the approximable functions beyond the RKHS to a larger function class in Hilbert space. Therefore, there exists potential for improving the expressivity. On the other hand, the weighting parameters v_i are still fixed once the neural network is trained, which leads to limitations similar to those of the kernel machines.

10.2.2 Frame Representation

Now, we review another class of function representation called a *frame* [1]. To understand the mathematical concept of a frame, we start with its simplified form—the *basis*.

In mathematics, a set $B = \{\boldsymbol{b}_i\}_{i=1}^m$ of elements (vectors) in a vector space V is called a basis, if every element of V may be written in a unique way as a linear combination of elements of B, that is, for every $\boldsymbol{f} \in V$, there exists unique coefficient $\{a_i\}$ such that

$$\boldsymbol{f} = \sum_{i=1}^{m} a_i \boldsymbol{b}_i. \tag{10.5}$$

Unlike the basis, which leads to the unique expansion, the frame is composed of redundant basis vectors, which allows multiple representation. Frames can also be extended to deal with function spaces, in which case the number of frame elements is infinite. Formally, a set of functions

$$\boldsymbol{\Phi} = [\boldsymbol{\phi}_k]_{k\in\Gamma} = \left[\cdots\ \boldsymbol{\phi}_{k-1}\ \boldsymbol{\phi}_k\ \cdots\right]$$

in a Hilbert space H is called a *frame* if it satisfies the following inequality [1]:

$$\alpha\|\boldsymbol{f}\|^2 \leq \sum_{k\in\Gamma} |\langle \boldsymbol{f}, \boldsymbol{\phi}_k\rangle|^2 \leq \beta\|\boldsymbol{f}\|^2, \quad \forall \boldsymbol{f} \in H, \tag{10.6}$$

where $\alpha, \beta > 0$ are called the frame bounds. If $\alpha = \beta$, then the frame is said to be tight. In fact, a basis is a special case of tight frames.

By writing $c_k := \langle \boldsymbol{f}, \boldsymbol{\phi}_k\rangle$ as the expansion coefficient with respect to the k-th frame vector $\boldsymbol{\phi}_k$ and defining the frame coefficient vector

$$\boldsymbol{c} = [c_k]_{k\in\Gamma} = \boldsymbol{\Phi}^\top \boldsymbol{f},$$

(10.6) can be equivalently represented by

$$\alpha\|\boldsymbol{f}\|^2 \leq \|\boldsymbol{c}\|^2 \leq \beta\|\boldsymbol{f}\|^2, \quad \forall \boldsymbol{f} \in H. \tag{10.7}$$

This implies that the energy of the expansion coefficients should be bounded by the original signal energy, and for the case of the tight frame, the expansion coefficient energy is the same as the original signal energy up to the scaling factor.

When the frame lower bound α is nonzero, then the recovery of the original signal can be done from the frame coefficient vector $\boldsymbol{c} = \boldsymbol{\Phi}^\top \boldsymbol{f}$ using the *dual frame operator* $\widetilde{\boldsymbol{\Phi}}$ given by

$$\widetilde{\boldsymbol{\Phi}} = \left[\cdots\ \widetilde{\boldsymbol{\phi}}_{k-1}\ \widetilde{\boldsymbol{\phi}}_k\ \cdots\right], \tag{10.8}$$

which satisfies the so-called *frame condition*:

$$\tilde{\boldsymbol{\Phi}}\boldsymbol{\Phi}^\top = \boldsymbol{I}, \tag{10.9}$$

because we have

$$\hat{\boldsymbol{f}} := \widetilde{\boldsymbol{\Phi}}\boldsymbol{c} = \boldsymbol{\Phi}\boldsymbol{\Phi}^\top \boldsymbol{f} = \boldsymbol{f},$$

or equivalently,

$$\boldsymbol{f} = \sum_{k\in\Gamma} c_k \widetilde{\boldsymbol{\phi}}_k = \sum_{k\in\Gamma} \langle \boldsymbol{f}, \boldsymbol{\phi}_k\rangle \widetilde{\boldsymbol{\phi}}_k. \tag{10.10}$$

Note that (10.10) is a linear signal expansion, so it is not useful for machine learning tasks. However, something more interesting occurs when it is combined with a nonlinear regularization. For example, consider a regression problem to estimate a noiseless signal from the noisy measurement $\boldsymbol{y}$:

$$\boldsymbol{y} = \boldsymbol{f} + \boldsymbol{w}, \tag{10.11}$$

where $\boldsymbol{w}$ is the additive noise and $\boldsymbol{f}$ is the unknown signal to estimate. If we formulate a loss function as follows:

$$\min_{\boldsymbol{f}} \frac{1}{2}\|\boldsymbol{y}-\boldsymbol{f}\|^2+\lambda\|\boldsymbol{\Phi}^\top \boldsymbol{f}\|_1, \tag{10.12}$$

where $\|\cdot\|_1$ is the l_1 norm, then the solution satisfies the following [106]:

$$\hat{\boldsymbol{f}}=\sum_{k\in\Gamma}\rho_\lambda\left(\langle \boldsymbol{y},\boldsymbol{\phi}_k\rangle\right)\widetilde{\boldsymbol{\phi}}_k, \tag{10.13}$$

where $\rho_\lambda(\cdot)$ is a nonlinear thresholding function that depends on the regularization parameter λ. This implies that the signal representation changes depending on the input $\boldsymbol{y}$, since only a small set of coefficients $\langle \boldsymbol{y},\boldsymbol{\phi}_k\rangle$ will be nonzero after processing with the nonlinear thresholding, and the signal is represented by only a small set of dual bases $\widetilde{\boldsymbol{\phi}}_k$ corresponding to the locations of the nonzero expansion coefficients.

For the last few decades, one of the most widely used frame representations in signal processing is the wavelet frame, or framelet [106], where its basis function captures the multi-resolution scale and shift dependent features. For example, Fig. 10.1 illustrates the Haar wavelet basis across different scale parameters j. As the scale increases, the support of the basis ϕ_k becomes narrow so that it can capture more localized behavior of the signal after applying the inner product. More specifically, Fig. 10.2 shows the noiseless original signal $\boldsymbol{f}$ and its noisy version $\boldsymbol{y}$, and their wavelet expansion coefficients. Here, $d_s(n)$ denotes the s-scale wavelet expansion coefficients. As shown in Fig. 10.2, for the smooth noiseless signal, most of the wavelet expansion coefficients are zero except a few expansion coefficients at lower scales. On the other hand, for the noisy signal, the small magnitude nonzero wavelet expansion coefficients are found across all scales. Therefore, the main idea of the wavelet shrinkage for signal denoising [107] is zeroing out the small-magnitude wavelet coefficients using a thresholding operation $\rho_\lambda(\cdot)$ and retaining large wavelet coefficients beyond the threshold values that have important signal characteristics. Accordingly, reconstruction using (10.13) can recover the underlying noiseless signals.

Extending this idea beyond the signal denoising, other successful tools in the signal processing theory are the compressed sensing or sparse recovery techniques [46]. In particular, compressed sensing theory is based on the observation that when images are represented via bases of frames, in many cases they can be represented as a sparse combination of bases or frames, as shown in Fig. 10.3. Thanks to the sparse representation, even when the measurements are very few below the classical limits such as Nyquist limit, one could obtain a stable solution of the inverse problem by searching for the sparse representation that generates an output consistent with the measured data, as shown in Fig. 10.3. As a result, the goal of the image reconstruction problem is to find an optimal set of sparse basis functions suitable for the given measurement data. This is why the classical method is often called the *basis pursuit* [46].

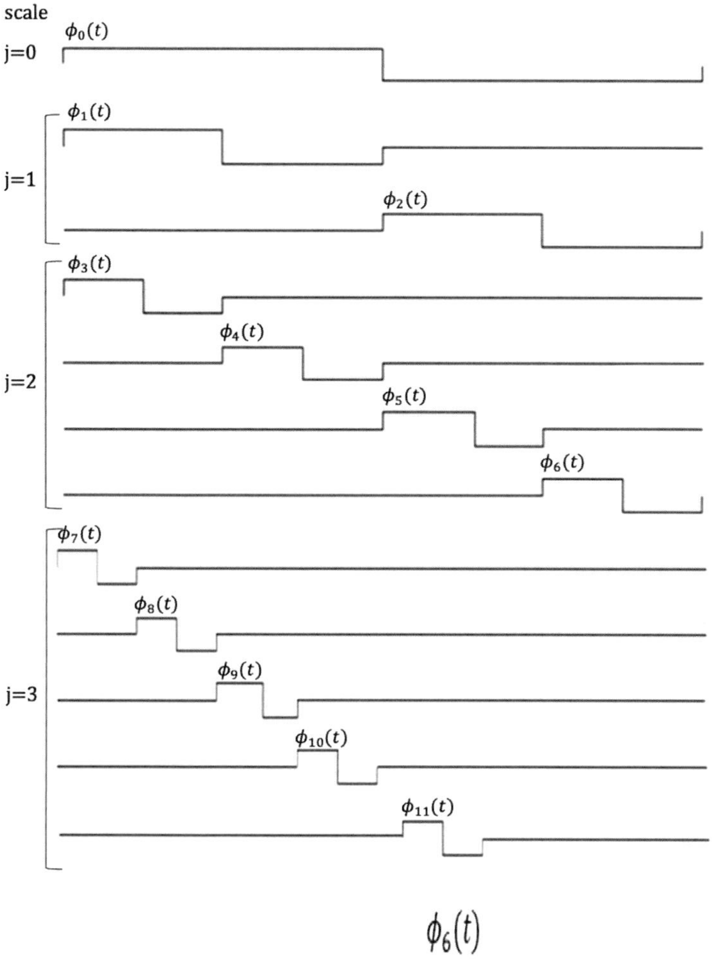

Fig. 10.1 Haar wavelet basis across scales

In contrast to the kernel machine in (10.2), the basis pursuit using the frame representation has several unique advantages. First, the function space that the basis pursuit can generate is often larger than the RKHS from (10.2). In fact, this space is often called the *union of subspaces* [108], which is a large subset of a Hilbert space. Second, among the given frames, the choice of *active* dual frame basis $\widetilde{\phi}_k$ is totally data-dependent. Therefore, the basis pursuit representation is an adaptive model. Moreover, the expansion coefficients $\rho_\lambda\left(\langle \boldsymbol{y}, \boldsymbol{\phi}_k\rangle\right)$ of the basis pursuit are also totally dependent on the input $\boldsymbol{y}$, thereby generating more diverse representation than the kernel machine with fixed expansion coefficients.

Having said this, one of the most fundamental limitations of the basis pursuit approach in (10.13) is that it is *transductive*, which does not allow *inductive* learning

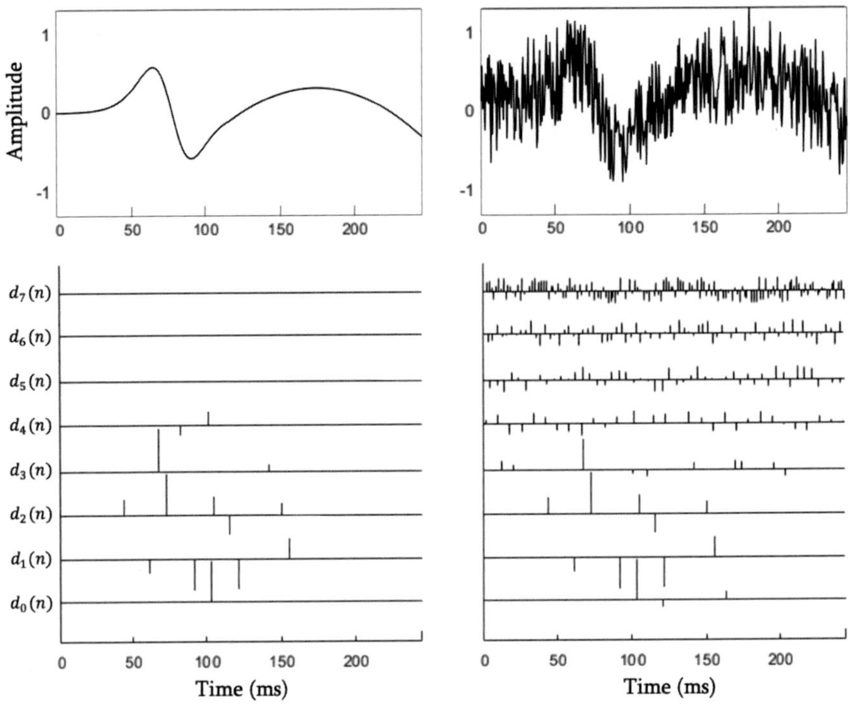

Fig. 10.2 Wavelet coefficients for two signals across scales

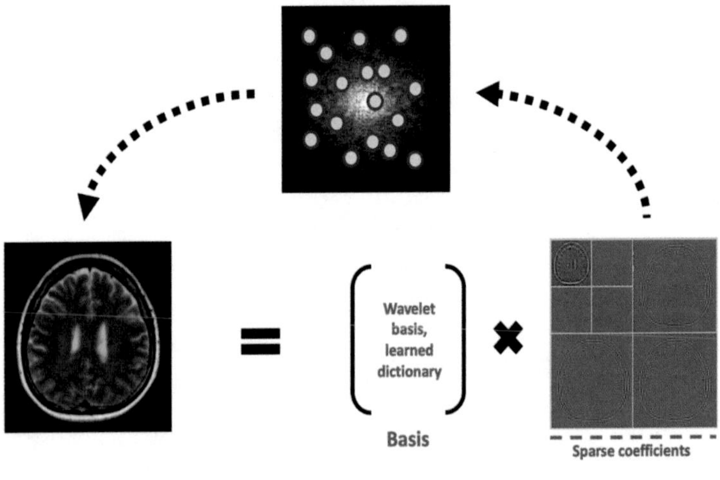

Fig. 10.3 Reconstruction principle of compressed sensing

from the training data. In general, the basis pursuit regression in (10.12) should be solved for each data set, since the nonlinear thresholding function should be found by an optimization method for each data set. Therefore, it is difficult to transfer the learning from one data set to another.

10.3 Convolution Framelets

Before we dive into the convolutional neural network, here we briefly review the theory of deep convolutional framelets [42], which is a linear frame expansion but turns out to be an important stepping stone to understand the geometry of CNN. For simplicity, we consider the 1-D version of the theory.

10.3.1 *Convolution and Hankel Matrix*

Let an n-dimensional signal $\boldsymbol{x} \in \mathbb{R}^n$ be represented by

$$\boldsymbol{x} = \begin{bmatrix} x[0] & \cdots & x[n-1] \end{bmatrix}^\top \in \mathbb{R}^n.$$

Then, the following results are standard in signal processing:

- Given two vectors $\boldsymbol{x}, \boldsymbol{h} \in \mathbb{R}^n$, the circular convolution is defined by

$$(\boldsymbol{x} \circledast \boldsymbol{h})[i] = \sum_{k=0}^{n-1} x[i-k]h[k], \tag{10.14}$$

  where appropriate periodic boundary conditions are imposed on $\boldsymbol{x}$.
- For any $\boldsymbol{v} \in \mathbb{R}^{n_1}$ and $\boldsymbol{w} \in \mathbb{R}^{n_2}$ with $n_1, n_2 \leq n$, define the convolution in $\mathbb{R}^n$ as

$$\boldsymbol{v} \circledast \boldsymbol{w} = \boldsymbol{v}^0 \circledast \boldsymbol{w}^0,$$

  where

$$\boldsymbol{v}^0 = \begin{bmatrix} \boldsymbol{v}^\top & \mathbf{0}_{n-n_1}^\top \end{bmatrix}^\top, \quad \boldsymbol{w}^0 = \begin{bmatrix} \boldsymbol{w}^\top & \mathbf{0}_{n-n_2}^\top \end{bmatrix}^\top.$$

- For any $\boldsymbol{v} \in \mathbb{R}^{n_1}$ with $n_1 \leq n$, define the *flip* of $\boldsymbol{v}$ as $\overline{v}[n] = v^0[-n]$, where we use the periodic boundary condition.

Using these notations, a single-input single-output (SISO) circular convolution of the input $\boldsymbol{f}$ and the filter $\overline{\boldsymbol{\psi}} \in \mathbb{R}^r$ with $r \leq n$ can be represented by:

$$y[i] = (\boldsymbol{x} \circledast \overline{\boldsymbol{\psi}})[i] = \sum_{k=0}^{n-1} x[i-k]\psi^0[-k]. \tag{10.15}$$

By defining a Hankel matrix $\mathbb{H}_r^n(\boldsymbol{x}) \in \mathbb{R}^{n\times r}$ as

$$\mathbb{H}_r^n(\boldsymbol{x}) = \begin{bmatrix} x[0] & x[1] & \cdots & x[r-1] \\ x[1] & x[2] & \cdots & x[r] \\ \vdots & \vdots & \ddots & \vdots \\ x[n-1] & x[n] & \cdots & x[r-2] \end{bmatrix} \tag{10.16}$$

the convolution in (10.15) can be compactly represented by

$$\boldsymbol{y} = \boldsymbol{x} \circledast \overline{\boldsymbol{\psi}} = \mathbb{H}_r^n(x)\boldsymbol{\psi}. \tag{10.17}$$

Then, we can obtain the following key equality [109], whose proof is repeated here for educational purposes:

Lemma 10.1 *For a given $\boldsymbol{f} \in \mathbb{R}^n$, let $\mathbb{H}_r^n(\boldsymbol{f}) \in \mathbb{R}^{n\times r}$ denote the associated Hankel matrix. Then, for any vectors $\boldsymbol{u} \in \mathbb{R}^n$ and $\boldsymbol{v} \in \mathbb{R}^r$ with $r \leq n$ and Hankel matrix $\boldsymbol{F} := \mathbb{H}_r^n(\boldsymbol{f})$, we have*

$$\boldsymbol{u}^\top \boldsymbol{F}\boldsymbol{v} = \boldsymbol{u}^\top(\boldsymbol{f} \circledast \overline{\boldsymbol{v}}) = \boldsymbol{f}^\top(\boldsymbol{u} \circledast \boldsymbol{v}) = \langle \boldsymbol{f}, \boldsymbol{u} \circledast \boldsymbol{v}\rangle, \tag{10.18}$$

where $\overline{v}[n] := v[-n]$ denotes the flipped version of the vector v.

Proof We only need to show the second equality. This can be shown as

$$\begin{aligned} \boldsymbol{f}^\top(\boldsymbol{u} \circledast \boldsymbol{v}) &= \boldsymbol{f}^\top\left(\boldsymbol{u} \circledast \boldsymbol{v}^0\right) \\ &= \sum_{i=0}^{n-1} f[i]\left(\sum_{k=0}^{n-1} u[k]v^0[i-k]\right) \\ &= \sum_{k=0}^{n-1} u[k]\left(\sum_{i=0}^{n-1} v^0[i-k]f[i]\right) \\ &= \sum_{k=0}^{n-1} u[k]\left(\sum_{i=0}^{n-1} v^0[-(k-i)]f[i]\right) \end{aligned}$$

$$
\begin{aligned}
&= \sum_{k=0}^{n-1} u[k](f \circledast \overline{v})[k] \\
&= \boldsymbol{u}^\top (\boldsymbol{f} \circledast \overline{\boldsymbol{v}}) .
\end{aligned}
$$

This concludes the proof. □

10.3.2 Convolution Framelet Expansion

Lemma 10.1 provides an important clue for the convolution framelet expansion. Specifically, for a given signal $\boldsymbol{f} \in \mathbb{R}^n$, consider the following two sets of matrices, $\widetilde{\boldsymbol{\Phi}}, \boldsymbol{\Phi} \in \mathbb{R}^{n\times n}$ and $\widetilde{\boldsymbol{\Psi}}, \boldsymbol{\Psi} \in \mathbb{R}^{r\times r}$, such that they satisfy the following *frame condition*[42]:

$$
\widetilde{\boldsymbol{\Phi}}\boldsymbol{\Phi}^\top = \boldsymbol{I}_n \quad \boldsymbol{\Psi}\widetilde{\boldsymbol{\Psi}}^\top = \boldsymbol{I}_r . \tag{10.19}
$$

Then, we have the following trivial equality:

$$
\mathbb{H}_r^n(\boldsymbol{f}) = \widetilde{\boldsymbol{\Phi}}\boldsymbol{\Phi}^\top \mathbb{H}_r^n(\boldsymbol{f})\boldsymbol{\Psi}\widetilde{\boldsymbol{\Psi}}^\top = \widetilde{\boldsymbol{\Phi}}\boldsymbol{C}\widetilde{\boldsymbol{\Psi}}^\top , \tag{10.20}
$$

where

$$
\boldsymbol{C} = \boldsymbol{\Phi}^\top \mathbb{H}_r^n(\boldsymbol{f})\boldsymbol{\Psi} \quad \in \mathbb{R}^{n\times r}, \tag{10.21}
$$

whose (i, j)-th element is given by

$$
c_{ij} = \boldsymbol{\phi}_i^\top \mathbb{H}_r^n(\boldsymbol{f})\boldsymbol{\psi}_j = \langle \boldsymbol{f}, \boldsymbol{\phi}_i \circledast \boldsymbol{\psi}_j \rangle , \tag{10.22}
$$

where $\boldsymbol{\phi}_i$ and $\boldsymbol{\psi}_j$ denote the i-th and the j-th column vector of $\boldsymbol{\Phi}$ and $\boldsymbol{\Psi}$, respectively, and the last equality of (10.22) comes from Lemma 10.1.

Now, we define an inverse Hankel operator $\mathbb{H}_r^{n(-)} : \mathbb{R}^{n\times r} \mapsto \mathbb{R}^n$ such that for any $\boldsymbol{f} \in \mathbb{R}^n$, the following equality satisfies

$$
\boldsymbol{f} = \mathbb{H}_r^{n(-)}\left(\mathbb{H}_r^n(\boldsymbol{f})\right) . \tag{10.23}
$$

Then, the following key equality can be obtained [42]:

$$
\begin{aligned}
\mathbb{H}_r^{n(-)}\left(\widetilde{\boldsymbol{\Phi}}\boldsymbol{C}\widetilde{\boldsymbol{\Psi}}^\top\right) &= \frac{1}{r}\sum_{j=1}^{r} \left(\widetilde{\boldsymbol{\Phi}}\boldsymbol{c}_j\right) \circledast \widetilde{\boldsymbol{\psi}}_j &\quad (10.24)\\
&= \frac{1}{r}\sum_{i,j} c_{ij}(\widetilde{\boldsymbol{\phi}}_i \circledast \widetilde{\boldsymbol{\psi}}_j) . &\quad (10.25)
\end{aligned}
$$

By combining (10.25) with (10.20) and (10.22), we have

$$f = \frac{1}{r}\sum_{i,j}\langle f, \boldsymbol{\phi}_i \circledast \boldsymbol{\psi}_j\rangle \left(\widetilde{\boldsymbol{\phi}}_i \circledast \widetilde{\boldsymbol{\psi}}_j\right). \tag{10.26}$$

This implies that $\{\boldsymbol{\phi}_i \circledast \boldsymbol{\psi}_j\}_{i,j}$ constitutes a frame for $\mathbb{R}^n$ and $\{\widetilde{\boldsymbol{\phi}}_i \circledast \widetilde{\boldsymbol{\psi}}_j\}_{i,j}$ corresponds to its dual frame. Furthermore, for many interesting signals f in real applications, the Hankel matrix $\mathbb{H}_r^n(f)$ has low-rank structures [110–112], which makes the expansion coefficients c_{ij} nonzero only at small index sets. Therefore, the convolution framelet expansion is a concise signal representation similar to the wavelet frames [42, 109].

In the convolution framelet, the functions $\boldsymbol{\phi}_i, \tilde{\boldsymbol{\phi}}_i$ correspond to the global basis, whereas $\boldsymbol{\psi}_i, \tilde{\boldsymbol{\psi}}_i$ are local basis functions. Therefore, by the convolution between the global and local basis to generate a new frame basis, convolution framelets can exploit both local and global structures of signals [42, 109], which is an important advance in signal representation theory.

10.3.3 Link to CNN

Although the convolution framelet is a linear representation, the reason we care about it so much is that it reveals the role of the pooling and convolution filters in CNNs. More specifically, using (10.17), we can show that the convolution framelet coefficient matrix $\boldsymbol{C}$ in (10.21) can be represented by

$$\begin{aligned} \boldsymbol{C} &= \begin{bmatrix} \boldsymbol{c}_1 & \cdots & \boldsymbol{c}_r \end{bmatrix} \\ &= \boldsymbol{\Phi}^\top \mathbb{H}_r^n(f)\boldsymbol{\Psi} = \boldsymbol{\Phi}^\top (f \circledast \overline{\boldsymbol{\Psi}}), \end{aligned} \tag{10.27}$$

where

$$f \circledast \overline{\boldsymbol{\Psi}} := \begin{bmatrix} f \circledast \overline{\boldsymbol{\psi}}_1 & \cdots & f \circledast \overline{\boldsymbol{\psi}}_r \end{bmatrix} \tag{10.28}$$

which corresponds to the single-input multi-output (SIMO) convolution. Note that the convolution operation is local since the filter weights are multiplied with the pixels within the receptive field. After the convolution operation, $\boldsymbol{\Phi}^\top$ is multiplied with all elements of the filtered output, which corresponds to the global operation.

On the other hand, by combining (10.24) with (10.20), we have

$$f = \frac{1}{r}\sum_{j=1}^{r}\left(\widetilde{\boldsymbol{\Phi}}\boldsymbol{c}_j\right) \circledast \widetilde{\boldsymbol{\psi}}_j, \tag{10.29}$$

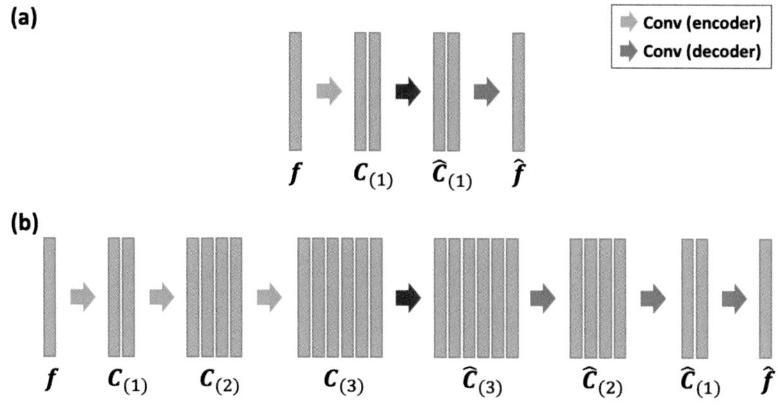

Fig. 10.4 Single-resolution encoder–decoder networks. (**a**) single-level convolutional framelet decomposition with identity pooling. (**b**) multi-level convolutional framelet deconvolution with identity pooling

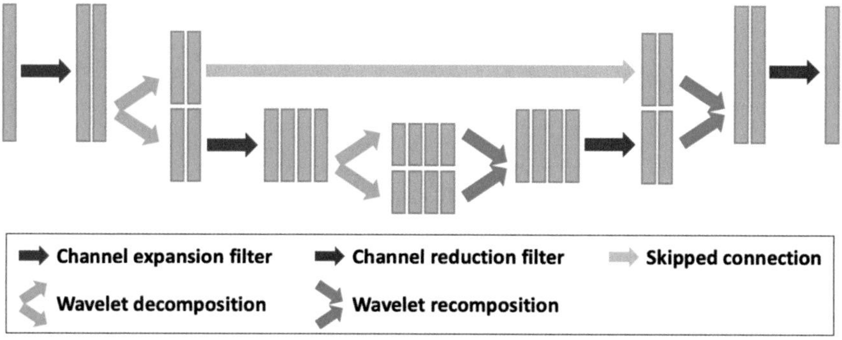

Fig. 10.5 Multi-resolution encoder–decoder networks

which shows the processing step of the framelet coefficient $\boldsymbol{C}$ at the decoder. More specifically, we apply the global operation $\widetilde{\boldsymbol{\Psi}}$ to $\boldsymbol{c}_j$ first, after which multi-input single-output (MISO) convolution operation is performed to obtain the final reconstruction.

In fact, the order of these signal processing operations is very similar to the two-layer encoder–decoder architecture, as shown in Figs. 10.4 and 10.5. At the encoder side, the SIMO convolution operation is performed first to generate multi-channel feature maps, after which the global pooling operation is performed. At the decoder side, the feature map is unpooled first, after which the MISO convolution is performed. Therefore, we can easily see the important analogy: the convolution framelet coefficients are similar to the feature maps in CNNs, and $\boldsymbol{\Phi}$, $\widetilde{\boldsymbol{\Phi}}$ work as a pooling and unpooling layers, respectively, whereas $\boldsymbol{\Psi}$, $\widetilde{\boldsymbol{\Psi}}$ correspond to the encoder and decoder filters, respectively. This implies that the pooling operation defines the global basis, whereas the convolution filters determine the local basis, and the CNN tries to exploit both global and local structure of the signal.

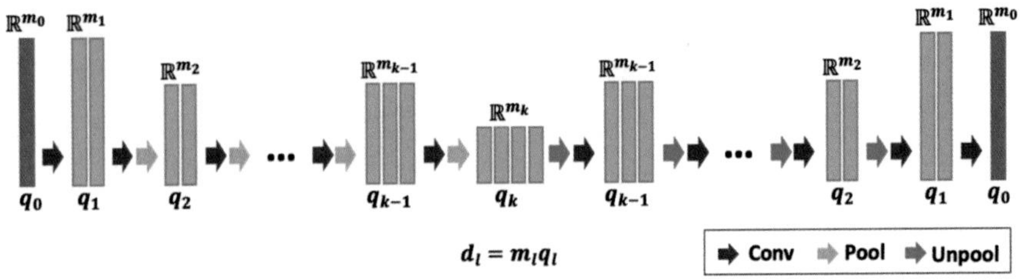

Fig. 10.6 Encoder–decoder CNNs without skip connection

Furthermore, by simply changing the global basis, we can obtain various network architectures. For example, in Fig. 10.4, we use $\boldsymbol{\Phi} = \tilde{\boldsymbol{\Phi}} = \boldsymbol{I}_n$, whereas we use the Haar wavevelet transform as global pooling for the case of Fig. 10.5.

10.3.4 Deep Convolutional Framelets

Now, we are ready to explain the multilayer convolution framelets, which we call deep convolutional framelets [42]. For simplicity, we consider encoder–decoder networks without skip connections, as shown in Fig. 10.6, although the analysis can be applied equally well when the skip connections are present. Furthermore, we assume symmetric configuration so that both encoder and decoder have the same number of layers, say κ; the input and output dimensions for the encoder layer $\mathcal{E}^l$ and the decoder layer $\mathcal{D}^l$ are symmetric:

$$\mathcal{E}^l : \mathbb{R}^{d_{l-1}} \mapsto \mathbb{R}^{d_l}, \quad \mathcal{D}^l : \mathbb{R}^{d_l} \mapsto \mathbb{R}^{d_{l-1}}, \quad l \in [\kappa], \tag{10.30}$$

where $[n]$ denotes the set $\{1, \cdots, n\}$. At the l-th layer, m_l and q_l denote the dimension of the signal, and the number of filter channels, respectively. The length of filter is assumed to be r.

We now define the l-th layer input signal for the encoder layer from q_{l-1}-input channels,

$$z^{l-1} := \left[z_1^{l-1\top} \cdots z_{q_{l-1}}^{l-1\top} \right]^\top \in \mathbb{R}^{d_{l-1}}, \tag{10.31}$$

where $^\top$ denotes the transpose, and $z_j^{l-1} \in \mathbb{R}^{m_{l-1}}$ refers to the j-th channel input with the dimension m_{l-1}. The l-th layer output signal z^l is similarly defined. Note that the filtered output is now stacked as a single column vector in (10.31), which is different from the former treatment at the convolution framelet where the filter output for each channel is stacked as an additional column. It turns out that the notation in (10.31) makes the mathematical derivation for multilayer convolutional neural networks much more trackable than the former notation, although the role of the global and local basis are clearly seen in the former notation.

Then, for the *linear* encoder–decoder CNN *without* skip connections, as shown in Fig. 10.6a, we have the following linear representation at the l-th encoder layer [35]:

$$z^l = \boldsymbol{E}^{l\top} z^{l-1}, \tag{10.32}$$

where

$$\boldsymbol{E}^l = \begin{bmatrix} \boldsymbol{\Phi}^l \circledast \boldsymbol{\psi}^l_{1,1} & \cdots & \boldsymbol{\Phi}^l \circledast \boldsymbol{\psi}^l_{q_l,1} \\ \vdots & \ddots & \vdots \\ \boldsymbol{\Phi}^l \circledast \boldsymbol{\psi}^l_{1,q_{l-1}} & \cdots & \boldsymbol{\Phi}^l \circledast \boldsymbol{\psi}^l_{q_l,q_{l-1}} \end{bmatrix}, \tag{10.33}$$

where $\boldsymbol{\Phi}^l$ denotes the $m_l \times m_l$ matrix that represents the pooling operation at the l-th layer, and $\boldsymbol{\psi}^l_{i,j} \in \mathbb{R}^r$ represents the l-th layer encoder filter to generate the i-th channel output from the contribution of the j-th channel input, and $\boldsymbol{\Phi}^l \circledast \boldsymbol{\psi}^l_{i,j}$ represents a single-input multi-output (SIMO) convolution [35]:

$$\boldsymbol{\Phi}^l \circledast \boldsymbol{\psi}^l_{i,j} = \left[\boldsymbol{\phi}^l_1 \circledast \boldsymbol{\psi}^l_{i,j} \cdots \boldsymbol{\phi}^l_n \circledast \boldsymbol{\psi}^l_{i,j}\right]. \tag{10.34}$$

Note that the inclusion of the bias can be readily done by including additional rows into $\boldsymbol{E}^l$ as the bias and augmenting the last element of z^{l-1} by 1.

Similarly, the l-th decoder layer can be represented by

$$\tilde{z}^{l-1} = \boldsymbol{D}^l \tilde{z}^l, \tag{10.35}$$

where

$$\boldsymbol{D}^l = \begin{bmatrix} \tilde{\boldsymbol{\Phi}}^l \circledast \tilde{\boldsymbol{\psi}}^l_{1,1} & \cdots & \tilde{\boldsymbol{\Phi}}^l \circledast \tilde{\boldsymbol{\psi}}^l_{1,q_l} \\ \vdots & \ddots & \vdots \\ \tilde{\boldsymbol{\Phi}}^l \circledast \tilde{\boldsymbol{\psi}}^l_{q_{l-1},1} & \cdots & \tilde{\boldsymbol{\Phi}}^l \circledast \tilde{\boldsymbol{\psi}}^l_{q_{l-1},q_l} \end{bmatrix}, \tag{10.36}$$

where $\tilde{\boldsymbol{\Phi}}^l$ denotes the $m_l \times m_l$ matrix that represents the unpooling operation at the l-th layer, and $\tilde{\boldsymbol{\psi}}^l_{i,j} \in \mathbb{R}^r$ represents the l-th layer decoder filter to generate the i-th channel output from the contribution of the j-th channel input.

Then, the output $\boldsymbol{v}$ of the encoder-decoder CNN with respect to input z can be represented by the following representation [35]:

$$\boldsymbol{v} = \mathcal{T}_{\boldsymbol{\Theta}}(z) = \sum_i \langle \boldsymbol{b}_i, z\rangle \tilde{\boldsymbol{b}}_i \tag{10.37}$$

where $\boldsymbol{\Theta}$ refers to all encoder and decoder convolution filters, and $\boldsymbol{b}_i$ and $\tilde{\boldsymbol{b}}_i$ denote the i-th column of the following matrices, respectively:

$$\boldsymbol{B} = \boldsymbol{E}^1\boldsymbol{E}^2\cdots\boldsymbol{E}^\kappa \ , \ \tilde{\boldsymbol{B}} = \boldsymbol{D}^1\boldsymbol{D}^2\cdots\boldsymbol{D}^\kappa \tag{10.38}$$

Note that this representation is completely linear, since the representation does not vary once the network parameters $\boldsymbol{\Theta}$ are trained. Furthermore, consider the following *multilayer frame conditions* for the pooling and filter layers:

$$\tilde{\boldsymbol{\Phi}}^l\boldsymbol{\Phi}^{l\top} = \alpha\boldsymbol{I}_{m_{l-1}} \ , \ \boldsymbol{\Psi}^l\tilde{\boldsymbol{\Psi}}^{l\top} = \frac{1}{r\alpha}\boldsymbol{I}_{rq_{l-1}}, \qquad \forall l, \tag{10.39}$$

where $\boldsymbol{I}_n$ denotes the $n \times n$ identity matrix and $\alpha > 0$ is a nonzero constant, and

$$\boldsymbol{\Psi}^l = \begin{bmatrix} \boldsymbol{\psi}^l_{1,1} & \cdots & \boldsymbol{\psi}^l_{q_l,1} \\ \vdots & \ddots & \vdots \\ \boldsymbol{\psi}^l_{1,q_{l-1}} & \cdots & \boldsymbol{\psi}^l_{q_l,q_{l-1}} \end{bmatrix}, \tag{10.40}$$

$$\tilde{\boldsymbol{\Psi}}^l = \begin{bmatrix} \tilde{\boldsymbol{\psi}}^l_{1,1} & \cdots & \tilde{\boldsymbol{\psi}}^l_{1,q_l} \\ \vdots & \ddots & \vdots \\ \tilde{\boldsymbol{\psi}}^l_{q_{l-1},1} & \cdots & \tilde{\boldsymbol{\psi}}^l_{q_{l-1},q_l} \end{bmatrix}, \tag{10.41}$$

Under these frame conditions, we showed in [35] that (10.37) satisfies the perfect reconstruction condition, i.e

$$z = \mathcal{L}_{\boldsymbol{\Theta}}(z) := \sum_i \langle \boldsymbol{b}_i, z\rangle \tilde{\boldsymbol{b}}_i, \tag{10.42}$$

hence the corresponding deep convolutional framelet is indeed a frame representation, similar to wavelet frames [113].

In the deep convolutional framelets, all the encoder and decoder filters can be estimated from the training data set; hence, it is a data-driven model. More specifically, for the given training data $\{\boldsymbol{x}_i, y_i\}_{i=1}^n$, the CNN parameter $\boldsymbol{\Theta}$ is estimated by solving the following optimization problem:

$$\min_{\boldsymbol{\Theta}} \sum_{i=1}^n \ell\left(y_i, \mathcal{L}_{\boldsymbol{\Theta}}(\boldsymbol{x}_i)\right) + \lambda R(\boldsymbol{\Theta}). \tag{10.43}$$

Once the parameter $\boldsymbol{\Theta}$ is learned, the encoder and decoder matrices $\boldsymbol{E}^l$ and $\boldsymbol{D}^l$ are determined. Therefore, the representations are entirely data-driven and dependent on the filter sets that are learned from the training data set, which is different from the classical kernel machine or basis pursuit approaches, where underlying kernels or frames are specified in a top-down manner.

That said, the deep convolutional framelet does not yet meet the desiderata of the machine learning, since once it is trained, the frame representation does not vary, hence the data-driven adaptation is not possible. In the next section, we will show that the last missing element is the nonlinearity such as ReLU, which plays key roles in machine learning.

10.4 Geometry of CNN

10.4.1 Role of Nonlinearity

In fact, the analysis of deep convolutional framelets with the ReLU nonlinearities turns out to be a simple modification, but it provides very fundamental insights on the geometry of the deep neural network.

Specifically, in [35] we showed that even with ReLU nonlinearities the expression (10.37) is still valid. The only change is that the basis matrices have additional ReLU pattern blocks in between encoder, decoder, and skipped blocks. For example, the expression in (10.38) is changed as follows:

$$\boldsymbol{B}(z) = \boldsymbol{E}^1\boldsymbol{\Lambda}^1(z)\boldsymbol{E}^2\boldsymbol{\Lambda}^2(z)\cdots\boldsymbol{\Lambda}^{\kappa-1}(z)\boldsymbol{E}^{\kappa}, \tag{10.44}$$

$$\tilde{\boldsymbol{B}}(z) = \boldsymbol{D}^1\tilde{\boldsymbol{\Lambda}}^1(z)\boldsymbol{D}^2\tilde{\boldsymbol{\Lambda}}^2(z)\cdots\tilde{\boldsymbol{\Lambda}}^{\kappa-1}(z)\boldsymbol{D}^{\kappa}, \tag{10.45}$$

where $\boldsymbol{\Lambda}^l(z)$ and $\tilde{\boldsymbol{\Lambda}}^l(z)$ are the diagonal matrices with 0 and 1 elements indicating the ReLU activation patterns.

Accordingly, the linear representation in (10.37) should be modified as a nonlinear representation:

$$\boldsymbol{v} = \mathcal{T}_{\boldsymbol{\Theta}}(z) = \sum_i \langle \boldsymbol{b}_i(z), z\rangle\, \tilde{\boldsymbol{b}}_i(z), \tag{10.46}$$

where we now have an explicit dependency on z for $\boldsymbol{b}_i(z)$ and $\tilde{\boldsymbol{b}}_i(z)$ due to the input-dependent ReLU activation patterns, which makes the representation nonlinear.

Again the filter parameter $\boldsymbol{\Theta}$ is estimated by solving the optimization problem in (10.43) by replacing $\mathcal{L}_{\Theta}(z)$ with $\mathcal{T}_{\Theta}(z)$ in (10.46). Therefore, the representations are entirely data-driven.

10.4.2 Nonlinearity Is the Key for Inductive Learning

In (10.44) and (10.45), the encoder and decoder basis matrices have an explicit dependence on the ReLU activation pattern on the input. Here we will show that

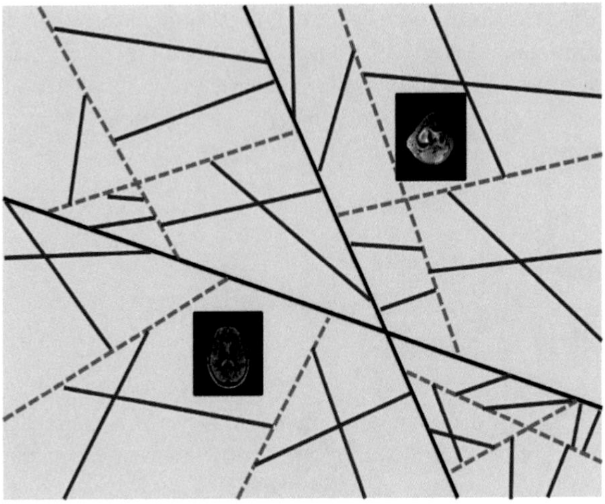

Fig. 10.7 Reconstruction principle of deep learning

this ReLU-activation-dependent diagonal matrix provides a key role in enabling inductive learning.

Specifically, the nonlinearity is applied after the convolution operation, so the on-and-off activation pattern of each ReLU determines a binary partition of the feature space at each layer across the hyperplane that is determined by the convolution. Accordingly, in deep neural networks, the input space is partitioned into multiple non-overlapping regions so that input images for each region share the same linear representation, but not across the partition. This implies that two different input images are automatically switched to two distinct linear representations that are different from each other, as shown in Fig. 10.7.

This leads to an important insight: although the CNN approach and the basis pursuit in Fig. 10.3 appear to be two completely different approaches, there exists a very close relationship between the two. Specifically, the CNN is indeed similar to the classical basis pursuit algorithm that searches for the distinct linear representation for each input, but in contrast to the basis pursuit, the CNN is *inductive* since it does not solve the optimization problem for a new input, rather it only switches to different frame representations by changing the ReLU activation patterns. This inductivity from the learned filter coefficients is an important advance over the classical signal processing approach.

10.4.3 Expressivity

Given the partition-dependent framelet geometry of CNN, we can easily expect that with a greater number of input space partitions, the nonlinear function approxima-

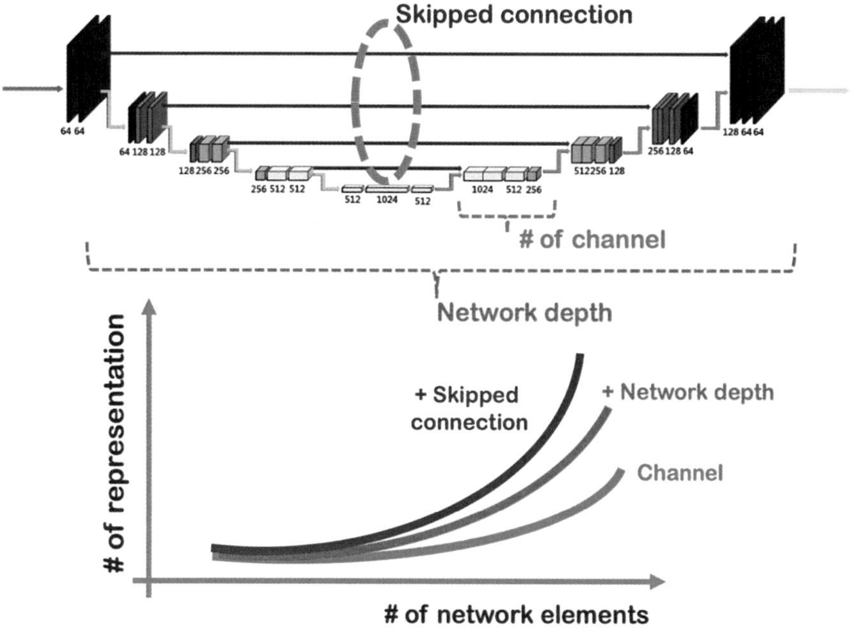

Fig. 10.8 Expressivity increases exponentially with channels, depth, and skip connections

tion by the piecewise linear frame representation becomes more accurate. Therefore, the number of piecewise linear regions is directly related to the expressivity or representation power of the neural network. If each ReLU activation pattern is independent of the others, then the number of distinct ReLU activation patterns is $2^{\# \text{ of neurons}}$, where the number of neurons is determined by the number of the entire features. Therefore, the number of distinct linear representation increases exponentially with the depth, width, and skip connection as shown in Fig. 10.8 [35]. This again confirms the expressive power of CNN thanks to the ReLU nonlinearities.

10.4.4 Geometric Meaning of Features

One of the interesting questions in neural networks is understanding the meaning of the intermediate features that are obtained as an output of each layer of neural network. Although these are largely regarded as latent variables, to our best knowledge the geometric understanding of each latent variable is still not complete. In this section, we show that this intermediate feature is directly related to the relative coordinates with respect to the hyperplanes that partition the product space of the previous layer features.

To understand the claim, let us first revisit the ReLU operation for each neuron at the encoder layer. Let $\boldsymbol{E}_i^l$ denote the i-th column of encoder matrix $\boldsymbol{E}^l$ and z_i^l be the i-th element of z^l. Then, the output of an *activated* neuron can be represented as:

$$z_i^l = \underbrace{\frac{|\langle \boldsymbol{E}_i^l, z^{l-1}\rangle|}{\|\boldsymbol{E}_i^l\|}}_{\text{distance to the hyperplane}} \times \|\boldsymbol{E}_i^l\|, \tag{10.47}$$

where the normal vector of the hyperplane can be identified as

$$\boldsymbol{n}^l = \boldsymbol{E}_i^l. \tag{10.48}$$

This implies that the output of the activated neuron is the scaled version of the distance to the hyperplane which partitions the space of feature vector z^{l-1} into active and non-active regions. Therefore, the role of the neural network can be understood as representing the input data with a coordinate vector using the relative distances with respect to multiple hyperplanes.

In fact, the aforementioned interpretation of the feature may not be novel, since a similar interpretation can be used to explain the geometrical meaning of the linear frame coefficients. Instead, one of the most important differences comes from the multilayer representation. To understand this, consider the following two layer neural network:

$$z_i^l = \sigma(\boldsymbol{E}_i^{l\top} z^{l-1}), \tag{10.49}$$

where

$$z^{l-1} = \sigma\left(\boldsymbol{E}^{(l-1)\top} z^{l-2}\right) = \boldsymbol{\Lambda}(z^{l-1})\boldsymbol{E}^{(l-1)\top} z^{l-2}, \tag{10.50}$$

where $\boldsymbol{\Lambda}(z^{l-1})$ again encodes the ReLU activation pattern. Using the property of the inner product and adjoint operator, we have

$$\begin{aligned} z_i^l &= \sigma(\boldsymbol{E}_i^{l\top} z^{l-1}) \\ &= \sigma\left(\left\langle \boldsymbol{E}_i^l, \boldsymbol{\Lambda}(z^{l-1})\boldsymbol{E}^{(l-1)\top} z^{l-2}\right\rangle\right) \\ &= \sigma\left(\left\langle \boldsymbol{\Lambda}(z^{l-1})\boldsymbol{E}_i^l, \boldsymbol{E}^{(l-1)\top} z^{l-2}\right\rangle\right). \end{aligned} \tag{10.51}$$

This indicates that on the space of the *unconstrained* feature vector from the previous layer (i.e. no ReLU is assumed), the hyperplane normal vector is now changed to

$$\boldsymbol{n}^l = \boldsymbol{\Lambda}(z^{l-l})\boldsymbol{E}_i^l. \tag{10.52}$$

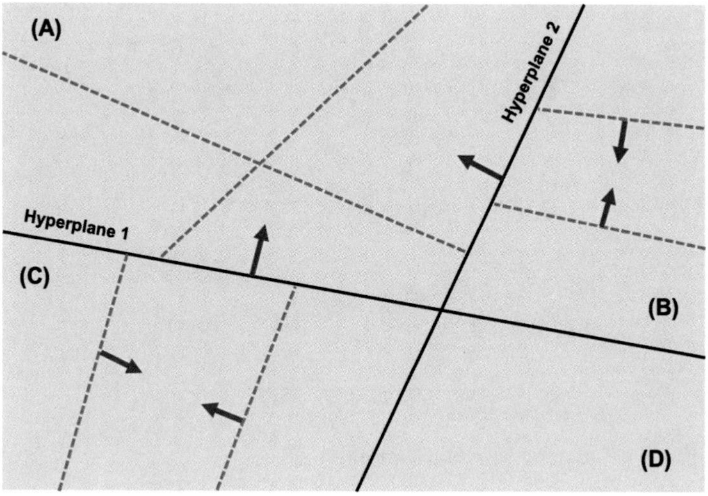

Fig. 10.9 Two-layer neural network with two neurons for each layer. Blue arrows indicate the normal direction of the hyperplane. The black lines are hyperplanes for the first layers, and the red lines correspond to the second layer hyperplanes

This implies that the hyperplane in the current layer is adaptively changed with respect to the input data, since the ReLU activation pattern in the previous layer, i.e. $\mathbf{\Lambda}(z^{l-l})$, can vary depending on inputs. This is an important difference over the linear multilayer frame representation, whose hyperplane structure is the same regardless of different inputs.

For example, Fig. 10.9 shows a partition geometry of $\mathbb{R}^2$ by a two-layer neural network with two neurons at each layer. The normal vector directions for the second layer hyperplanes are determined by the ReLU activation patterns such that the coordinate values at the inactive neuron become degenerate. More specifically, for the (A) quadrant where two neurons at the first layers are active, we can obtain two hyperplanes in any normal direction determined by the filter coefficients. However, for the (B) quadrant where the second neuron is inactive, the situation is different. Specifically, due to (10.52), the second coordinate of the normal vector, which corresponds to the inactive neuron, becomes degenerate. This leads to the two parallel hyperplanes that are distinct only by the bias term. A similar phenomenon occurs for the quadrant (C) where the first neuron is inactive. For the (D) quadrant where two neurons are inactive, the normal vector becomes zero and there exists no partitioning. Therefore, we can conclude that the hyperplane geometry is adaptively determined by the feature vectors in the previous layer.

In the following, we provide several toy examples in which the partition geometry can be easily calculated.

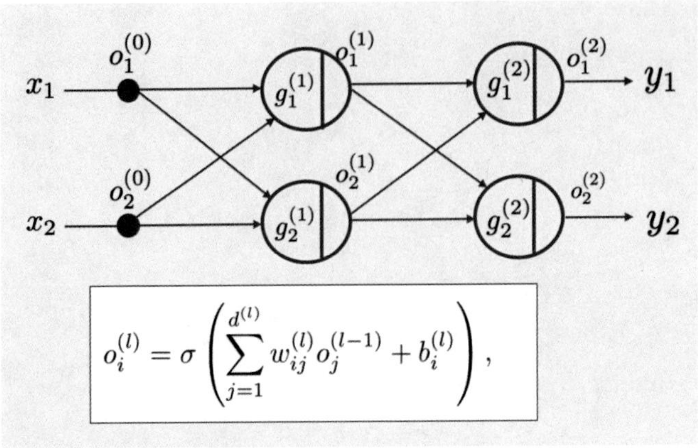

Fig. 10.10 An example two-layer neural network

Problem 10.1 (Partition Geometry of Two-Layer Neural Network in $\mathbb{R}^2$) Consider a two layer fully connected network $f_\Theta : \mathbb{R}^2 \to \mathbb{R}^2$ with ReLU nonlinearity, as shown in Fig. 10.10.

(a) Suppose the weight matrices and biases are given by

$$W^{(0)} = \begin{bmatrix} 2 & -1 \\ 1 & 1 \end{bmatrix}, \quad b^{(0)} = \begin{bmatrix} 1 \\ -1 \end{bmatrix},$$
$$W^{(1)} = \begin{bmatrix} 1 & 2 \\ -1 & 1 \end{bmatrix}, \quad b^{(1)} = \begin{bmatrix} -9 \\ -2 \end{bmatrix}.$$

Draw the corresponding input space partition, and compute the output mapping with respect to an input vector (x, y) in each input partition. Please derive all the steps explicitly.

(b) In problem (a), suppose that the bias terms are zero. Compute the input space partition and the output mapping. What do you observe compared to the one with bias?

(c) In problem (a), suppose that the second layer weight and bias are changed as

$$W^{(1)} = \begin{bmatrix} 1 & 2 \\ 0 & 1 \end{bmatrix}, \quad b^{(1)} = \begin{bmatrix} 0 \\ 1 \end{bmatrix}.$$

Draw the corresponding input space partition, and compute the output mapping with respect to an input vector (x, y) in each input partition. Compared to the original problem in (a), what do you observe?

Solution 10.1

(a) Let $\boldsymbol{x} = [x, y]^\top \in \mathbb{R}^2$. At the first layer, the output signal is given by

$$\boldsymbol{o}^{(1)} = \sigma\left(W^{(0)}\mathbf{x} + b^{(0)}\right) = \begin{bmatrix} \sigma(2x - y + 1) \\ \sigma(x + y - 1) \end{bmatrix},$$

where σ is the ReLU. Now, at the second layer, we need to consider all cases where each ReLU is active or inactive.

(i) If $2x - y + 1 < 0$ and $x + y - 1 < 0$, then $\boldsymbol{o}^{(1)} = [0, 0]^\top, \boldsymbol{o}^{(2)} = \sigma\left(W^{(1)}\boldsymbol{o}^{(1)} + b^{(1)}\right) = \sigma[-9, -2]^\top = [0, 0]^\top$.

(ii) If $2x - y + 1 \geq 0$ and $x + y - 1 < 0$, then $\boldsymbol{o}^{(1)} = [2x - y + 1, 0]^\top$. Hence, $\boldsymbol{o}^{(2)} = \sigma\left(W^{(1)}\boldsymbol{o}^{(1)} + b^{(1)}\right) = \sigma\left([2x - y - 8, -2x + y - 3]\right)^\top$. Therefore,

$$\boldsymbol{o}^{(2)} = \begin{cases} [0, 0]^\top, & 2x - y - 8 < 0, \\ [2x - y - 8, 0]^\top, & \text{otherwise}. \end{cases}$$

(iii) If $2x - y + 1 < 0$ and $x + y - 1 \geq 0$, then $\boldsymbol{o}^{(1)} = [0, x + y - 1]^\top$ and $\boldsymbol{o}^{(2)} = \sigma\left(W^{(1)}\boldsymbol{o}^{(1)} + b^{(1)}\right) = \sigma\left([2x + 2y - 11, x + y - 3]\right)^\top$. Therefore,

$$\boldsymbol{o}^{(2)} = \begin{cases} [0, 0]^\top, & x + y - 3 < 0, \\ [0, x + y - 3]^\top, & 2x + 2y - 11 < 0, x + y - 3 \geq 0, \\ [2x + 2y - 11, x + y - 3]^\top, & \text{otherwise}. \end{cases}$$

(iv) If $2x - y + 1 \geq 0$ and $x + y - 1 \geq 0$, then $\boldsymbol{o}^{(1)} = [2x - y + 1, x + y - 1]^\top$ and $\boldsymbol{o}^{(2)} = \sigma\left(W^{(1)}\boldsymbol{o}^{(1)} + b^{(1)}\right) = \sigma\left([4x + y - 10, -x + 2y - 4]\right)^\top$. Therefore,

$$\boldsymbol{o}^{(2)} = \begin{cases} [0, 0]^\top, & 4x + y - 10 < 0, -x + 2y - 4 < 0, \\ [4x + y - 10, 0]^\top, & 4x + y - 10 < 0, -x + 2y - 4 \geq 0, \\ [0, -x + 2y - 4]^\top, & 4x + y - 10 \geq 0, -x + 2y - 4 < 0, \\ [4x + y - 10, -x + 2y - 4]^\top, & \text{otherwise}. \end{cases}$$

The resulting input space partition is shown in Fig. 10.11, where the corresponding linear mapping and its rank are illustrated. Note that around the two full rank partitions, there exist rank-1 mapping partitions, which join with the rank-0 mapping partition.

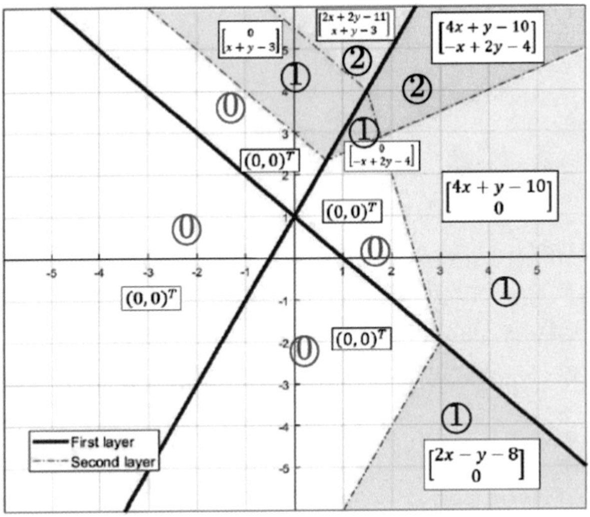

Fig. 10.11 Input space partitioning for the problem (a) case

(b) At the first layer, the output signal is given by

$$\boldsymbol{o}^{(1)} = \sigma\left(W^{(0)}\mathbf{x} + b^{(0)}\right) = \begin{bmatrix} \sigma(2x - y) \\ \sigma(x + y) \end{bmatrix},$$

where σ is the ReLU. At the second layer, we again consider all cases where each ReLU is active or inactive.

(i) If $2x - y < 0$ and $x + y < 0$, then $\boldsymbol{o}^{(1)} = [0, 0]^\top$, $\boldsymbol{o}^{(2)} = \sigma\left(W^{(1)}\boldsymbol{o}^{(1)}\right) = [0, 0]^\top$.

(ii) If $2x - y \geq 0$ and $x + y < 0$, then $\boldsymbol{o}^{(1)} = [2x - y, 0]^\top$. Hence,

$$\boldsymbol{o}^{(2)} = \sigma\left(W^{(1)}\boldsymbol{o}^{(1)} + b^{(1)}\right) = \sigma\left([2x - y, -2x + y]\right)^\top = [2x - y, 0]^\top.$$

(iii) If $2x - y < 0$ and $x + y \geq 0$, then $\boldsymbol{o}^{(1)} = [0, x + y]]^\top$ and

$$\boldsymbol{o}^{(2)} = \sigma\left(W^{(1)}\boldsymbol{o}^{(1)}\right) = \sigma\left([2x + 2y, x + y]\right)^\top = [2x + 2y, x + y]^\top.$$

(iv) If $2x - y \geq 0$ and $x + y \geq 0$, then $\boldsymbol{o}^{(1)} = [2x - y, x + y]^\top$ and $\boldsymbol{o}^{(2)} = \sigma\left(W^{(1)}\boldsymbol{o}^{(1)} + b^{(1)}\right) = \sigma\left([4x + y, -x + 2y]\right)^\top$. Therefore,

$$\boldsymbol{o}^{(2)} = \begin{cases} [4x + y, 0]^\top, & -x + 2y < 0, \\ [4x + y, -x + 2y]^\top, & \text{otherwise.} \end{cases}$$

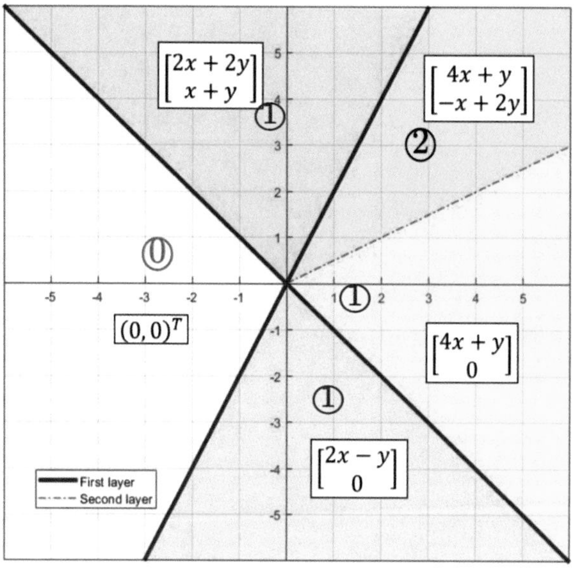

Fig. 10.12 Input space partitioning for the problem (b) case

The resulting input space partition is shown in Fig. 10.12, where the corresponding linear mapping and its rank are illustrated. Similar to problem (a), around the two full rank partitions, there exist rank-1 mapping partitions, which join with the rank-0 mapping partition. Since there is no bias term, all the hyperplanes should contain the origin. Also, there are no hyperplane with same normal vector, since parallel hyperplanes cannot be formed without bias terms. As a result, the input space partition becomes simpler compared to (a).

(c) At the first layer, the output signal is given by

$$\boldsymbol{o}^{(1)} = \sigma\left(W^{(0)}\mathbf{x} + b^{(0)}\right) = \begin{bmatrix}\sigma(2x - y + 1)\\ \sigma(x + y - 1)\end{bmatrix}$$

where σ is the ReLU. Now, at the second layer, we need to consider all cases where each ReLU is active or inactive.

(i) If $2x - y + 1 < 0$ and $x + y - 1 < 0$, then $\boldsymbol{o}^{(1)} = [0, 0]^\top$, $\boldsymbol{o}^{(2)} = \sigma\left(W^{(1)}\boldsymbol{o}^{(1)} + b^{(1)}\right) = \sigma[0, 1]^\top = [0, 1]^\top$.

(ii) If $2x - y + 1 \geq 0$ and $x + y - 1 < 0$, then $\boldsymbol{o}^{(1)} = [2x - y + 1, 0]^\top$. Hence, $\boldsymbol{o}^{(2)} = \sigma\left(W^{(1)}\boldsymbol{o}^{(1)} + b^{(1)}\right) = \sigma\left([2x - y + 1, 1]\right)^\top = [2x - y + 1, 1]^\top$.

(iii) If $2x - y + 1 < 0$ and $x + y - 1 \geq 0$, then $\boldsymbol{o}^{(1)} = [0, x + y - 1]^\top$ and $\boldsymbol{o}^{(2)} = \sigma\left(W^{(1)}\boldsymbol{o}^{(1)} + b^{(1)}\right) = \sigma\left([2x + 2y - 2, x + y]\right)^\top = [2x+2y-2, x+y]^\top$.

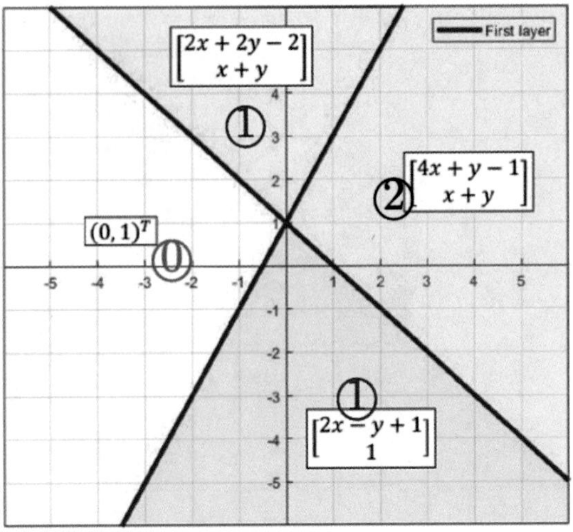

Fig. 10.13 Input space partitioning for the problem (c) case

(iv) If $2x - y + 1 \geq 0$ and $x + y - 1 \geq 0$, then $\boldsymbol{o}^{(1)} = [2x - y + 1, x + y - 1]^\top$ and $\boldsymbol{o}^{(2)} = \sigma\left(W^{(1)}\boldsymbol{o}^{(1)} + b^{(1)}\right) = \sigma\left([4x + y - 1, x + y]\right)^\top = [4x + y - 1, x+y]^\top$. The resulting input space partition is shown in Fig. 10.13, where the corresponding linear mapping and its rank are illustrated. There is no hyperplane formed by the second layer. This shows how weight and bias change the complexity of the input partition.

10.4.5 Geometric Understanding of Autoencoder

We are now interested in providing a more in-depth discussion on geometry of deep neural networks for regression problems, in particular, autoencoder. Autoencoders have the same input and output domains, and are commonly used for low-level computer vision problems, such as image denoising [90, 98], super-resolution [99, 100] and so on. Although we provide a discussion on the autoencoder, similar geometric understanding can be applied to other regression problems, where the input and output domains are different. Later we will show that the geometric understanding of the autoencoder also gives a clear insight on the geometry of classifiers.

Based on the discussion so far, we now understand that the deep neural network with ReLU nonlinearities partitions the input data space into piecewise linear regions. In fact, this view is directly related to the manifold structure of the data, and we believe that the main fundamental principle to explain the success of deep learning is its efficient use of the manifold structure in the data.

First, we provide some differential geometric definition.

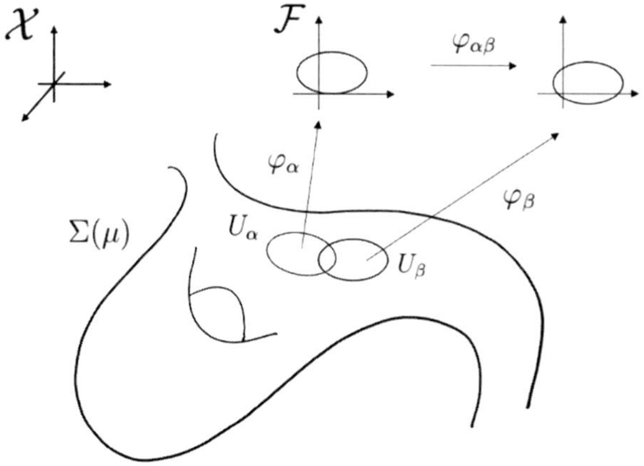

Fig. 10.14 Manifold geometry of autoencoder [114]

Definition 10.1 An n-dimensional manifold is a topological space, covered by a set of open sets $\Sigma \subset \cup_\alpha U_\alpha$. For each open set U_α, there exists a homeomorphism $\varphi_\alpha : U_\alpha \mapsto \mathbb{R}^n$, and the pair $(U_\alpha, \varphi_\alpha)$ form a chart. The union of the chart forms an atlas $\mathcal{A} = \{(U_\alpha, \varphi_\alpha)\}$.

As shown in Fig. 10.14, suppose $\mathcal{X}$ is the ambient space, μ is a probability distribution defined on $\mathcal{X}$. The support of μ,

$$\Sigma(\mu) := \{\boldsymbol{x} \in \mathcal{X} \ : \ \mu(\boldsymbol{x}) > 0\} \tag{10.53}$$

is a low-dimensional manifold in $\mathcal{X}$. For a given local chart $(U_\alpha, \varphi_\alpha)$, $\varphi_\alpha : U_\alpha \mapsto \mathcal{F}$ is called an *encoder*, where $\mathcal{F}$ is called the *latent space* or *feature space*. A point $\boldsymbol{x} \in \Sigma$ is called a *sample*; its image $\varphi_\alpha(\boldsymbol{x})$ is the corresponding feature of $\boldsymbol{x}$. The inverse map $\psi_\alpha := \varphi_\alpha^{-1} : \mathcal{F} \mapsto \Sigma$ is called the *decoder* [114].

Then, an autoencoder consists of two parts, the encoder and the decoder. The encoder takes a sample $\boldsymbol{x} \in \mathcal{X}$ and maps it to the feature map $z \in \mathcal{F}$, $z = \varphi(\boldsymbol{x})$. The encoder $\varphi : \mathcal{X} \mapsto \mathcal{F}$ maps Σ to its latent representation $D = \varphi(\Sigma)$ homomorphically. After that, the decoder $\psi : \mathcal{F} \mapsto \mathcal{X}$ maps z to the reconstruction $\boldsymbol{x}$ of the same shape as $\boldsymbol{x}$,

$$\widehat{\boldsymbol{x}} = \psi(z) = \psi \circ \varphi(\boldsymbol{x}).$$

This relation can be seen in the following commutative diagram [114]:

$$\begin{array}{ccc} \{(\mathcal{X}, \boldsymbol{x}), \mu, \Sigma\} & \xrightarrow{\ \varphi\ } & \{(\mathcal{F}, z), D\} \\ & \searrow^{\psi\circ\varphi} & \downarrow \psi \\ & & \{(\mathcal{X}, \widehat{\boldsymbol{x}}), \mu, \widehat{\Sigma}\} \end{array}$$

In practice, both encoder and decoder are parameterized with the parameter $\boldsymbol{\Theta}$, so that the autoencoder is described by

$$\widehat{\boldsymbol{x}} = \mathcal{T}_{\boldsymbol{\Theta}}(\boldsymbol{x}) = \psi_{\boldsymbol{\Theta}} \circ \varphi_{\boldsymbol{\Theta}}(\boldsymbol{x})$$

and the parameter estimation problem can be solved by

$$\min_{\boldsymbol{\Theta}} \sum_{i=1}^{n} \ell\left(y_i, \mathcal{T}_{\boldsymbol{\Theta}}(\boldsymbol{x}_i)\right) + \lambda R(\boldsymbol{\Theta}), \tag{10.54}$$

which is the same as the CNN training.

Figure 10.14 shows an example geometry of each step of the autoencoder with ReLU nonlinearities. Here, the ambient space $\mathcal{X}$ is $\mathbb{R}^3$ and the feature space is two-dimensional, i.e. $\mathcal{F} \subset \mathbb{R}^2$. The sample $\boldsymbol{x}$ is a 3-D point so that the input manifold $M := \Sigma(\mu) \subset \mathcal{X}$ is a two-dimensional surface within $\mathbb{R}^3$, which is low-dimensional (see Fig. 10.14). The input samples are mapped to the feature space manifold in Fig. 10.14 using the parameterized encoder $\varphi_{\boldsymbol{\Theta}}$. Then, this feature manifold is mapped back to the original ambient space using the parameterized decoder $\psi_{\boldsymbol{\Theta}}$ as Fig. 10.14. Due to the ReLU nonlinearities, the input manifold M is partitioned into piecewise linear regions $\mathcal{D}(\varphi_{\boldsymbol{\Theta}})$.

The specific operation on each piecewise linear region is then defined during the training phase. For example, in Fig. 10.15, the input manifold is a noisy point cloud, whereas the label data at the output are the noiseless 3D surfaces. During the training, the specific operation of the neural network is guided as a low-rank mapping on the reconstruction manifold, as discussed in Problem 10.1. Therefore, the noisy outliers from the input manifold are projected into the reconstruction manifold via a trained neural network, which is piecewise linear at each cell but globally nonlinear [114].

10.4.6 Geometric Understanding of Classifier

The geometric understanding of the autoencoder now gives a clear picture of what happens in the deep neural network classifier. In this case, we only have an encoder to map to the latent space, which leads to a simplified commutative diagram:

$$\{(\mathcal{X}, \boldsymbol{x}), \mu, \Sigma\} \xrightarrow{\varphi} \{(\mathcal{F}, z), D\}.$$

Since the encoder is also parameterized with $\boldsymbol{\Theta}$ and equipped with a ReLU, the input manifold is also partitioned into piecewise linear regions, as shown in Fig. 10.14d. Then, the linear layer followed by softmax assigns to the class probability for each piecewise linear cell.

Fig. 10.15 Denoising as a piecewise linear projection on the reconstruction manifold [114]

10.5 Open Problems

Our discussion so far reveals that the deep neural network is indeed trained to partition the input data manifold such that the linear mapping at each piecewise linear region can effectively perform machine learning tasks, such as classification, regression, etc. Therefore, we strongly believe that the clue to unveil the mystery of deep neural networks comes from the understanding of the high-dimensional manifold structure and its piecewise linear partition, and how the partitions can be controlled.

In fact, many machine learning theoreticians have been focusing on this, thereby generating many intriguing theoretical and empirical observations [115–118]. For example, although we mentioned that the number of linear regions can potentially increase exponentially with the network complexity, they observed that the actual number of piecewise linear representation for specific tasks is much smaller. For example, Fig. 10.16 shows that the number of linear regions indeed converges to a smaller value compared to the initialization as the number of epochs increases [115, 116].

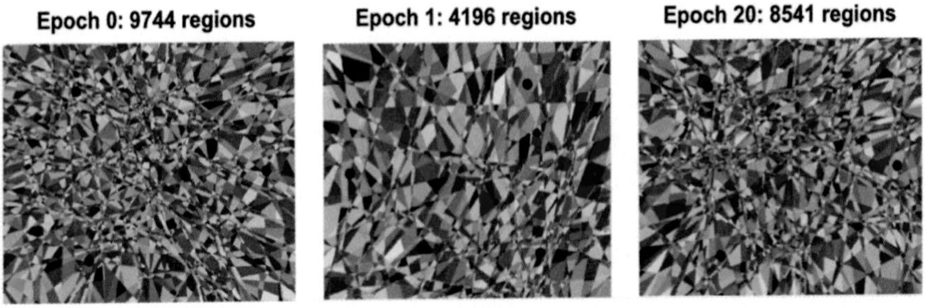

Fig. 10.16 Here the authors [115, 116] show the linear regions that intersect a 2D plane through input space for a network of depth 3 and width 64 trained on MNIST

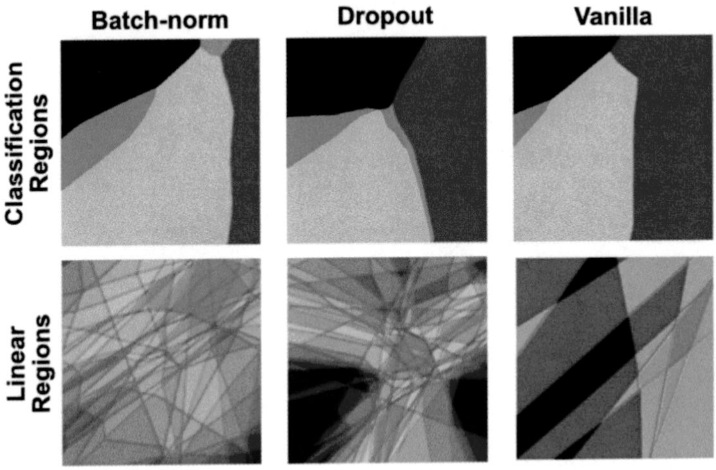

Fig. 10.17 Linear regions and classification regions of models trained with different optimization techniques [117]

Note that only the number of epochs determines the number of piecewise linear regions, but also, depending on the choice of the optimization algorithms, the number of linear regions varies. For example, Fig. 10.17 shows that the number of linear regions varies depending on the optimization algorithms, which leads to the different classification boundaries. Here, the gray curves in the bottom row are transition boundaries separating different linear regions, and the color represents the activation rate of the corresponding linear region. In the top row, different colors represent different classification regions, separated by the decision boundaries. The models were trained on the vectorized MNIST data set, and this figure shows a two-dimensional slice of the input space.

In fact, this phenomenon can be understood as a data-driven adaptation to eliminate the unnecessary partitions for machine learning tasks. Note that the partition boundary can collapse, resulting in a smaller number of partitions, as discussed in Problem 10.1(c). It is believed that there is a compromise between the approximation error and the robustness of the neural network in terms of the number of piecewise linear areas. Many of these questions remain unanswered, and many research efforts need to clearly understand the partition geometry of the neural network.

Finally, while it was largely disregarded during our discussion, for the case of CNNs, the choice of the hyperplanes becomes further constrained due to the convolutional relationship. For example, to encode the data manifold in $\mathbb{R}^3$ with the $r = 2$ convolution filter with the filter coefficient of $[1, 2]$, the following three vectors determine the normal direction of the three hyperplanes:

$$\boldsymbol{n}_1^l = \begin{bmatrix}1 \ 2 \ 0\end{bmatrix}, \quad \boldsymbol{n}_2^l = \begin{bmatrix}0 \ 1 \ 2\end{bmatrix}, \quad \boldsymbol{n}_3^l = \begin{bmatrix}2 \ 0 \ 1\end{bmatrix}, \tag{10.55}$$

where we assume the circular convolution and no pooling operation (i.e. $\boldsymbol{\Phi}^l = \boldsymbol{I}_3$). This implies that each channel of the convolution filter determines an orthant of the underlying feature space, and the feature vectors are directly related to the coordinate on the resultant orthant. Therefore, understanding piecewise linear regions in CNNs requires a much more in-depth understanding of the high-dimensional geometry, which may be another very exciting research topic.

10.6 Exercises

1. Prove (10.24).
2. Prove the equality (10.25).
3. Fill in the missing step in (10.26).
4. Show (10.29).
5. Our goal is to derive the input–output relation in (10.32) at the encoder.

   (a) Show that

$$(\boldsymbol{\Phi}^l \circledast \boldsymbol{\psi}_{j,k}^l)^\top \boldsymbol{z}_k^{l-1} = \boldsymbol{\Phi}^{l\top}(\boldsymbol{z}_k^{l-1} \circledast \overline{\boldsymbol{\psi}}_{j,k}^l). \tag{10.56}$$

   (b) Using (10.56), prove (10.32).

6. Our goal is to derive the input–output relation in (10.35) at the decoder.

   (a) Show that

$$(\tilde{\boldsymbol{\Phi}}^l \circledast \tilde{\boldsymbol{\psi}}_{j,k}^l)\tilde{z}_k^l = \tilde{\boldsymbol{\Phi}}^l \tilde{z}_k^l \circledast \tilde{\boldsymbol{\psi}}_{j,k}^l. \tag{10.57}$$

   (b) Using (10.57), prove (10.35).

7. Under the frame condition (10.39), derive the perfect reconstruction condition in (10.42).

8. Consider a three-layer fully connected network $f_\Theta : \mathbb{R}^2 \rightarrow \mathbb{R}^2$ with ReLU nonlinearity.

(a) Suppose the weight matrices and biases are given by

$$\boldsymbol{W}^{(0)} = \begin{bmatrix} 1 & -1 \\ 1 & 1 \end{bmatrix}, \quad \boldsymbol{b}^{(0)} = \begin{bmatrix} 1 \\ -1 \end{bmatrix},$$

$$\boldsymbol{W}^{(1)} = \begin{bmatrix} 2 & 2 \\ 1 & 1 \end{bmatrix}, \quad \boldsymbol{b}^{(1)} = \begin{bmatrix} 0 \\ 1 \end{bmatrix},$$

$$\boldsymbol{W}^{(2)} = \begin{bmatrix} 1 & 2 \\ -1 & 1 \end{bmatrix}, \quad \boldsymbol{b}^{(2)} = \begin{bmatrix} -1 \\ -1 \end{bmatrix}.$$

Draw the corresponding input space partition, and compute the output mapping with respect to an input vector (x, y) in each input partition. Please derive all the steps explicitly.

(b) In problem (a), suppose that the bias terms are zero. Compute the input space partition and the output mapping. What do you observe compared to the one with bias?

(c) In problem (a), the last layer weight $\boldsymbol{W}^{(2)}$ and bias $\boldsymbol{b}^{(2)}$ are changed due to the fine tuning. Please give an example of $\boldsymbol{W}^{(2)}$ and bias $\boldsymbol{b}^{(2)}$ that gives the smallest number of partitions.

Chapter 11
Deep Learning Optimization

11.1 Introduction

In Chap. 6, we discussed various optimization methods for deep neural network training. Although they are in various forms, these algorithms are basically gradient-based local update schemes. However, the biggest obstacle recognized by the entire community is that the loss surfaces of deep neural networks are extremely non-convex and not even smooth. This non-convexity and non-smoothness make the optimization unaffordable to analyze, and the main concern was whether popular gradient-based approaches might fall into local minimizers.

Surprisingly, the success of modern deep learning may be due to the remarkable effectiveness of gradient-based optimization methods despite its highly non-convex nature of the optimization problem. Extensive research has been carried out in recent years to provide a theoretical understanding of this phenomenon. In particular, several recent works [119–121] have noted the importance of the over-parameterization. In fact, it was shown that when hidden layers of a deep network have a large number of neurons compared to the number of training samples, the gradient descent or stochastic gradient converges to a global minimum with zero training errors. While these results are intriguing and provide important clues for understanding the geometry of deep learning optimization, it is still unclear why simple local search algorithms can be successful for deep neural network training.

Indeed, the area of deep learning optimization is a rapidly evolving area of intense research, and there are too many different approaches to cover in a single chapter. Rather than treating a variety of techniques in a disorganized way, this chapter explains two different lines of research just for food for thought: one is based on the geometric structure of the loss function and the other is based on the results of Lyapunov stability. Although the two approaches are closely related, they have different advantages and disadvantages. By explaining these two approaches, we can cover some of the key topics of research exploration such as optimization landscape [122–124], over-parameterization [119, 125–129], and neural tangent kernel (NTK)

J. C. Ye, *Geometry of Deep Learning*, Mathematics in Industry 37,
https://doi.org/10.1007/978-981-16-6046-7_11

[130–132] that have been used extensively to analyze the convergence properties of local deep learning search methods.

11.2 Problem Formulation

In Chap. 6, we pointed out that the basic optimization problem in neural network training can be formulated as

$$\min_{\boldsymbol{\theta} \in \mathbb{R}^n} \ell(\boldsymbol{\theta}), \tag{11.1}$$

where $\boldsymbol{\theta}$ refers to the network parameters and $\ell : \mathbb{R}^n \mapsto \mathbb{R}$ is the loss function. In the case of supervised learning with the mean square error (MSE) loss, the loss function is defined by

$$\ell(\boldsymbol{\theta}) := \frac{1}{2}\|\boldsymbol{y} - \boldsymbol{f}_{\boldsymbol{\theta}}(\boldsymbol{x})\|^2, \tag{11.2}$$

where $\boldsymbol{x}, \boldsymbol{y}$ denotes the pair of the network input and the label, and $\boldsymbol{f}_{\boldsymbol{\theta}}(\cdot)$ is a neural network parameterized by trainable parameters $\boldsymbol{\theta}$. For the case of an L-layer feed-forward neural network, the regression function $\boldsymbol{f}_{\Theta}(\boldsymbol{x})$ can be represented by

$$\boldsymbol{f}_{\boldsymbol{\theta}}(\boldsymbol{x}) := \left(\boldsymbol{\sigma} \circ \boldsymbol{g}^{(L)} \circ \boldsymbol{\sigma} \circ \boldsymbol{g}^{(L-1)} \cdots \circ \boldsymbol{g}^{(1)}\right)(\boldsymbol{x}), \tag{11.3}$$

where $\boldsymbol{\sigma}(\cdot)$ denotes the element-wise nonlinearity and

$$\boldsymbol{g}^{(l)} = \boldsymbol{W}^{(l)}\boldsymbol{o}^{(l-1)} + \boldsymbol{b}^{(l-1)}, \tag{11.4}$$

$$\boldsymbol{o}^{(l)} = \boldsymbol{\sigma}(\boldsymbol{g}^{(l)}), \tag{11.5}$$

$$\boldsymbol{o}^{(0)} = \boldsymbol{x}, \tag{11.6}$$

for $l = 1, \cdots, L$. Here, the number of the l-th layer hidden neurons, often referred to as the width, is denoted by $d^{(l)}$, so that $\boldsymbol{g}^{(l)}, \boldsymbol{o}^{(l)} \in \mathbb{R}^{d^{(l)}}$ and $\boldsymbol{W}^{(l)} \in \mathbb{R}^{d^{(l)} \times d^{(l-1)}}$.

The popular local search approaches using the gradient descent use the following update rule:

$$\boldsymbol{\theta}[k+1] = \boldsymbol{\theta}[k] - \eta_k \left.\frac{\partial \ell(\boldsymbol{\theta})}{\partial \boldsymbol{\theta}}\right|_{\boldsymbol{\theta}=\boldsymbol{\theta}[k]}, \tag{11.7}$$

where η_k denotes the k-th iteration step size. In a differential equation form, the update rule can be represented by

$$\dot{\boldsymbol{\theta}}[t] = -\frac{\partial \ell(\boldsymbol{\theta}[k])}{\partial \boldsymbol{\theta}}, \tag{11.8}$$

where $\dot{\boldsymbol{\theta}}[t] = \partial \boldsymbol{\theta}[t]/\partial t$.

As previously explained, the optimization problem (11.1) is strongly non-convex, and it is known that the gradient-based local search schemes using (11.7) and (11.8) may get stuck in the local minima. Interestingly, many deep learning optimization algorithms appear to avoid the local minima and even result in zero training errors, indicating that the algorithms are reaching the global minima. In the following, we present two different approaches to explain this fascinating behavior of gradient descent approaches.

11.3 Polyak–Łojasiewicz-Type Convergence Analysis

The loss function ℓ is said to be strongly convex (SC) if

$$\ell(\boldsymbol{\theta}') \geq \ell(\boldsymbol{\theta}) + \langle \nabla \ell(\boldsymbol{\theta}), \boldsymbol{\theta}' - \boldsymbol{\theta} \rangle + \frac{\mu}{2}\|\boldsymbol{\theta}' - \boldsymbol{\theta}\|^2, \quad \forall \boldsymbol{\theta}, \boldsymbol{\theta}'. \tag{11.9}$$

It is known that if ℓ is SC, then gradient descent achieves a global linear convergence rate for this problem [133]. Note that SC in (11.9) is a stronger condition than the convexity in Proposition 1.1, which is given as

$$\ell(\boldsymbol{\theta}') \geq \ell(\boldsymbol{\theta}) + \langle \nabla \ell(\boldsymbol{\theta}), \boldsymbol{\theta}' - \boldsymbol{\theta} \rangle, \quad \forall \boldsymbol{\theta}, \boldsymbol{\theta}'. \tag{11.10}$$

Our starting point is the observation that the convex analysis mentioned above is not the right approach to analyzing a deep neural network. The non-convexity is essential for the analysis. This situation has motivated a variety of alternatives to the convexity to prove the convergence. One of the oldest of these conditions is the error bounds (EB) of Luo and Tseng [134], but other conditions have been recent considered, which include essential strong convexity (ESC) [135], weak strong convexity (WSC) [136], and the restricted secant inequality (RSI) [137]. See their specific forms of conditions in Table 11.1. On the other hand, there is a much older condition called the Polyak–Łojasiewicz (PL) condition, which was originally introduced by Polyak [138] and found to be a special case of the inequality of Łojasiewicz [139]. Specifically, we will say that a function satisfies the PL inequality if the following holds for some $\mu > 0$:

$$\frac{1}{2}\|\nabla \ell(\boldsymbol{\theta})\|^2 \geq \mu(\ell(\boldsymbol{\theta}) - \ell^*), \quad \forall \boldsymbol{\theta}. \tag{11.11}$$

Table 11.1 Examples of conditions for gradient descent (GD) convergence. All of these definitions involve some constant $\mu > 0$ (which may not be the same across conditions). $\boldsymbol{\theta}_p$ denotes the projection of $\boldsymbol{\theta}$ onto the solution set $\mathcal{X}^*$, and ℓ^* refers to the minimum cost

Name	Conditions	
Strong convexity (SC)	$\ell(\boldsymbol{\theta}') \geq \ell(\boldsymbol{\theta}) + \langle \nabla \ell(\boldsymbol{\theta}), \boldsymbol{\theta}' - \boldsymbol{\theta} \rangle + \frac{\mu}{2} \|\boldsymbol{\theta}' - \boldsymbol{\theta}\|^2,$	$\forall \boldsymbol{\theta}, \boldsymbol{\theta}'$
Essential strong convexity (ESC)	$\ell(\boldsymbol{\theta}') \geq \ell(\boldsymbol{\theta}) + \langle \nabla \ell(\boldsymbol{\theta}), \boldsymbol{\theta}' - \boldsymbol{\theta} \rangle + \frac{\mu}{2} \|\boldsymbol{\theta}' - \boldsymbol{\theta}\|^2,$	$\forall \boldsymbol{\theta}, \boldsymbol{\theta}' \; s.t. \; \boldsymbol{\theta}_p = \boldsymbol{\theta}'_p$
Weak strong convexity (WSC)	$\ell^* \geq \ell(\boldsymbol{\theta}) + \langle \nabla \ell(\boldsymbol{\theta}), \boldsymbol{\theta}_p - \boldsymbol{\theta} \rangle + \frac{\mu}{2} \|\boldsymbol{\theta}_p - \boldsymbol{\theta}\|^2,$	$\forall \boldsymbol{\theta}$
Restricted secant inequality (RSI)	$\langle \nabla \ell(\boldsymbol{\theta}), \boldsymbol{\theta} - \boldsymbol{\theta}_p \rangle \geq \mu \|\boldsymbol{\theta}_p - \boldsymbol{\theta}\|^2,$	$\forall \boldsymbol{\theta}$
Error bound (EB)	$\|\nabla \ell(\boldsymbol{\theta})\| \geq \mu \|\boldsymbol{\theta}_p - \boldsymbol{\theta}\|^2,$	$\forall \boldsymbol{\theta}$
Polyak–Łojasiewicz (PL)	$\frac{1}{2} \|\nabla \ell(\boldsymbol{\theta})\|^2 \geq \mu(\ell(\boldsymbol{\theta}) - \ell^*),$	$\forall \boldsymbol{\theta}$

Note that this inequality implies that every stationary point is a global minimum. But unlike SC, it does not imply that there is a unique solution. We will revisit this issue later.

Similar to other conditions in Table 11.1, PL is a sufficient condition for gradient descent to achieve a linear convergence rate [122]. In fact, PL is the mildest condition among them. Specifically, the following relationship between the conditions holds [122]:

$$(SC) \rightarrow (ESC) \rightarrow (WSC) \rightarrow (RSI) \rightarrow (EB) \equiv (PL),$$

if ℓ have a Lipschitz continuous gradient, i.e. there exists $L > 0$ such that

$$\|\nabla \ell(\boldsymbol{\theta}) - \nabla \ell(\boldsymbol{\theta}')\| \leq L\|\boldsymbol{\theta} - \boldsymbol{\theta}'\|, \quad \forall \boldsymbol{\theta}, \boldsymbol{\theta}'. \tag{11.12}$$

In the following, we provide a convergence proof of the gradient descent method using the PL condition, which turns out to be an important tool for non-convex deep learning optimization problems.

Theorem 11.1 (Karimi et al. [122]) *Consider problem* (11.1), *where* ℓ *has an* L*-Lipschitz continuous gradient, a non-empty solution set, and satisfies the PL inequality* (11.11). *Then the gradient method with a step-size of* $1/L$*:*

$$\boldsymbol{\theta}[k+1] = \boldsymbol{\theta}[k] - \frac{1}{L}\nabla \ell(\boldsymbol{\theta}[k]) \tag{11.13}$$

has a global convergence rate

$$\ell(\boldsymbol{\theta}[k]) - \ell^* \leq \left(1 - \frac{\mu}{L}\right)^k \left(\ell(\boldsymbol{\theta}[0]) - \ell^*\right).$$

Proof Using Lemma 11.1 (see next section), L-Lipschitz continuous gradient of the loss function ℓ implies that the function

$$g(\boldsymbol{\theta}) = \frac{L}{2}\|\boldsymbol{\theta}\|^2 - \ell(\boldsymbol{\theta})$$

is convex. Thus, the first-order equivalence of convexity in Proposition 1.1 leads to the following:

$$\begin{aligned}
\frac{L}{2}\|\boldsymbol{\theta}'\|^2 - \ell(\boldsymbol{\theta}') &\geq \frac{L}{2}\|\boldsymbol{\theta}\|^2 - \ell(\boldsymbol{\theta}) + \langle \boldsymbol{\theta}' - \boldsymbol{\theta}, L\boldsymbol{\theta} - \nabla \ell(\boldsymbol{\theta})\rangle \\
&= -\frac{L}{2}\|\boldsymbol{\theta}\|^2 - \ell(\boldsymbol{\theta}) + L\langle \boldsymbol{\theta}', \boldsymbol{\theta}\rangle - \langle \boldsymbol{\theta}' - \boldsymbol{\theta}, \nabla \ell(\boldsymbol{\theta})\rangle.
\end{aligned}$$

By arranging terms, we have

$$\ell(\boldsymbol{\theta}') \leq \ell(\boldsymbol{\theta}) + \langle \nabla \ell(\boldsymbol{\theta}), \boldsymbol{\theta}' - \boldsymbol{\theta} \rangle + \frac{L}{2} \|\boldsymbol{\theta}' - \boldsymbol{\theta}\|^2, \quad \forall \boldsymbol{\theta}, \boldsymbol{\theta}'.$$

By setting $\boldsymbol{\theta}' = \boldsymbol{\theta}[k+1]$ and $\boldsymbol{\theta} = \boldsymbol{\theta}[k]$ and using the update rule (11.13), we have

$$\ell(\boldsymbol{\theta}[k+1]) - \ell(\boldsymbol{\theta}[k]) \leq -\frac{1}{2L} \|\nabla \ell(\boldsymbol{\theta}[k])\|^2. \tag{11.14}$$

Using the PL inequality (11.11), we get

$$\ell(\boldsymbol{\theta}[k+1]) - \ell(\boldsymbol{\theta}[k]) \leq -\frac{\mu}{L} \left(\ell(\boldsymbol{\theta}[k]) - \ell^* \right).$$

Rearranging and subtracting ℓ^* from both sides gives us

$$\ell(\boldsymbol{\theta}[k+1]) - \ell^* \leq \left(1 - \frac{\mu}{L}\right) \left(\ell(\boldsymbol{\theta}[k]) - \ell^* \right).$$

Applying this inequality recursively gives the result. □

The beauty of this proof is that we can replace the long and complicated proofs from other conditions with simpler proofs based on the PL inequality [122].

11.3.1 Loss Landscape and Over-Parameterization

In Theorem 11.1, we use the two conditions for the loss function: (1) ℓ satisfies the PL condition and (2) the gradient of ℓ is Lipschitz continuous. Although these conditions are much weaker than the convexity of the loss function, they still impose the geometric constraint for the loss function, which deserves further discussion.

Lemma 11.1 *If the gradient of $\ell(\boldsymbol{\theta})$ satisfies the L-Lipschitz condition in* (11.12), *then the transformed function $g : \mathbb{R}^n \mapsto \mathbb{R}$ defined by*

$$g(\boldsymbol{\theta}) := \frac{L}{2} \boldsymbol{\theta}^\top \boldsymbol{\theta} - \ell(\boldsymbol{\theta}) \tag{11.15}$$

is convex.

Proof Using the Cauchy–Schwarz inequality, (11.12) implies

$$\langle \nabla \ell(\boldsymbol{\theta}) - \nabla \ell(\boldsymbol{\theta}'), \boldsymbol{\theta} - \boldsymbol{\theta}' \rangle \leq L \|\boldsymbol{\theta} - \boldsymbol{\theta}'\|^2, \quad \forall \boldsymbol{\theta}, \boldsymbol{\theta}'.$$

This is equivalent to the following condition:

$$\langle \boldsymbol{\theta}' - \boldsymbol{\theta}, \nabla g(\boldsymbol{\theta}') - \nabla g(\boldsymbol{\theta}) \rangle \geq 0, \quad \forall \boldsymbol{\theta}, \boldsymbol{\theta}', \tag{11.16}$$

where

$$g(\boldsymbol{\theta}) = \frac{L}{2}\|\boldsymbol{\theta}\|^2 - \ell(\boldsymbol{\theta}).$$

Thus, using the monotonicity of gradient equivalence in Proposition 1.1, we can show that $g(\boldsymbol{\theta})$ is convex. □

Lemma 11.1 implies that although ℓ is not convex, its transformed function by (11.15) can be convex. Figure 11.1a shows an example of such case. Another important geometric consideration for the loss landscape comes from the PL condition. More specifically, the PL condition in (11.11) implies that every stationary point is a global minimizer, although the global minimizers may not be unique, as shown in Fig. 11.1b,c. While the PL inequality does not imply convexity of ℓ, it does imply the weaker condition of *invexity* [122]. A function is invex if it is differentiable and there exists a vector-valued function η such that for any $\boldsymbol{\theta}$ and $\boldsymbol{\theta}'$ in $\mathbb{R}^n$, the following inequality holds:

$$\ell(\boldsymbol{\theta}') \geq \ell(\boldsymbol{\theta}) + \langle \nabla \ell(\boldsymbol{\theta}), \eta(\boldsymbol{\theta}, \boldsymbol{\theta}') \rangle. \tag{11.17}$$

A convex function is a special case of invex functions since (11.17) holds when we set $\eta(\boldsymbol{\theta}, \boldsymbol{\theta}') = \boldsymbol{\theta}' - \boldsymbol{\theta}$. It was shown that a smooth ℓ is invex if and only if every stationary point of ℓ is a global minimum [140]. As the PL condition implies that every stationary point is a global minimizer, a function satisfying PL is an invex function. The inclusion relationship between convex, invex, and PL functions is illustrated in Fig. 11.2.

The loss landscape, where every stationary point is a global minimizer, implies that that there are no spurious local minimizers. This is often called the *benign* optimization landscape. Finding the conditions for a benign optimization landscape

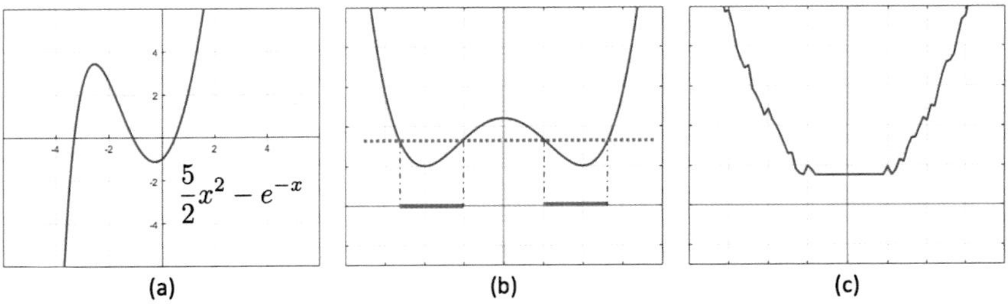

Fig. 11.1 Loss landscape for the function $\ell(x)$ with (**a**) (11.15) is convex, and (**b**, **c**) PL conditions

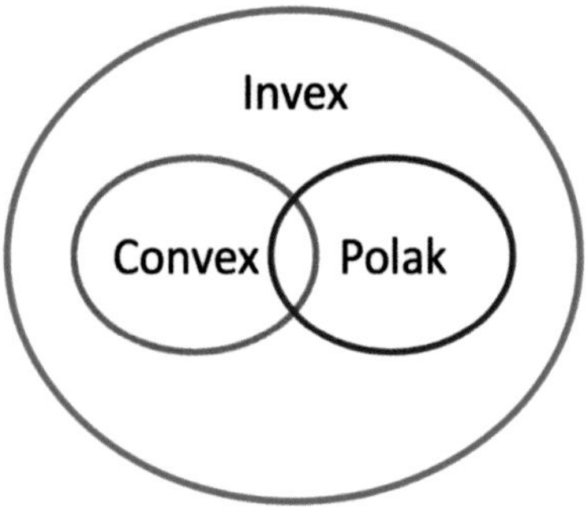

Fig. 11.2 Inclusion relationship between invex, convex and PL-type functions

of neural networks was an important theoretical interest of the theorists in machine learning. Originally observed by Kawaguch [141], Lu and Kawaguchi [142] and Zhou and Liang [143] have proven that the loss surfaces of linear neural networks, whose activation functions are all linear functions, do not have any spurious local minima under some conditions and all local minima are equally good.

Unfortunately, this good property no longer stands when the activations are nonlinear. Zhou and Liang [143] show that ReLU neural networks with one hidden layer have spurious local minima. Yun et al. [144] prove that ReLU neural networks with one hidden layer have infinitely many spurious local minima when the outputs are one-dimensional.

These somewhat negative results were surprising and seemed to contradict the empirical success of optimization in neural networks. Indeed, it was later shown that if the activation functions are continuous, and the loss functions are convex and differentiable, over-parameterized fully-connected deep neural networks do not have any spurious local minima [145].

The reason for the benign optimization landscape for an over-parameterized neural network was analyzed by examining the geometry of the global minimum. Nguyen [123] discovered that the global minima are interconnected and concentrated on a unique valley if the neural networks are sufficiently over-parameterized. Similar results were obtained by Liu et al. [124]. In fact, they found that the set of solutions of an over-parameterized system is generically a manifold of positive dimensions, with the Hessian matrices of the loss function being positive semidefinite but not positive definite. Such a landscape is incompatible with convexity unless the set of solutions is a linear manifold. However, the linear manifold with zero curvature of the curve of global minima is unlikely to occur due to the essential non-convexity of the underlying optimization problem. Hence, gradient type algorithms can converge to any of the global minimum, although the exact point of the convergence depends on a specific optimization algorithm. This *implicit bias* of an optimization algorithm is another important theoretical topic in deep learning, which will be covered in a later chapter. In contrast, an under-parameterized landscape generally has several isolated local minima with a positive definite Hessian of the loss, the function being locally convex. This is illustrated in Fig. 11.3.

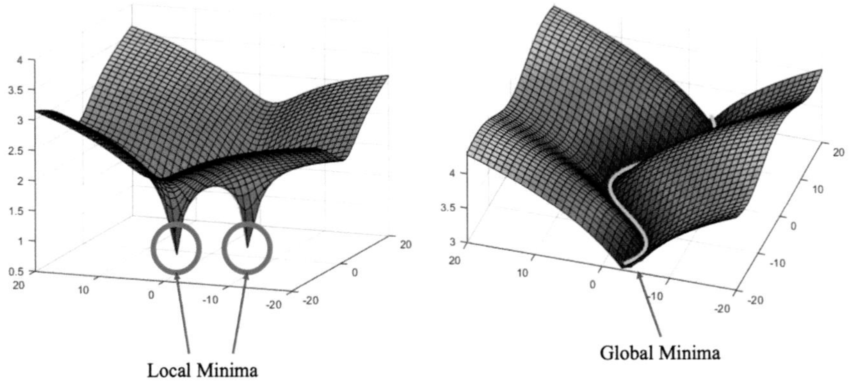

Fig. 11.3 Loss landscapes of (**a**) under-parameterized models and (**b**) over-parameterized models

11.4 Lyapunov-Type Convergence Analysis

Now let us introduce a different type of convergence analysis with a different mathematical flavor. In contrast to the methods discussed above, the analysis of the global loss landscape is not required here. Rather, a local loss geometry along the solution trajectory is the key to this analysis.

In fact, this type of convergence analysis is based on Lyapunov stability analysis [146] for the solution dynamics described by (11.8). Specifically, for a given nonlinear system,

$$\dot{\boldsymbol{\theta}}[t] = \boldsymbol{g}(\boldsymbol{\theta}[t]), \tag{11.18}$$

the Lyapunov stability analysis is concerned with checking whether the solution trajectory $\boldsymbol{\theta}[t]$ converges to zero as $t \to \infty$. To provide a general solution for this, we first define the Lyapunov function $V(z)$, which satisfies the following properties:

Definition 11.1 A function $V : \mathbb{R}^n \mapsto \mathbb{R}$ is positive definite (PD) if

- $V(z) \geq 0$ for all z.
- $V(z) = 0$ if and only if $z = \mathbf{0}$.
- All sublevel sets of V are bounded.

The Lyapunov function V has an analogy to the potential function of classical dynamics, and $-\dot{V}$ can be considered the associated generalized dissipation function. Furthermore, if we set $z := \boldsymbol{\theta}[t]$ to analyze the nonlinear dynamic system in (11.18), then $\dot{V} : z \in \mathbb{R}^n \mapsto \mathbb{R}$ is computed by

$$\dot{V}(z) = \left(\frac{\partial V}{\partial z}\right)^{\top} \dot{z} = \left(\frac{\partial V}{\partial z}\right)^{\top} \boldsymbol{g}(z). \tag{11.19}$$

The following Lyapunov global asymptotic stability theorem is one of the keys to the stability analysis of dynamic systems:

Theorem 11.2 (Lyapunov Global Asymptotic Stability [146]) *Suppose there is a function V such that 1) V is positive definite, and 2) $\dot{V}(z) < 0$ for all $z \neq \mathbf{0}$ and $\dot{V}(\mathbf{0}) = \mathbf{0}$. Then, every trajectory $\boldsymbol{\theta}[t]$ of $\dot{\boldsymbol{\theta}} = \mathbf{g}(\boldsymbol{\theta})$ converges to zero as $t \to \infty$. (i.e., the system is globally asymptotically stable).*

Example: 1-D Differential Equation
Consider the following ordinary differential equation:

$$\dot{\theta} = -\theta.$$

We can easily show that the system is globally asymptotically stable since the solution is $\theta[t] = C\exp(-t)$ for some constant C, and $\theta[t] \to 0$ as $t \to \infty$. Now, we want to prove this using Theorem 11.2 without ever solving the differential equation. First, choose a Lyapunov function

$$V(z) = \frac{z^2}{2},$$

where $z = \theta[t]$. We can easily show that $V(z)$ is positive definite. Furthermore, we have

$$\dot{V} = z\dot{z} = -(\theta[t])^2 < 0, \quad \forall\theta[t] \neq 0.$$

Therefore, using Theorem 11.2 we can show that $\theta[t]$ converges to zero as $t \to \infty$.

One of the beauties of Lyapunov stability analysis is that we do not need an explicit knowledge of the loss landscape to prove convergence. Instead, we just need to know the local dynamics along the solution path. To understand this claim, here we apply Lyapunov analysis to the convergence analysis of our gradient descent dynamics:

$$\dot{\boldsymbol{\theta}}[t] = -\frac{\partial \ell}{\partial \boldsymbol{\theta}}(\boldsymbol{\theta}[t]).$$

For the MSE loss, this leads to

$$\dot{\boldsymbol{\theta}}[t] = -\frac{\partial \boldsymbol{f}_{\boldsymbol{\theta}[t]}(\boldsymbol{x})}{\partial \boldsymbol{\theta}}\left(\boldsymbol{y} - \boldsymbol{f}_{\boldsymbol{\theta}[t]}(\boldsymbol{x})\right). \tag{11.20}$$

Now let

$$\boldsymbol{e}[t] := \boldsymbol{f}_{\boldsymbol{\theta}[t]}(\boldsymbol{x}) - \boldsymbol{y} ,$$

and consider the following positive definite Lyapunov function

$$V(z) = \frac{1}{2} z^\top z,$$

where $z = \boldsymbol{e}[t]$. Then, we have

$$\dot{V}(z) = \left(\frac{\partial V}{\partial z}\right)^\top \dot{z} = z^\top \dot{z}. \tag{11.21}$$

Using the chain rule, we have

$$\dot{z} = \dot{\boldsymbol{e}}[t] = \left(\frac{\partial \boldsymbol{f}}{\partial \boldsymbol{\theta}}\right)^\top \dot{\boldsymbol{\theta}}[t] = -\boldsymbol{K}_t \boldsymbol{e}[t],$$

where

$$\boldsymbol{K}_t = \boldsymbol{K}_{\boldsymbol{\theta}[t]} := \left(\frac{\partial \boldsymbol{f}_{\boldsymbol{\theta}}}{\partial \boldsymbol{\theta}}\right)^\top \frac{\partial \boldsymbol{f}_{\boldsymbol{\theta}}}{\partial \boldsymbol{\theta}}\bigg|_{\boldsymbol{\theta}=\boldsymbol{\theta}[t]} \tag{11.22}$$

is often called the *neural tangent kernel (NTK)* [130–132]. By plugging this into (11.21), we have

$$\dot{V} = -\eta \boldsymbol{e}[t]^\top \boldsymbol{K}_t \boldsymbol{e}[t]. \tag{11.23}$$

Accordingly, if the NTK is positive definite for all t, then $\dot{V}(z) < 0$. Therefore, $\boldsymbol{e}[t] \to \boldsymbol{0}$ so that $\boldsymbol{f}(\boldsymbol{\theta}[t]) \to \boldsymbol{y}$ as $t \to \infty$. This proves the convergence of gradient descent approach.

11.4.1 The Neural Tangent Kernel (NTK)

In the previous discussion we showed that the Lyapunov analysis only requires a positive-definiteness of the NTK along the solution trajectory. While this is a great advantage over PL-type analysis, which requires knowledge of the global loss landscape, the NTK is a function of time, so it is important to obtain the conditions for the positive-definiteness of NTK along the solution trajectory.

To understand this, here we are interested in deriving the explicit form of the NTK to understand the convergence behavior of the gradient descent methods.

Using the backpropagation in Chap. 6, we can obtain the weight update as follows:

$$\frac{\partial \boldsymbol{f}_{\boldsymbol{\theta}}}{\partial \text{VEC}(\boldsymbol{W}^{(l)})} = \frac{\partial \boldsymbol{g}_n^{(l)}}{\partial \text{VEC}(\boldsymbol{W}^{(l)})} \frac{\partial \boldsymbol{o}_n^{(l)}}{\partial \boldsymbol{g}_n^{(l)}} \frac{\partial \boldsymbol{g}_n^{(l+1)}}{\partial \boldsymbol{o}_n^{(l)}} \cdots \frac{\partial \boldsymbol{o}_n^{(L)}}{\partial \boldsymbol{g}_n^{(L)}}$$
$$= (\boldsymbol{o}^{(l)} \otimes \boldsymbol{I}_{d^{(l)}}) \boldsymbol{\Lambda}_n^{(l)} \boldsymbol{W}^{(l+1)\top} \boldsymbol{\Lambda}_n^{(l+1)} \boldsymbol{W}^{(l+2)\top} \cdots \boldsymbol{W}^{(L)\top} \boldsymbol{\Lambda}_n^{(L)}.$$

Similarly, we have

$$\frac{\partial \boldsymbol{f}_{\boldsymbol{\theta}}}{\partial \boldsymbol{b}^{(l)}} = \frac{\partial \boldsymbol{g}_n^{(l)}}{\partial \boldsymbol{b}^{(l)}} \frac{\partial \boldsymbol{o}_n^{(l)}}{\partial \boldsymbol{g}_n^{(l)}} \frac{\partial \boldsymbol{g}_n^{(l+1)}}{\partial \boldsymbol{o}_n^{(l)}} \cdots \frac{\partial \boldsymbol{o}_n^{(L)}}{\partial \boldsymbol{g}_n^{(L)}}$$
$$= \boldsymbol{\Lambda}_n^{(l)} \boldsymbol{W}^{(l+1)\top} \boldsymbol{\Lambda}_n^{(l+1)} \boldsymbol{W}^{(l+2)\top} \cdots \boldsymbol{W}^{(L)\top} \boldsymbol{\Lambda}_n^{(L)}.$$

Therefore, the NTK can be computed by

$$\boldsymbol{K}_t^{(L)} := \left(\frac{\partial \boldsymbol{f}_{\boldsymbol{\theta}}}{\partial \boldsymbol{\theta}}\right)^\top \frac{\partial \boldsymbol{f}_{\boldsymbol{\theta}}}{\partial \boldsymbol{\theta}}\Bigg|_{\boldsymbol{\theta}=\boldsymbol{\theta}[t]}$$
$$= \sum_{l=1}^{L} \left(\frac{\partial \boldsymbol{f}_{\boldsymbol{\theta}}}{\partial \text{VEC}(\boldsymbol{W}^{(l)})}\right)^\top \frac{\partial \boldsymbol{f}_{\boldsymbol{\theta}}}{\partial \text{VEC}(\boldsymbol{W}^{(l)})} + \left(\frac{\partial \boldsymbol{f}_{\boldsymbol{\theta}}}{\partial \boldsymbol{b}^{(l)}}\right)^\top \frac{\partial \boldsymbol{f}_{\boldsymbol{\theta}}}{\partial \boldsymbol{b}^{(l)}}$$
$$= \sum_{l=1}^{L} (\|\boldsymbol{o}^{(l)}[t]\|^2 + 1) \boldsymbol{M}^{(l)}[t],$$

where

$$\boldsymbol{M}^{(l)}[t] = \boldsymbol{\Lambda}^{(L)} \boldsymbol{W}^{(L)}[t] \cdots \boldsymbol{W}^{(l+1)}[t] \boldsymbol{\Lambda}^{(l)} \boldsymbol{\Lambda}^{(l)} \boldsymbol{W}^{(l+1)\top}[t] \cdots \boldsymbol{W}^{(L)\top}[t] \boldsymbol{\Lambda}^{(L)}. \tag{11.24}$$

Therefore, the positive definiteness of the NTK comes from the properties of $\boldsymbol{M}^{(l)}[t]$. In particular, if $\boldsymbol{M}^{(l)}[t]$ is positive definite for any l, the resulting NTK is positive definite. Moreover, the positive-definiteness of $\boldsymbol{M}^{(l)}[t]$ can be readily shown if the following sensitivity matrix is full row ranked:

$$\boldsymbol{S}^{(l)} := \boldsymbol{\Lambda}^{(L)} \boldsymbol{W}^{(L)}[t] \cdots \boldsymbol{W}^{(l+1)}[t] \boldsymbol{\Lambda}^{(l)}.$$

11.4.2 NTK at Infinite Width Limit

Although we derived the explicit form of the NTK using backpropagation, still the component matrix in (11.24) is difficult to analyze due to the stochastic nature of the weights and ReLU activation patterns.

To address this problem, the authors in [130] calculated the NTK at the infinite width limit and showed that it satisfies the positive definiteness. Specifically, they considered the following normalized form of the neural network update:

$$\boldsymbol{o}_n^{(0)} = \boldsymbol{x}, \tag{11.25}$$

$$\boldsymbol{g}^{(l)} = \frac{1}{\sqrt{d^{(l)}}} \boldsymbol{W}^{(l)} \boldsymbol{o}_n^{(l-1)} + \beta \boldsymbol{b}^{(l-1)}, \tag{11.26}$$

$$\boldsymbol{o}^{(l)} = \boldsymbol{\sigma}(\boldsymbol{g}^{(l)}), \tag{11.27}$$

for $l = 1, \cdots, L$, and $d^{(l)}$ denotes the width of the l-th layer. Furthermore, they considered what is sometimes called LeCun initialization, taking $W_{ij}^{(l)} \sim \mathcal{N}\left(0, \frac{1}{d^{(l)}}\right)$ and $b_j^{(l)} \sim \mathcal{N}(0, 1)$. Then, the following asymptotic form of the NTK can be obtained.

Theorem 11.3 (Jacot et al. [130]) *For a network of depth L at initialization, with a Lipschitz nonlinearity σ, and in the limit as the layers width $d^{(1)} \cdots, d^{(L-1)} \to \infty$, the neural tangent kernel $\boldsymbol{K}^{(L)}$ converges in probability to a deterministic limiting kernel:*

$$\boldsymbol{K}^{(L)} \to \kappa_\infty^{(L)} \otimes \boldsymbol{I}_{d_L}. \tag{11.28}$$

Here, the scalar kernel $\kappa_\infty^{(L)} : \mathbb{R}^{d^{(0)} \times d^{(0)}} \mapsto \mathbb{R}$ is defined recursively by

$$\kappa_\infty^{(1)}(\boldsymbol{x}, \boldsymbol{x}') = \frac{1}{d^{(0)}} \boldsymbol{x}^\top \boldsymbol{x}' + \beta^2, \tag{11.29}$$

$$\kappa_\infty^{(l+1)}(\boldsymbol{x}, \boldsymbol{x}') = \kappa_\infty^{(l)}(\boldsymbol{x}, \boldsymbol{x}')\dot{\nu}^{(l+1)}(\boldsymbol{x}, \boldsymbol{x}') + \nu^{(l+1)}(\boldsymbol{x}, \boldsymbol{x}'), \tag{11.30}$$

where

$$\nu^{(l+1)}(\boldsymbol{x}, \boldsymbol{x}') = E_g\left[\sigma(g(\boldsymbol{x}))\sigma(g(\boldsymbol{x}'))\right] + \beta^2, \tag{11.31}$$

$$\dot{\nu}^{(l+1)}(\boldsymbol{x}, \boldsymbol{x}') = E_g\left[\dot{\sigma}(g(\boldsymbol{x}))\dot{\sigma}(g(\boldsymbol{x}'))\right], \tag{11.32}$$

where the expectation is with respect to a centered Gaussian process g of covariance $\nu^{(l)}$, and where $\dot{\sigma}$ denotes the derivative of σ.

Note that the symptotic form of the NTK is positive definite since $\kappa_\infty^{(L)} > 0$. Therefore, the gradient descent using the infinite width NTK converges to the global minima. Again, we can clearly see the benefit of the over-parameterization in terms of large network width.

11.4.3 NTK for General Loss Function

Now, we are interested in extending the example above to the general loss function with multiple training data sets. For a given training data set $\{\boldsymbol{x}_n\}_{n=1}^N$, the gradient dynamics in (11.7) can be extended to

$$\dot{\boldsymbol{\theta}} = -\sum_{n=1}^{N} \frac{\partial \ell(\boldsymbol{f}_{\boldsymbol{\theta}}(\boldsymbol{x}_n))}{\partial \boldsymbol{\theta}} = -\sum_{n=1}^{N} \frac{\partial \boldsymbol{f}_{\boldsymbol{\theta}}(\boldsymbol{x}_n)}{\partial \boldsymbol{\theta}} \frac{\partial \ell(\boldsymbol{x}_n)}{\partial \boldsymbol{f}_{\boldsymbol{\theta}}(\boldsymbol{x}_n)},$$

where $\ell(\boldsymbol{x}_n) := \ell(\boldsymbol{f}(\boldsymbol{x}_n))$ with a slight abuse of notation. This leads to

$$\begin{aligned}
\dot{\boldsymbol{f}}_{\boldsymbol{\theta}}(\boldsymbol{x}_m) &= \left(\frac{\partial \boldsymbol{f}_{\boldsymbol{\theta}}(\boldsymbol{x}_m)}{\partial \boldsymbol{\theta}}\right)^\top \dot{\boldsymbol{\theta}} \\
&= -\sum_{n=1}^{N} \left(\frac{\partial \boldsymbol{f}_{\boldsymbol{\theta}}(\boldsymbol{x}_m)}{\partial \boldsymbol{\theta}}\right)^\top \frac{\partial \boldsymbol{f}_{\boldsymbol{\theta}}(\boldsymbol{x}_n)}{\partial \boldsymbol{\theta}} \frac{\partial \ell(\boldsymbol{x}_n)}{\partial \boldsymbol{f}_{\boldsymbol{\theta}}(\boldsymbol{x}_n)} \\
&= -\sum_{n=1}^{N} \boldsymbol{K}_t(\boldsymbol{x}_m, \boldsymbol{x}_n) \frac{\partial \ell(\boldsymbol{x}_n)}{\partial \boldsymbol{f}_{\boldsymbol{\theta}}(\boldsymbol{x}_n)},
\end{aligned}$$

where $\boldsymbol{K}_t(\boldsymbol{x}_m, \boldsymbol{x}_n)$ denotes the (m, n)-th block NTK defined by

$$\boldsymbol{K}_t(\boldsymbol{x}_m, \boldsymbol{x}_n) := \left(\frac{\partial \boldsymbol{f}_{\boldsymbol{\theta}}(\boldsymbol{x}_m)}{\partial \boldsymbol{\theta}}\right)^\top \frac{\partial \boldsymbol{f}_{\boldsymbol{\theta}}(\boldsymbol{x}_n)}{\partial \boldsymbol{\theta}}\Bigg|_{\boldsymbol{\theta}=\boldsymbol{\theta}[t]}.$$

Now, consider the following Lyapunov function candidate:

$$V(\boldsymbol{z}) = \sum_{m=1}^{N} \ell(\boldsymbol{f}_{\boldsymbol{\theta}}(\boldsymbol{x}_m)) = \sum_{m=1}^{N} \ell(\boldsymbol{z}_m + \boldsymbol{f}_m^*),$$

where

$$\boldsymbol{z} = \begin{bmatrix} \boldsymbol{z}_1 \\ \boldsymbol{z}_2 \\ \vdots \\ \boldsymbol{z}_N \end{bmatrix} = \begin{bmatrix} \boldsymbol{f}_{\boldsymbol{\theta}}(\boldsymbol{x}_1) - \boldsymbol{f}^*(\boldsymbol{x}_1) \\ \boldsymbol{f}_{\boldsymbol{\theta}}(\boldsymbol{x}_2) - \boldsymbol{f}^*(\boldsymbol{x}_2) \\ \vdots \\ \boldsymbol{f}_{\boldsymbol{\theta}}(\boldsymbol{x}_N) - \boldsymbol{f}^*(\boldsymbol{x}_N) \end{bmatrix},$$

and $\boldsymbol{f}^*(\boldsymbol{x}_m)$ refers to $\boldsymbol{f}_{\boldsymbol{\theta}^*}(\boldsymbol{x}_m)$ with $\boldsymbol{\theta}^*$ being the global minimizer. We further assume that the loss function satisfies the property that

$$\forall n, \quad \ell(\boldsymbol{f}_{\boldsymbol{\theta}}(\boldsymbol{x}_n)) > 0, \quad \text{if } \boldsymbol{f}_{\boldsymbol{\theta}}(\boldsymbol{x}_n) \neq \boldsymbol{f}_n^*, \quad \ell(\boldsymbol{f}_n*) = 0,$$

so that $V(z)$ is a positive definite function. Under this assumption, we have

$$\begin{aligned}
\dot{V}(z) &= \sum_{m=1}^{N} \left(\frac{\partial \ell(\boldsymbol{f}_{\boldsymbol{\theta}}(\boldsymbol{x}_m))}{\partial z_m} \right)^{\top} \dot{z}_m = \sum_{m=1}^{N} \left(\frac{\partial \ell(\boldsymbol{x}_m)}{\partial \boldsymbol{f}_{\boldsymbol{\theta}}(\boldsymbol{x}_m)} \right)^{\top} \dot{\boldsymbol{f}}_{\boldsymbol{\theta}}(\boldsymbol{x}_m) \Bigg|_{\boldsymbol{\theta}=\boldsymbol{\theta}[t]} \\
&= - \sum_{m=1}^{N} \sum_{n=1}^{N} \left(\frac{\partial \ell(\boldsymbol{f}_{\boldsymbol{\theta}}(\boldsymbol{x}_m))}{\partial \boldsymbol{f}_{\boldsymbol{\theta}}(\boldsymbol{x}_m)} \right)^{\top} \boldsymbol{K}_t(\boldsymbol{x}_m, \boldsymbol{x}_n) \frac{\partial \ell(\boldsymbol{f}_{\boldsymbol{\theta}}(\boldsymbol{x}_n))}{\partial \boldsymbol{f}_{\boldsymbol{\theta}}(\boldsymbol{x}_n)} \Bigg|_{\boldsymbol{\theta}=\boldsymbol{\theta}[t]} \\
&= -\boldsymbol{e}[t]^{\top} \mathcal{K}[t] \boldsymbol{e}[t],
\end{aligned}$$

where

$$\boldsymbol{e}[t] = \begin{bmatrix} \frac{\partial \ell(\boldsymbol{f}_{\boldsymbol{\theta}}(\boldsymbol{x}_1))}{\partial \boldsymbol{f}_{\boldsymbol{\theta}}(\boldsymbol{x}_1)} \\ \vdots \\ \frac{\partial \ell(\boldsymbol{f}_{\boldsymbol{\theta}}(\boldsymbol{x}_N))}{\partial \boldsymbol{f}_{\boldsymbol{\theta}}(\boldsymbol{x}_N)} \end{bmatrix}_{\boldsymbol{\theta}=\boldsymbol{\theta}[t]}, \quad \mathcal{K}[t] = \begin{bmatrix} \boldsymbol{K}_t(\boldsymbol{x}_1, \boldsymbol{x}_1) & \cdots & \boldsymbol{K}_t(\boldsymbol{x}_1, \boldsymbol{x}_N) \\ \vdots & \ddots & \vdots \\ \boldsymbol{K}_t(\boldsymbol{x}_N, \boldsymbol{x}_1) & \cdots & \boldsymbol{K}_t(\boldsymbol{x}_N, \boldsymbol{x}_N). \end{bmatrix}$$

Therefore, if the NTK $\mathcal{K}[t]$ is positive definite for all t, then Lyapunov stability theory guarantees that the gradient dynamics converge to the global minima.

11.5 Exercises

1. Show that a smooth $\ell(\boldsymbol{\theta})$ is invex if and only if every stationary point of $\ell(\boldsymbol{\theta})$ is a global minimum.
2. Show that a convex function is invex.
3. Let $a > 0$. Show that $V(x, y) = x^2 + 2y^2$ is a Lyapunov function for the system

$$\dot{x} = ay^2 - x \ , \ \dot{y} = -y - ax^2.$$

4. Show that $V(x, y) = \ln(1 + x^2) + y^2$ is a Lyapunov function for the system

$$\dot{x} = x(y - 1) \ , \ \dot{y} = -\frac{x^2}{1 + x^2}.$$

5. Consider a two-layer fully connected network $f_{\Theta} : \mathbb{R}^2 \to \mathbb{R}^2$ with ReLU nonlinearity, as shown in Fig. 10.10.

(a) Suppose the weight matrices and biases are given by

$$\boldsymbol{W}^{(0)} = \begin{bmatrix} 2 & -1 \\ 1 & 1 \end{bmatrix}, \quad \boldsymbol{b}^{(0)} = \begin{bmatrix} 1 \\ -1 \end{bmatrix}$$

$$\boldsymbol{W}^{(1)} = \begin{bmatrix} 1 & 2 \\ -1 & 1 \end{bmatrix}, \quad \boldsymbol{b}^{(1)} = \begin{bmatrix} -9 \\ -2 \end{bmatrix}.$$

Given the corresponding input space partition in Fig. 10.11, compute the neural tangent kernel for each partition. Are they positive definite?

(b) In problem (a), suppose that the second layer weight and bias are changed to

$$\boldsymbol{W}^{(1)} = \begin{bmatrix} 1 & 2 \\ 0 & 1 \end{bmatrix}, \quad \boldsymbol{b}^{(1)} = \begin{bmatrix} 0 \\ 1 \end{bmatrix}.$$

Given the corresponding input space partition, compute the neural tangent kernel for each partition. Are they positive definite?

Chapter 12
Generalization Capability of Deep Learning

12.1 Introduction

One of the main reasons for the enormous success of deep neural networks is their amazing ability to generalize, which seems mysterious from the perspective of classic machine learning. In particular, the number of trainable parameters in deep neural networks is often greater than the training data set, this situation being notorious for overfitting from the point of view of classical statistical learning theory. However, empirical results have shown that a deep neural network generalizes well at the test phase, resulting in high performance for the unseen data.

This apparent contradiction has raised questions about the mathematical foundations of machine learning and their relevance to practitioners. A number of theoretical papers have been published to understand the intriguing generalization phenomenon in deep learning models [147–153]. The simplest approach to studying generalization in deep learning is to prove a generalization bound, which is typically an upper limit for test error. A key component in these generalization bounds is the notion of complexity measure: a quantity that monotonically relates to some aspect of generalization. Unfortunately, it is difficult to find tight bounds for a deep neural network that can explain the fascinating ability to generalize.

Recently, the authors in [154, 155] have delivered groundbreaking work that can reconcile classical understanding and modern practice in a unified framework. The so-called "double descent" curve extends the classical U-shaped bias-variance trade-off curve by showing that increasing the model capacity beyond the interpolation point leads to improved performance in the test phase. Particularly, the induced bias by optimization algorithms such as the stochastic gradient descent (SGD) offers simpler solutions that improve generalization in the over-parameterized regime. This relationship between the algorithms and structure of machine learning models describes the limits of classical analysis and has implications for the theory and practice of machine learning.

J. C. Ye, *Geometry of Deep Learning*, Mathematics in Industry 37,
https://doi.org/10.1007/978-981-16-6046-7_12

This chapter also presents new results showing that a generalization bound based on the robustness of the algorithm can be a promising tool to understand the generalization ability of the ReLU network. In particular, we claim that it can potentially offer a tight generalization bound that depends on the piecewise linear nature of the deep neural network and the inductive bias of the optimization algorithms.

12.2 Mathematical Preliminaries

Let Q be an arbitrary distribution over $z := (\boldsymbol{x}, \boldsymbol{y})$, where $\boldsymbol{x} \in \mathcal{X}$ and $\boldsymbol{y} \in \mathcal{Y}$ denote the input and output of the learning algorithm, and $\mathcal{Z} := \mathcal{X} \times \mathcal{Y}$ refer to the sample space. Let $\mathcal{F}$ be a hypothesis class and let $\ell(\boldsymbol{f}, z)$ be a loss function. For the case of regression with MSE loss, the loss can be defined as

$$\ell(\boldsymbol{f}, z) = \frac{1}{2}\|\boldsymbol{y} - \boldsymbol{f}(\boldsymbol{x})\|^2.$$

Over the choice of an i.i.d. training set $\mathcal{S} := \{z_n\}_{n=1}^N$, which is sampled according to Q, an algorithm $\mathcal{A}$ returns the estimated hypothesis

$$\boldsymbol{f}_{\mathcal{S}} = \mathcal{A}(\mathcal{S}). \tag{12.1}$$

For example, the estimated hypothesis from the popular empirical risk minimization (ERM) principle [10] is given by

$$\boldsymbol{f}_{ERM} = \arg\min_{\boldsymbol{f}\in\mathcal{F}} \hat{R}_N(\boldsymbol{f}), \tag{12.2}$$

where the empirical risk $\hat{R}_N(\boldsymbol{f})$ is defined by

$$\hat{R}_N(\boldsymbol{f}) := \frac{1}{N}\sum_{n=1}^{N} \ell(\boldsymbol{f}, z_n), \tag{12.3}$$

which is assumed to uniformly converge to the population (or expected) risk defined by:

$$R(\boldsymbol{f}) = \mathbb{E}_{z\sim Q}\ell(\boldsymbol{f}, z). \tag{12.4}$$

If uniform convergence holds, then the empirical risk minimizer (ERM) is consistent, that is, the population risk of the ERM converges to the optimal population risk, and the problem is said to be learnable using the ERM [10].

In fact, learning algorithms that satisfy such performance guarantees are called the probably approximately correct (PAC) learning [156]. Formally, PAC learnability is defined as follows.

Definition 12.1 (PAC Learnability [156]) A concept class C is PAC learnable if there exist some algorithm $\mathcal{A}$ and a polynomial function $poly(\cdot)$ such that the following holds. Pick any target concept $\boldsymbol{c} \in C$. Pick any input distribution $\mathcal{P}$ over $\mathcal{X}$. Pick any $\epsilon, \delta \in [0, 1]$. Define $\mathcal{S} := \{\boldsymbol{x}_n, \boldsymbol{c}(\boldsymbol{x}_n)\}_{n=1}^N$ where $\boldsymbol{x}_n \sim \mathcal{P}$ are i.i.d samples. Given $N \geq poly(1/\epsilon, 1/\delta, \dim(\mathcal{X}), \text{size}(\boldsymbol{c}))$, where $\dim(\mathcal{X})$, $\text{size}(\boldsymbol{c})$ denote the computational costs of representing inputs $\boldsymbol{x} \in \mathcal{X}$ and target $\boldsymbol{c}$, the generalization error is bounded as

$$\mathbb{P}_{\boldsymbol{x}\sim Q}\{\mathcal{A}_{\mathcal{S}}(\boldsymbol{x}) \neq \boldsymbol{c}(\boldsymbol{x})\} \leq \epsilon, \tag{12.5}$$

where $\boldsymbol{A}_{\mathcal{S}}$ denotes the learned hypothesis by the algorithm $\mathcal{A}$ using the training data $\mathcal{S}$.

The PAC learnability is closely related to the generalization bounds. More specifically, the ERM could only be considered a solution to a machine learning problem or PAC-learnable if the difference between the training error and the generalization error, called the generalization gap, is small enough. This implies that the following probability should be sufficiently small:

$$\mathbb{P}\left\{\sup_{f\in\mathcal{F}} |R(\boldsymbol{f}) - \hat{R}_N(\boldsymbol{f})| > \epsilon\right\}. \tag{12.6}$$

Note that this is the worst-case probability, so even in the worst-case scenario, we try to minimize the difference between the empirical risk and the expected risk.

A standard trick to bound the probability in (12.6) is based on concentration inequalities. For example, Hoeffding's inequality is useful.

Theorem 12.1 (Hoeffding's Inequality [157]) *If $x_1, x_2, \cdots, x_N$ are N i.i.d. samples of a random variable X distributed by $\mathcal{P}$, and $a \leq x_n \leq b$ for every n, then for a small positive nonzero value ϵ:*

$$\mathbb{P}\left\{\left|\mathbb{E}[X] - \frac{1}{N}\sum_{n=1}^{N} x_n\right| > \epsilon\right\} \leq 2\exp\left(\frac{-2N\epsilon^2}{(b-a)^2}\right). \tag{12.7}$$

Assuming that our loss is bounded between 0 and 1 using a 0/1 loss function or by squashing any other loss between 0 and 1, (12.6) can be bounded as follows using Hoeffding's inequality:

$$\begin{aligned}\mathbb{P}\left\{\sup_{f\in\mathcal{F}} |R(f) - \hat{R}_N(f)| > \epsilon\right\} &= \mathbb{P}\left\{\bigcup_{f\in\mathcal{F}} |R(f) - \hat{R}_N(f)| > \epsilon\right\} \\ &\overset{(a)}{\leq} \sum_{f\in\mathcal{F}} \mathbb{P}\left\{|R(f) - \hat{R}_N(f)| > \epsilon\right\} \\ &= 2|\mathcal{F}|\exp(-2N\epsilon^2),\end{aligned} \tag{12.8}$$

where $|\mathcal{F}|$ is the size of the hypothesis space and we use the union bound in (a) to obtain the inequality. By denoting the right hand side of the above inequality by δ, we can say that with probability at least $1 - \delta$, we have

$$R(f) \leq \hat{R}_N(f) + \sqrt{\frac{\ln|\mathcal{F}| + \ln\frac{2}{\delta}}{2N}}. \tag{12.9}$$

Indeed, (12.9) is one of the simplest forms of the generalization bound, but still reveals the fundamental bias–variance trade-off in classical statistical learning theory. For example, the ERM for a given function class $\mathcal{F}$ results in the minimum empirical loss:

$$\hat{R}_N(f_{ERM}) = \min_{f\in\mathcal{F}} \hat{R}_N(f), \tag{12.10}$$

which goes to zero as the hypothesis class $\mathcal{F}$ becomes bigger. On the other hand, the second term in (12.9) grows with increasing $|\mathcal{F}|$. This trade-off in the generalization bound with respect to the hypothesis class size $|\mathcal{F}|$ is illustrated in Fig. 12.1.

Although the expression in (12.9) looks very nice, it turns out that the bound is very loose. This is due to the term $|\mathcal{F}|$ which originates from the union bound of all elements in the hypothesis class $\mathcal{F}$. In the following, we discuss some representative classical approaches to obtain tighter generation bounds.

12.2.1 Vapnik–Chervonenkis (VC) Bounds

One of the key ideas of the work of Vapnik and Chervonenkis [10] is to replace the union bound for all hypothesis class in (12.8) with the union bound of simpler empirical distributions. This idea is historically important, so we will review it here.

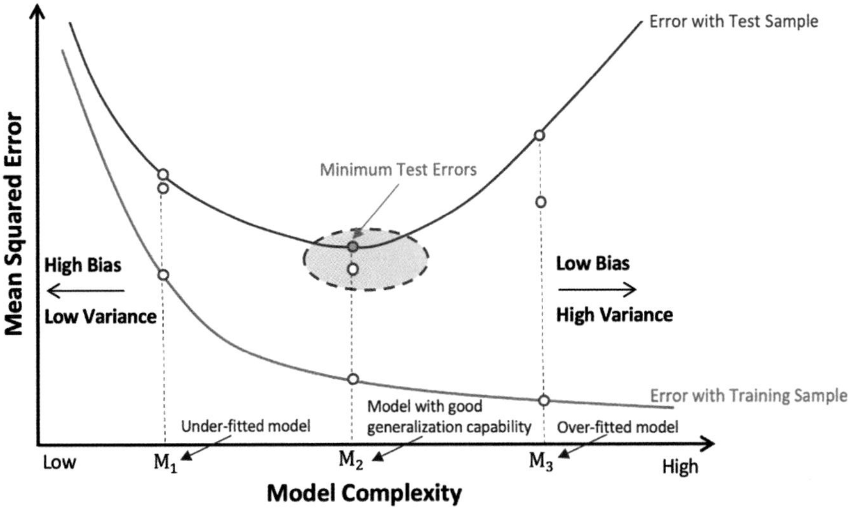

Fig. 12.1 Generation bound behavior according to the hypothesis class size $|\mathcal{F}|$

More specifically, consider independent samples $z'_n := (\boldsymbol{x}'_n, \boldsymbol{y}'_n)$ for $n = 1, \cdots, N$, which are often called "ghost" samples. The associated empirical risk is given by

$$\hat{R}'_N(f) = \frac{1}{N}\sum_{n=1}^{N} \ell\left(f, z'_n\right). \tag{12.11}$$

Then, we have the following symmetrization lemma.

Lemma 12.1 (Symmetrization[10]) *For a given sample set* $\mathcal{S} := \{\boldsymbol{x}_n, \boldsymbol{y}_n\}_{n=1}^N$ *and its ghost samples set* $\mathcal{S}' := \{\boldsymbol{x}'_n, \boldsymbol{y}'_n\}_{n=1}^N$ *from a distribution* Q *and for any* $\epsilon > 0$ *such that* $\epsilon \geq \sqrt{2/N}$, *we have*

$$\mathbb{P}\left\{\sup_{f\in\mathcal{F}} |R(f) - \hat{R}_N(f)| > \epsilon\right\} \leq 2\mathbb{P}\left\{\sup_{f\in\mathcal{F}} |\hat{R}'_N(f) - \hat{R}_N(f)| > \frac{\epsilon}{2}\right\}. \tag{12.12}$$

Vapnik and Chervonenkis [10] used the symmetrization lemma to obtain a much tighter generalization bound:

$$\begin{aligned}\mathbb{P}\left\{\sup_{f\in\mathcal{F}} |R(f) - \hat{R}_N(f)| > \epsilon\right\} &\leq 2\mathbb{P}\left\{\sup_{f\in\mathcal{F}_{\mathcal{S},\mathcal{S}'}} |\hat{R}'_N(f) - \hat{R}_N(f)| > \frac{\epsilon}{2}\right\} \\ &= 2\mathbb{P}\left\{\bigcup_{f\in\mathcal{F}_{\mathcal{S},\mathcal{S}'}} |\hat{R}'_N(f) - \hat{R}_N(f)| > \epsilon\right\}\end{aligned}$$

$$\leq 2G_{\mathcal{F}}(2N) \cdot \mathbb{P}\left\{|\hat{R}'_N(f) - \hat{R}_N(f)| > \epsilon\right\}$$
$$\leq 2G_{\mathcal{F}}(2N) \exp(-N\epsilon^2/8),$$

where the last inequality is obtained by Hoeffding's inequality and $\mathcal{F}_{\mathcal{S},\mathcal{S}'}$ denotes the restriction of the hypothesis class to the empirical distribution for $\mathcal{S}, \mathcal{S}'$. Here, $G_{\mathcal{F}}(\cdot)$ is called the *growth function* defined by

$$G_{\mathcal{F}}(2N) := |\mathcal{F}_{\mathcal{S},\mathcal{S}'}|, \tag{12.13}$$

which represents the number of the most possible sets of dichotomies using the hypothesis class $\mathcal{F}$ on any $2N$ points from $\mathcal{S}$ and $\mathcal{S}'$.

The discovery of the growth function is one of the important contributions of Vapnik and Chervonenkis [10]. This is closely related to the concept of *shattering*, which is formally defined as follows.

Definition 12.2 (Shattering) We say $\mathcal{F}$ shatters $\mathcal{S}$ if $|\mathcal{F}| = 2^{|\mathcal{S}|}$.

In fact, the growth function $G_{\mathcal{F}}(N)$ is often called the *shattering number*: the number of the most possible sets of dichotomies using the hypothesis class $\mathcal{F}$ on any N points. Below, we show several facts for the growth function:

- By definition, the shattering number satisfies $G_{\mathcal{F}}(N) \leq 2^N$.
- When $\mathcal{F}$ is finite, we always have $G_{\mathcal{F}}(N) = |\mathcal{F}|$.
- If $G_{\mathcal{F}}(N) = 2^N$, then there is a set of N points such that the class of functions $\mathcal{F}$ can generate any possible classification result on these points. Figure 12.2 shows such a case where $\mathcal{F}$ is the class of linear classifiers.

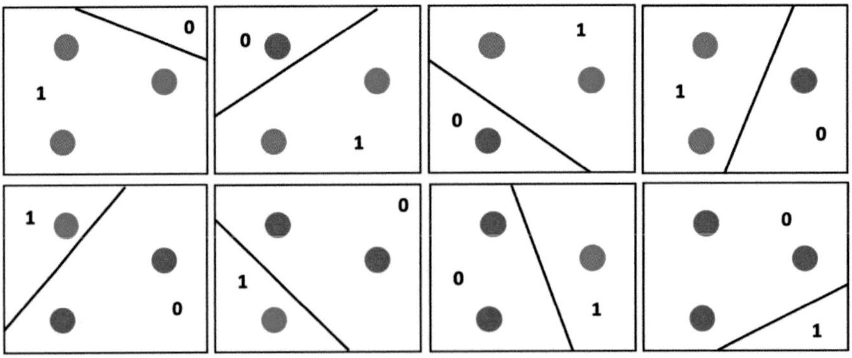

Fig. 12.2 Most possible sets of dichotomies using linear classifier on any three points. The resulting shattering number is $G_{\mathcal{F}}(3) = 8$

Accordingly, we arrive at the following classical VC bound [10]:

Theorem 12.2 (VC Bound) *For any $\delta > 0$, with probability at least $1-\delta$, we have*

$$R(f) \leq \hat{R}_N(f) + \sqrt{\frac{8 \ln G_{\mathcal{F}}(2N) + 8 \ln \frac{2}{\delta}}{N}}. \tag{12.14}$$

Another important contribution of the work by Vapnik and Chervonenkis [10] is that the growth function can be bounded by the so-called *VC dimension*, and the number of data points for which we cannot get all possible dichotomies (=VC dimension +1) is called the *break point*.

Definition 12.3 (VC Dimension) The VC dimension of a hypothesis class $\mathcal{F}$ is the largest $N = d_{VC}(\mathcal{F})$ such that

$$G_{\mathcal{F}}(N) = 2^N.$$

In other words, the VC dimension of a function class $\mathcal{F}$ is the cardinality of the largest set that it can shatter.

This means that the VC dimension is a measure of the capacity (complexity, expressiveness, richness, or flexibility) of a set of functions that can be learned from a statistical binary classification algorithm. It is defined as the cardinality of the largest number of points that the algorithm can classify with zero training error. In the following, we show several examples where we can explicitly calculate the VC dimensions.

Example: Half-Sided Interval
Consider any function of the form $\mathcal{F} = \{f(x) = \chi(x \leq \theta), \theta \in \mathbb{R}\}$. It can shatter two points, but any three points cannot be shattered. Therefore, $d_{VC}(\mathcal{F}) = 2$.

Example: Half Plane
Consider a hypothesis class $\mathcal{F}$ composed of half planes in $\mathbb{R}^d$. It can shatter $d+1$ points, but any $d+2$ points cannot be shattered. Therefore, $d_{VC}(\mathcal{F}) = d+1$.

Example: Sinusoids
f is a single-parametric sine classifier, i.e, for a certain parameter θ, the classifier f_θ returns 1 if the input number x is larger than $\sin(\theta x)$ and 0 otherwise. The VC dimension of f is infinite, since it can shatter any finite subset of the set $\{2^{-m} \mid m \in \mathbb{N}\}$.

Finally, we can derive the generalization bound using the VC dimension. For this, the following lemma by Sauer is the key element.

Lemma 12.2 (Sauer's Lemma[158]) *Suppose that $\mathcal{F}$ has a finite VC dimension d_{VC}. Then*

$$G_{\mathcal{F}}(n) \le \sum_{i=1}^{d_{VC}} \binom{n}{i} \tag{12.15}$$

and for all $n \ge d_{VC}$,

$$G_{\mathcal{F}}(n) \le \left(\frac{en}{d_{VC}}\right)^{d_{VC}}. \tag{12.16}$$

Corollary 12.1 (VC Bound Using VC Dimension) *Let $d_{VC} \ge N$. Then, for any $\delta > 0$, with probability at least $1 - \delta$, we have*

$$R(f) \le \hat{R}_N(f) + \sqrt{\frac{8 d_{VC} \ln \frac{2eN}{d_{VC}} + 8 \ln \frac{2}{\delta}}{N}}. \tag{12.17}$$

Proof This is a direct consequence of Theorem 12.2 and Lemma 12.2. □

The VC dimension has been studied for deep neural networks to understand their generalization behaviors [159]. Bartlett et al. [160] proves bounds on the VC dimension of piece-wise linear networks with potential weight sharing. Although this measure could be predictive when the architecture changes, which happens only in depth and width hyperparameter types, the authors in [159] also found that it is negatively correlated with the generalization gap, which contradicts the widely known empirical observation that over-parametrization improves generalization in deep learning [159].

12.2.2 Rademacher Complexity Bounds

Another important classical approach for the generalization error bound is Rademacher complexity [161]. To understand this concept, consider the following toy example. Let $\mathcal{S} := \{(\boldsymbol{x}_n, y_n)\}_{n=1}^N$ denote the training sample set, where $y_n \in \{-1, 1\}$. Then, the training error can be computed by

$$err_N(f) = \frac{1}{N}\sum_{n=1}^{N} \mathbf{1}\left[f(\boldsymbol{x}_n) \neq y_n\right], \tag{12.18}$$

where $\mathbf{1}[\cdot]$ is an indicator function computed by

$$\mathbf{1}\left[f(\boldsymbol{x}_n) \neq y_n\right] = \begin{cases} 1, & \{f(\boldsymbol{x}_n), y_n\} = \{1, -1\}, \{-1, 1\} \\ 0, & \{f(\boldsymbol{x}_n), y_n\} = \{1, 1\}, \{-1, -1\} \end{cases}. \tag{12.19}$$

Then, (12.18) can be equivalently represented by

$$\begin{aligned} err_N(f) &= \frac{1}{N}\sum_{n=1}^{N} \frac{1 - y_n f(\boldsymbol{x}_n)}{2} \\ &= \frac{1}{2} - \underbrace{\frac{1}{N}\sum_{i=1}^{N} y_n f(\boldsymbol{x}_n)}_{\text{correlation}}. \end{aligned} \tag{12.20}$$

Therefore, minimizing the training error is equivalent to maximizing the correlation. Now, the core idea of the Rademacher complexity is to consider a game where a player generates random targets $\{y_n\}_{n=1}^N$ and another player provides the hypothesis that maximize the correlation:

$$\sup_{f \in \mathcal{F}} \frac{1}{N}\sum_{n=1}^{N} y_n f(\boldsymbol{x}_n). \tag{12.21}$$

Note that the idea is closely related to the shattering in VC analysis. Specifically, if the hypothesis class $\mathcal{F}$ shatters $\boldsymbol{S} = \{\boldsymbol{x}_n, y_n\}_{n=1}^N$, then the correlation becomes a maximum. However, in contrast to the VC analysis that considers the worst-case scenario, Rademacher complexity analysis deals with average-case analysis. Formally, we define the so-called Rademacher complexity [161].

Definition 12.4 (Rademacher Complexity[161]) Let $\sigma_1 \cdots, \sigma_N$ be independent random variables $\mathbb{P}\{\sigma_n = 1\} = \mathbb{P}\{\sigma_= - 1\} = \frac{1}{2}$. Then, the empirical Rademacher complexity of $\mathcal{F}$ is defined by

$$Rad_N(\mathcal{F}, \mathcal{S}) = \mathbb{E}_{\boldsymbol{\sigma}} \left[\sup_{f \in \mathcal{F}} \frac{1}{N} \sum_{n=1}^{N} \sigma_n f(\boldsymbol{x}_n) \right], \tag{12.22}$$

where $\boldsymbol{\sigma} = [\sigma_1, \cdots, \sigma_N]^\top$. In addition, the general notion of Rademacher complexity is computed by

$$Rad_N(\mathcal{F}) := \mathbb{E}_{\mathcal{S}} \left[Rad_N(\mathcal{F}, \mathcal{S}) \right]. \tag{12.23}$$

Another important advantage of Rademacher complexity is that it can be easily generalized to the regression problem for the vector target. For example, (12.23) can be generalized as follows:

$$Rad_N(\mathcal{F}) = \mathbb{E} \left[\sup_{f \in \mathcal{F}} \frac{1}{N} \sum_{n=1}^{N} \langle \boldsymbol{\sigma}_n, \boldsymbol{f}(\boldsymbol{x}_n) \rangle \right], \tag{12.24}$$

where $\{\boldsymbol{\sigma}_n\}_{n=1}^N$ refers to the independent random vectors. In the following, we provide some examples where the Rademacher complexity can be explicitly calculated.

Example: Minimum Rademacher Complexity
When the hypothesis class has one element, i.e. $|\mathcal{F}| = 1$, we have

$$Rad(\mathcal{F}) = \mathbb{E} \left[\sup_{f \in \mathcal{F}} \frac{1}{N} \sum_{n=1}^{N} \sigma_n f(\boldsymbol{x}_n) \right] = f(\boldsymbol{x}_1) \cdot \mathbb{E} \left[\frac{1}{N} \sum_{n=1}^{N} \sigma_n \right] = 0,$$

where the second equality comes from the fact that $f(\boldsymbol{x}_n) = f(\boldsymbol{x}_1)$ for all n when $|\mathcal{F}| = 1$. The final equation comes from the definition of the random variable σ_n.

Example: Maximum Rademacher Complexity
When $|\mathcal{F}| = 2^N$, we have

$$Rad(\mathcal{F}) = \mathbb{E} \left[\sup_{f \in \mathcal{F}} \frac{1}{N} \sum_{n=1}^{N} \sigma_n f(\boldsymbol{x}_n) \right] = \mathbb{E} \left[\frac{1}{N} \sum_{n=1}^{N} \sigma_n^2 \right] = 1,$$

(continued)

where the second equality comes from the fact that we can find a hypothesis such that $f(\boldsymbol{x}_n) = \sigma_n$ for all n. The final equation comes from the definition of the random variable σ_n.

Although the Rademacher complexity was originally derived above for the binary classifiers, it can also be used to evaluate the complexity of the regression. The following example shows that a closed form Rademacher complexity can be obtained for ridge regression.

Example: Ridge Regression
Let $\mathcal{F}$ be the class of linear predictors given by $y = \boldsymbol{w}^\top \boldsymbol{x}$ with the restriction of $\|\boldsymbol{w}\| \le W$ and $\|\boldsymbol{x}\| \le X$. Then, we have

$$\begin{aligned}
Rad(\mathcal{F}, \mathcal{S}) &= \mathbb{E}_\sigma \left[\sup_{\boldsymbol{w}:\|\boldsymbol{w}\|\le W} \frac{1}{N} \sum_{n=1}^{N} \sigma_n \boldsymbol{w}^\top \boldsymbol{x}_n \right] \\
&= \frac{1}{N} \mathbb{E}_\sigma \left[\sup_{\boldsymbol{w}:\|\boldsymbol{w}\|\le W} \boldsymbol{w}^\top \left(\sum_{n=1}^{N} \sigma_n \boldsymbol{x}_n \right) \right] \\
&\overset{(a)}{=} \frac{W}{N} \mathbb{E}_\sigma \left\| \sum_{n=1}^{N} \sigma_n \boldsymbol{x}_n \right\| \overset{(b)}{\le} \frac{W}{N} \sqrt{\sum_{n=1}^{N} \mathbb{E}_\sigma \|\sigma_n \boldsymbol{x}_n\|^2} \\
&= \frac{W}{N} \sqrt{\sum_{n=1}^{N} \|\boldsymbol{x}_n\|^2} \le \frac{WX}{\sqrt{N}},
\end{aligned}$$

where (a) comes from the definition of the l_1 norm, and (b) comes from Jensen's inequality.

Using the Rademacher complexity, we can now derive a new type of generalization bound. First, we need the following concentration inequality.

Lemma 12.3 (McDiarmid's Inequality[161]) *Let $x_1, \cdots, x_N$ be independent random variables taking on values in a set $\mathcal{X}$ and let $c_1, \cdots, c_n$ be positive real constants. If $\varphi : \mathcal{X}^N \mapsto \mathbb{R}$ satisfies*

$$\sup_{x_1,\cdots,x_N,x_n' \in \mathcal{A}} |\varphi(x_1, \cdots, x_n, \cdots, x_N) - \varphi(x_1, \cdots, x_n', \cdots, x_N)| \le c_n,$$

for $1 \leq n \leq N$, *then*

$$\mathbb{P}\{|\varphi(x_1, \cdots, x_N) - \mathbb{E}\varphi(x_1, \cdots, x_N)| \geq \epsilon\} \leq 2\exp\left(-\frac{2\epsilon^2}{\sum_{n=1}^{N} c_n^2}\right). \tag{12.25}$$

In particular, if $\varphi(x_1, \cdots, x_N) = \sum_{n=1}^{N} x_n/N$, *the inequality* (12.25) *reduces to Hoeffding's inequality.*

Using McDiarmid's inequality and symmetrization using "ghost samples", we can obtain the following generalization bound.

Theorem 12.3 (Rademacher Bound) *Let* $\mathcal{S} := \{\boldsymbol{x}_n, y_n\}_{n=1}^{N}$ *denote the training set and* $f(\boldsymbol{x}) \in [a, b]$. *For any* $\delta > 0$, *with probability at least* $1 - \delta$, *we have*

$$R(f) \leq \hat{R}_N(f) + 2Rad_N(\mathcal{F}) + (b-a)\sqrt{\frac{\ln 1/\delta}{2N}}, \tag{12.26}$$

and

$$R(f) \leq \hat{R}_N(f) + 2Rad_N(\mathcal{F}, \mathcal{S}) + 3(b-a)\sqrt{\frac{\ln 2/\delta}{2N}}. \tag{12.27}$$

Unfortunately, many theoretical efforts using the Rademacher complexity to understand the deep neural network were not successful [159], which often resulted in a vacuous bound similar to the attempts using VC bounds. Therefore, the need to obtain a tighter bound has been increasing.

12.2.3 PAC–Bayes Bounds

So far, we have discussed performance guarantees which hold whenever the training and test data are drawn independently from an identical distribution. In fact, learning algorithms that satisfy such performance guarantees are called the probably approximately correct (PAC) learning [156]. It was shown that the concept class C is PAC learnable if and only if the VC dimension of C is finite [162].

In addition to PAC learning, there is another important area of modern learning theory—Bayesian inference. Bayesian inferences apply whenever the training and test data are generated according to the specified prior. However, there is no guarantee of an experimental environment in which training and test data are generated according to a different probability distribution than the previous one. In fact, much of modern learning theory can be broken down into Bayesian inference and PAC learning. Both areas investigate learning algorithms that use training data as the input and generate a concept or model as the output, which can then be tested on test data.

The difference between the two approaches can be seen as a trade-off between generality and performance. We define an "experimental setting" as a probability distribution over training and test data. A PAC performance guarantee applies to a wide class of experimental settings. A Bayesian correctness theorem applies only to experimental settings that match those previously used in the algorithm. In this restricted class of settings, however, the Bayesian learning algorithm can be optimal and generally outperforms the PAC learning algorithms.

The PAC–Bayesian theory combines Bayesian and frequentist approaches [163]. The PAC–Bayesian theory is based on a prior probability distribution concerning the "situation" occurring in nature, and a "rule" expresses a learner's preference for some rules over others. There is no supposed relationship between the learner's bias for rules and the nature distribution. This differs from the Bayesian inference, where the starting point is a common distribution of rules and situations, which induces a conditional distribution of rules in certain situations.

Under this set-up, the following PAC–Bayes generalization bound can be obtained.

Theorem 12.4 (PAC–Bayes Generalization Bound) *[163] Let $\mathcal{Q}$ be an arbitrary distribution over $z := (\boldsymbol{x}, \boldsymbol{y}) \in \mathcal{Z} := \mathcal{X} \times \mathcal{Y}$. Let $\mathcal{F}$ be a hypothesis class and let ℓ be a loss function such that for all $\boldsymbol{f}$ and z we have $\ell(\boldsymbol{f}, z) \in [0, 1]$. Let $\mathcal{P}$ be a prior distribution over $\mathcal{F}$ and let $\delta \in (0, 1)$. Then, with probability of at least $1 - \delta$ over the choice of an i.i.d. training set $\mathcal{S} := \{z_n\}_{n=1}^N$ sampled according to $\mathcal{Q}$, for all distributions Q over $\mathcal{F}$ (even such that depend on $\mathcal{S}$), we have*

$$\mathbb{E}_{f\sim Q}[R(f)] \leq \mathbb{E}_{f\sim Q}\left[\hat{R}_N(f)\right] + \sqrt{\frac{KL(Q||\mathcal{P}) + \ln N/\delta}{2(N-1)}}, \tag{12.28}$$

where

$$KL(Q||\mathcal{P}) := \mathbb{E}_{f\sim Q}[\ln Q(f)/\mathcal{P}(f)] \tag{12.29}$$

is the Kullback–Leibler divergence.

Recently, PAC–Bayes approaches have been studied extensively to explain the generalization capability of neural networks [149, 153, 164]. According to a recent large scale experiment to test the correlation of different measures with the generalization of deep models [159], the authors confirmed the effectiveness of the PAC–Bayesian bounds and corroborate them as a promising direction for cracking the generalization puzzle. Another nice application of PAC–Bayes bounds is that it provides a mean to find the optimal distribution Q^* by minimizing the upper bounds. This technique has been successfully used for the linear classifier design [164], etc.

12.3 Reconciling the Generalization Gap via Double Descent Model

Recall that the following error bound can be obtained for the ERM estimate in (12.2):

$$R(f^*_{ERM}) \leq \underbrace{\hat{R}_N(f^*_{ERM})}_{\text{empirical risk (training error)}} + \underbrace{\mathcal{O}\left(\sqrt{\frac{c}{N}}\right)}_{\text{complexity penalty}} , \tag{12.30}$$

where $\mathcal{O}(\cdot)$ denotes the "big O" notation and c refers to the model complexity such as VC dimension, Rademacher complexity, etc.

In (12.30), with increasing hypothesis class size $|\mathcal{F}|$, the empirical risk or training error decreases, whereas the complexity penalty increases. The control of the functional class capacity can be therefore done explicitly by choosing $\mathcal{F}$ (e.g. selection of the neural network architecture). This is summarized in the classic U-shaped risk curve, which is shown in Fig. 12.3a and was often used as a guide for model selection. A widely accepted view from this curve is that a model with zero training error is overfitted to the training data and will typically generalize poorly [10]. Classical thinking therefore deals with the search for the "sweet spot" between underfitting and overfitting.

Lately, this view has been challenged by empirical results that seem mysterious. For example, in [165] the authors trained several standard architectures on a copy of the data, with the true labels being replaced by random labels. Their central finding can be summarized as follows: deep neural networks easily fit random labels. More precisely, neural networks achieve zero training errors if they are trained on a completely random labeling of the true data. While this observation is easy to formulate, it has profound implications from a statistical learning perspective: the effective capacity of neural networks is sufficient to store the entire data set. Despite the high capacity of the functional classes and the almost perfect fit to training data, these predictors often give very accurate predictions for new data in the test phase.

These observations rule out VC dimension, Rademacher complexity, etc. from describing the generalization behavior. In particular, the Rademacher complexity for the interpolation regime, which leads to a training error of 0, assumes the maximum value of 1, as previously explained in an example. Therefore, the classic generalization bounds are vacuous and cannot explain the amazing generalization ability of the neural network.

The recent breakthrough in Belkin et al.'s "double descent" risk curve [154, 155] reconciles the classic bias–variance trade-off with behaviors that have been observed in over-parameterized regimes for a large number of machine learning models. In particular, when the functional class capacity is below the "interpolation threshold", learned predictors show the classic U-shaped curve from Fig. 12.3a, where the function class capacity is identified with the number of parameters needed to specify a function within the class. The bottom of the U-shaped risk can be achieved at

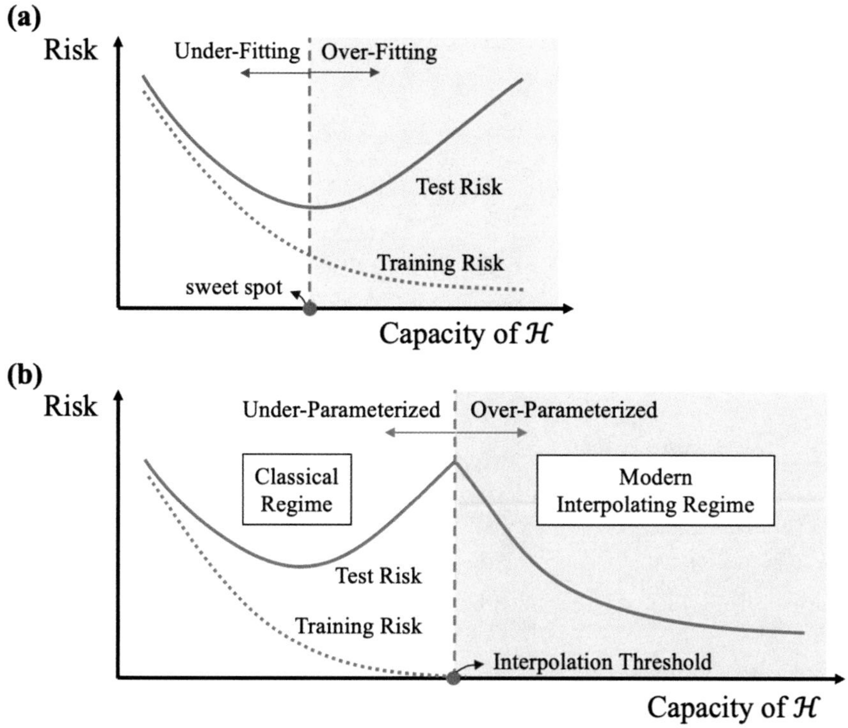

Fig. 12.3 Curves for training risk (dashed line) and test risk (solid line). (**a**) The classical U-shaped risk curve arising from the bias–variance trade-off. (**b**) The double descent risk curve, which incorporates the U-shaped risk curve (i.e., the "classical" regime) together with the observed behavior from using high-capacity function classes (i.e., the "modern" interpolating regime), separated by the interpolation threshold. The predictors to the right of the interpolation threshold have zero training risk

the sweet spot which balances the fit to the training data and the susceptibility to over-fitting. When we increase the function class capacity high enough by increasing the size of the neural network architecture, the learned predictors achieve (almost) perfect fits to the training data. Although the learned predictors obtained at the interpolation threshold typically have high risk, increasing the function class capacity beyond this point leads to decreasing risk, which typically falls below the risk achieved at the sweet spot in the "classic" regime (see Fig. 12.3b).

In the following example we provide concrete and explicit evidence for the double descent behavior in the context of simple linear regression models. The analysis shows the transition from under- to over-parameterized regimes. It also allows us to compare the risks at any point on the curve and explain how the risk in the over-parameterized regime can be lower than any risk in the under-parameterized regime.

Example: Double Descent in Regression [155]
We consider the following linear regression problem:

$$y = \boldsymbol{x}^\top \boldsymbol{\beta} + \epsilon, \tag{12.31}$$

where $\boldsymbol{\beta} \in \mathbb{R}^D$ and $\boldsymbol{x}$ and ϵ are a normal random vector and a variable, where $\boldsymbol{x} \sim \mathcal{N}(\boldsymbol{0}, \boldsymbol{I}_D)$ and $\epsilon \sim \mathcal{N}(0, \sigma^2)$. Given training data $\{\boldsymbol{x}_n, y_n\}_{n=1}^N$, we fit a linear model to the data using only a subset $T \subset [D]$ of cardinality of p, where $[D] := \{0, \cdots, D\}$. Let $\boldsymbol{X} = [\boldsymbol{x}_1, \cdots, \boldsymbol{x}_N] \in \mathbb{R}^{D\times N}$ be the design matrix, $\boldsymbol{y} = [y_1, \cdots, y_N]^\top$ be the vector of response. For a subset T, we use $\boldsymbol{\beta}_T$ to denote its $|T|$-dimensional subvector of entries from T; we also use $\boldsymbol{X}_T$ to denote an $N \times p$ sub-matrix of $\boldsymbol{X}$ composed of columns in T. Then, the risk of $\hat{\boldsymbol{\beta}}$, where $\hat{\boldsymbol{\beta}}_T = \boldsymbol{X}_T^\dagger \boldsymbol{y}$ and $\hat{\boldsymbol{\beta}}_{T^c} = 0$, is given by

$$\mathbb{E}\left[(y - \boldsymbol{x}^\top \hat{\boldsymbol{\beta}})^2\right] = \begin{cases} (\|\boldsymbol{\beta}_{T^c}\|^2 + \sigma^2)\left(1 + \frac{p}{N-p-1}\right); & if\ p \le N-2 \\ \infty; & if\ N-1 \le p \le N+1 \\ \|\boldsymbol{\beta}_T\|^2\left(1 - \frac{N}{p}\right) & \\ \quad +(\|\boldsymbol{\beta}_{T^c}\|^2 + \sigma^2)\left(1 + \frac{N}{p-N-1}\right); & if\ p \ge N+2. \end{cases} \tag{12.32}$$

Proof Recall that $\boldsymbol{x}$ is assumed to be a Gaussian distribution with zero mean and identity covariance, so that the mean squared prediction error can be written as

$$\begin{aligned} \mathbb{E}\left[(y - \boldsymbol{x}^\top \hat{\boldsymbol{\beta}})^2\right] &= \mathbb{E}\left[(\boldsymbol{x}^\top \boldsymbol{\beta} + \sigma\epsilon - \boldsymbol{x}^\top \hat{\boldsymbol{\beta}})^2\right] = \sigma^2 + \mathbb{E}\|\boldsymbol{\beta} - \hat{\boldsymbol{\beta}}\|^2 \\ &= \sigma^2 + \|\boldsymbol{\beta}_{T^c}\|^2 + \mathbb{E}\|\boldsymbol{\beta}_T - \hat{\boldsymbol{\beta}}_T\|^2, \end{aligned}$$

where $\boldsymbol{\beta}$ denotes the ground-truth regression parameter and we use the independency of the test phase regressor $\boldsymbol{x}$ and the training phase design matrix $\boldsymbol{X}$. Our goal is now to derive the closed form expression for the second term.

(Classical regime) For the given training data set, we have

$$\begin{aligned} \hat{\boldsymbol{\beta}}_T &= (\boldsymbol{X}_T \boldsymbol{X}_T^\top)^{-1} \boldsymbol{X}_T \boldsymbol{y} = (\boldsymbol{X}_T \boldsymbol{X}_T^\top)^{-1} \boldsymbol{X}_T \boldsymbol{X}_T^\top \boldsymbol{\beta}_T + (\boldsymbol{X}_T \boldsymbol{X}_T^\top)^{-1} \boldsymbol{X}_T \boldsymbol{\eta} \\ &= \boldsymbol{\beta}_T + (\boldsymbol{X}_T \boldsymbol{X}_T^\top)^{-1} \boldsymbol{X}_T \boldsymbol{\eta}, \end{aligned}$$

(continued)

where

$$\eta := y - X_T^\top \beta_T = \epsilon + X_{T^c}^\top \beta_{T^c}.$$

By plugging this into the second term, we have

$$\mathbb{E}\|\beta_T - \hat{\beta}_T\|^2 = \mathbb{E}\left[\eta^\top P_{\mathcal{R}(X_T)}\eta\right] = \mathrm{Tr}\left(\mathbb{E}\left[P_{\mathcal{R}(X_T)}\right]\mathbb{E}\left[\eta\eta^\top\right]\right).$$

In addition, we have

$$\begin{aligned}\mathbb{E}\left[\eta\eta^\top\right] &= \mathbb{E}\left[\epsilon\epsilon^\top\right] + \mathbb{E}\left[X_{T^c}^\top \beta_{T^c}\left(X_{T^c}^\top \beta_{T^c}\right)^\top\right] \\ &= (\sigma^2 + \|\beta_{T^c}\|^2)I_N,\end{aligned}$$

where $\mathcal{R}(X_T)$ denotes the range space of X_T and $P_{\mathcal{R}(X_T)}$ denotes the projection to the range space of X_T. Furthermore, $P_{\mathcal{R}(X_T)}$ is Hotelling's T-squared distribution with parameter p and $N - p + 1$ so that

$$\mathrm{Tr}\mathbb{E}\left[P_{\mathcal{R}(X_T)}\right] = \begin{cases}\frac{p}{N-p-1}, & \text{if } p \le N-2 \\ +\infty, & \text{if } p = N-1\end{cases}. \tag{12.33}$$

Therefore, by putting them together we conclude the proof for the classical regime.

(Modern interpolating regime) We consider $p \ge N$. Then, we have

$$\begin{aligned}\hat{\beta}_T &= X_T^\top(X_T X_T^\top)^{-1} y = X_T^\top(X_T X_T^\top)^{-1} X_T^\top \beta_T + X_T^\top(X_T X_T^\top)^{-1}\eta \\ &= X_T^\top(X_T X_T^\top)^{-1} X_T^\top \beta_T + X_T^\top(X_T X_T^\top)^{-1}\eta \\ &= \mathcal{P}_{\mathcal{R}(X_T^\top)}\beta_T + X_T^\top(X_T X_T^\top)^{-1}\eta,\end{aligned}$$

where

$$\eta := y - X_T^\top \beta_T = \epsilon + X_{T^c}^\top \beta_{T^c}.$$

Therefore,

$$\mathbb{E}\left[\|\beta_T - \hat{\beta}_T\|^2\right] = \mathbb{E}\left[\|\mathcal{P}^\perp_{\mathcal{R}(X_T^\top)}\beta_T\|^2\right] + \mathbb{E}\left[\eta^\top(X_T X_T^\top)^{-1}\eta\right].$$

(continued)

Furthermore, we have

$$\mathbb{E}\left[\|\mathcal{P}^{\perp}_{\mathcal{R}(X_T^\top)}\boldsymbol{\beta}_T\|^2\right] = \left(1-\frac{n}{p}\right)\|\boldsymbol{\beta}_T\|^2$$
$$\mathbb{E}\left[\boldsymbol{\eta}^\top(X_TX_T^\top)^{-1}\boldsymbol{\eta}\right] = \mathrm{Tr}\left(\mathbb{E}(X_TX_T^\top)^{-1}\mathbb{E}\left[\boldsymbol{\eta}\boldsymbol{\eta}^\top\right]\right),$$

where we use the independency between X_T and X_{T^c} and $\boldsymbol{\epsilon}$ for the second equality. In addition, we have

$$\begin{aligned}\mathbb{E}\left[\boldsymbol{\eta}\boldsymbol{\eta}^\top\right] &= \mathbb{E}\left[\boldsymbol{\epsilon}\boldsymbol{\epsilon}^\top\right] + \mathbb{E}\left[X_{T^c}^\top\boldsymbol{\beta}_{T^c}\left(X_{T^c}^\top\boldsymbol{\beta}_{T^c}\right)^\top\right] \\ &= (\sigma^2 + \|\boldsymbol{\beta}_{T^c}\|^2)\boldsymbol{I}_N.\end{aligned}$$

Finally, the distribution of $\boldsymbol{P} := (X_TX_T^\top)^{-1}$ is inverse-Wishart with identity scale matrix $\boldsymbol{I}_N$ with p degrees of freedom. Accordingly, we have

$$\mathrm{Tr}\left(\mathbb{E}(X_TX_T^\top)^{-1}\right) = \begin{cases}\frac{N}{p-N-1}, & \text{if } p \geq N+2 \\ +\infty, & \text{if } p = N, N+1\end{cases}.$$

By putting them together, we have

$$\mathbb{E}\left[(y - \boldsymbol{x}^\top\hat{\boldsymbol{\beta}})^2\right] = \left(1-\frac{N}{p}\right)\|\boldsymbol{\beta}_T\|^2 + (\sigma^2 + \|\boldsymbol{\beta}_{T^c}\|^2)\left(1 + \frac{N}{p-N-1}\right),$$

for $p \geq N$ and $\mathbb{E}\left[(y - \boldsymbol{x}^\top\hat{\boldsymbol{\beta}})^2\right] = \infty$ for $p = N, N+1$. This concludes the proof. □

Figure 12.4 illustrates an example plot for the linear regression problem analyzed above for a particular parameter set.

12.4 Inductive Bias of Optimization

All learned predictors to the right of the interpolation threshold fit perfectly with the training data and have no empirical risk. Then, why should some—especially those from larger functional classes—have a lower test risk than others so that they generalize better? The answer is that the functional class capacity, such as VC dimension, or Rademacher complexity, does not necessarily reflect the inductive bias of the predictor appropriate for the problem at hand. Indeed, one

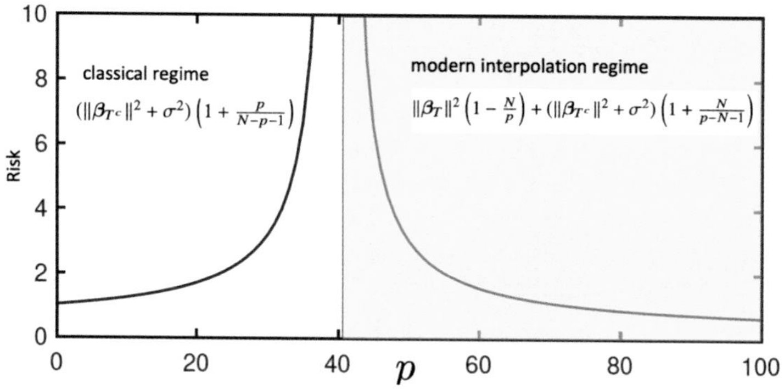

Fig. 12.4 Plot of the risk in (12.32) as a function of p under the random selection of T. Here $\|\boldsymbol{\beta}\|^2 = 1$, $\sigma^2 = 1/25$ and $N = 40$

of the underlying reasons for the appearance of the double descent model in the previous linear regression problem is that we impose an *inductive bias* to choose the minimum norm solution $\hat{\boldsymbol{\beta}}_T = \boldsymbol{X}_T(\boldsymbol{X}_T^\top \boldsymbol{X}_T)^{-1}\boldsymbol{y}$ for the over-parameterized regime, which leads to the smooth solution.

Among the various interpolation solutions, choosing the smooth or simple function that perfectly fits the observed data is a form of Occam's razor: the simplest explanation compatible with the observations should be preferred. By considering larger functional classes that contain more candidate predictors that are compatible with the data, we can find interpolation functions that are "simpler". Increasing the functional class capacity thus improves the performance of classifiers. One of the important advantages of choosing a simpler solution is that it is easy to generalize by avoiding unnecessary glitches in the data. Increasing the functional class capacity to the over-parameterized area thus improves the performance of the resulting classifiers.

Then, one of the remaining questions is: what is the underlying mechanism by which a trained network becomes smooth or simple? This is closely related to the inductive bias (or implicit bias) of an optimization algorithm such as gradient descent, stochastic gradient descent (SGD), etc. [166–171]. Indeed, this is an active area of research. For example, the authors in [168] show that the gradient descent for the linear classifier for specific loss function leads to the maximum margin SVM classifier. Other researchers have shown that the gradient descent in deep neural network training leads to a simple solution [169–171].

12.5 Generalization Bounds via Algorithm Robustness

Another important question is how we can quantify the inductive bias of the algorithm in terms of a generalization error bound. In this section, we introduce a notion of *algorithmic robustness* for quantifying the generalization error, which was originally proposed in [172], but has been largely neglected in deep learning research. It turns out that the generalization bound based on algorithmic robustness has all the ingredients to quantify the fascinating generalization behavior of the deep neural network, so it can be a useful tool for studying generalization.

Recall that the underlying assumption for the classical generalization bounds is the uniform convergence of empirical quantities to their mean [10], which provides ways to bound the gap between the expected risk and the empirical risk by the complexity of the hypothesis set. On the other hand, robustness requires that a prediction rule has comparable performance if tested on a sample close to a training sample. This is formally defined as follows.

Definition 12.5 (Algorithm Robustness [172]) Algorithm $\mathcal{A}$ is said to be $(K, \epsilon(\cdot))$-robust for $K \in \mathbb{N}$ and $\epsilon(\cdot) : \mathcal{Z} \mapsto \mathbb{R}$, if $\mathcal{Z} := \mathcal{X} \times \mathcal{Y}$ can be partitioned into K disjoint sets, denoted by $\{C_i\}_{i=1}^K$ such that the following holds for all training sets $\mathcal{S} \subset \mathcal{Z}$:

$$\forall \boldsymbol{s} \in \mathcal{S}, \forall \boldsymbol{z} \in \mathcal{Z}; \text{if } \boldsymbol{s}, \boldsymbol{z} \in C_i, \text{ then } \quad |\ell(\mathcal{A}_\mathcal{S}, \boldsymbol{s}) - \ell(\mathcal{A}_\mathcal{S}, \boldsymbol{z})| \leq \epsilon(\mathcal{S}) \tag{12.34}$$

for all $i = 1, \cdots, K$, where $\mathcal{A}_\mathcal{S}$ denotes the algorithm $\mathcal{A}$ trained with the data set $\mathcal{S}$.

Then, we can obtain the generalization bound based on algorithmic robustness. First, we need the following concentration inequality.

Lemma 12.4 (Breteganolle–Huber–Carol Inequality [173]) *If the random vector $(N_1, \cdots, N_k)$ is multinomially distributed with parameters N and $(p_1, \cdots, p_k)$, then*

$$\mathbb{P}\left\{\sum_{i=1}^{k} |N_i - Np_i| \geq 2\sqrt{N}\lambda\right\} \leq 2^k \exp(-2\lambda^2), \quad \lambda > 0. \tag{12.35}$$

Theorem 12.5 *If a learning algorithm $\mathcal{A}$ is $(K, \epsilon(\cdot))$-robust, and the training sample set $\mathcal{S}$ is generated by N i.i.d samples from the probability measure μ, then for any $\delta > 0$, with probability at least $1 - \delta$ we have*

$$|R(\mathcal{A}_\mathcal{S}) - \hat{R}_N(\mathcal{A}_\mathcal{S})| \leq \epsilon(\mathcal{S}) + M\sqrt{\frac{2K \ln 2 + 2\ln(1/\delta)}{N}}, \tag{12.36}$$

where

$$M := \max_{z \in \mathcal{Z}} |\ell(\mathcal{A}_\mathcal{S}, z)|.$$

Proof Let N_i be the set of indices of points of $\mathcal{S}$ that fall into the C_i. Note that $(|N_1|, \cdots, |N_K|)$ is an i.i.d. multinomial random variable with parameters N and $(\mu(C_i), \cdots, \mu(C_K))$. Then, the following holds by Lemma 12.4.

$$\mathbb{P}\left\{\sum_{i=1}^{K}\left|\frac{|N_i|}{N} - \mu(C_i)\right| \geq \lambda\right\} \leq 2^K \exp\left(-\frac{N\lambda^2}{2}\right). \tag{12.37}$$

Hence, the following holds with probability at least $1 - \delta$,

$$\sum_{i=1}^{K}\left|\frac{|N_i|}{N} - \mu(C_i)\right| \leq \sqrt{\frac{2K\ln 2 + 2\ln(1/\delta)}{N}}. \tag{12.38}$$

The generalization error is then given by

$$\begin{aligned}
|R(\mathcal{A}_\mathcal{S}) - \hat{R}_N(\mathcal{A}_\mathcal{S})| &\leq \left|\sum_{i=1}^{K}\mathbb{E}_{z\sim\mu}\ell(\mathcal{A}_\mathcal{S}, z|z \in C_i)\mu(C_i) - \frac{1}{N}\sum_{n=1}^{N}\ell(\mathcal{A}_\mathcal{S}, s_i)\right| \\
&\overset{(a)}{\leq} \left|\sum_{i=1}^{K}\mathbb{E}_{z\sim\mu}\ell(\mathcal{A}_\mathcal{S}, z|z \in C_i)\frac{|N_i|}{N} - \frac{1}{N}\sum_{n=1}^{N}\ell(\mathcal{A}_\mathcal{S}, s_i)\right| \\
&\quad + \left|\sum_{i=1}^{K}\mathbb{E}_{z\sim\mu}\ell(\mathcal{A}_\mathcal{S}, z|z \in C_i)\mu(C_i) - \sum_{n=1}^{N}\mathbb{E}_{z\sim\mu}\ell(\mathcal{A}_\mathcal{S}, z|z \in C_i)\frac{|N_i|}{N}\right| \\
&\overset{(b)}{\leq} \frac{1}{N}\left|\sum_{i=1}^{K}\sum_{j\in N_i}\max_{z_2\in C_j}|\ell(\mathcal{A}_\mathcal{S}, s_j) - \ell(\mathcal{A}_\mathcal{S}, z_2)|\right| \\
&\quad + \max_{z\in\mathcal{Z}}|\ell(\mathcal{A}_\mathcal{S}, z)|\sum_{i=1}^{K}\left|\frac{|N_i|}{N} - \mu(C_i)\right| \\
&\overset{(c)}{\leq} \epsilon(\mathcal{S}) + M\sum_{i=1}^{K}\left|\frac{|N_i|}{N} - \mu(C_i)\right| \\
&\overset{(d)}{\leq} \epsilon(\mathcal{S}) + M\sqrt{\frac{2K\ln 2 + 2\ln(1/\delta)}{N}},
\end{aligned}$$

where (a), (b), and (c) are due to the triangle inequality, the definition of N_i, and the definition of $\epsilon(\mathcal{S})$ and M, respectively. □

Note that the definition of robustness requires that (12.34) holds for every training sample. The parameters K and $\epsilon(\cdot)$ quantify the robustness of an algorithm. Since $\epsilon(\cdot)$ is a function of training samples, an algorithm can have different robustness properties for different training patterns. For example, a classification algorithm is more robust to a training set with a larger margin. Since (12.34) includes both the trained solution $\mathcal{A}_\mathcal{S}$ and the training set $\mathcal{S}$, robustness is a property of the learning algorithm, rather than the property of the "effective hypothesis space". This is why the robustness-based generalization bound can account for the inductive bias from the algorithm.

For example, for the case of a single-layer ReLU neural network $f_\Theta : \mathbb{R}^2 \to \mathbb{R}^2$ with the following weight matrix and bias:

$$W^{(0)} = \begin{bmatrix} 2 & -1 \\ 1 & 1 \end{bmatrix}, \; b^{(0)} = \begin{bmatrix} 1 \\ -1 \end{bmatrix}$$

the corresponding neural network output is given by

$$\boldsymbol{o}^{(1)} = \begin{cases} [0, 0]^\top, & 2x - y + 1 < 0, x + y - 1 < 0, \\ [2x - y + 1, 0]^\top, & 2x - y + 1 \geq 0, x + y - 1 < 0, \\ [0, x + y - 1]^\top, & 2x - y + 1 < 0, x + y - 1 \geq 0, \\ [2x - y + 1, x + y - 1]^\top, & 2x - y + 1 \geq 0, x + y - 1 \geq 0. \end{cases}$$

Here, the number of partitions is $K = 4$.

On the other hand, consider a two-layer ReLU network with the weight matrices and biases given by

$$W^{(0)} = \begin{bmatrix} 2 & -1 \\ 1 & 1 \end{bmatrix}, \; b^{(0)} = \begin{bmatrix} 1 \\ -1 \end{bmatrix},$$

$$W^{(1)} = \begin{bmatrix} 1 & 2 \\ 0 & 1 \end{bmatrix}, \; b^{(1)} = \begin{bmatrix} 0 \\ 1 \end{bmatrix}.$$

The corresponding neural network output is given by

$$\boldsymbol{o}^{(2)} = \begin{cases} [0, 1]^\top, & 2x - y + 1 < 0, x + y - 1 < 0, \\ [2x - y + 1, 1]^\top, & 2x - y + 1 \geq 0, x + y - 1 < 0, \\ [2x + 2y - 2, x + y]^\top, & 2x - y + 1 < 0, x + y - 1 \geq 0, \\ [4x + y - 1, x + y]^\top, & 2x - y + 1 \geq 0, x + y - 1 \geq 0. \end{cases}$$

Therefore, in spite of the twice larger parameter sizes, the number of partitions is $K = 4$, which is the same as the single-layer neural network. Therefore, in terms of the generalization bounds, the two algorithms have same upper bound up to the parameter $\epsilon(\mathcal{S})$. This example clearly confirms that generalization is a property of the learning algorithm, rather than the property of the effective hypothesis space or the number of parameters.

12.6 Exercises

1. Compute the VC dimension of the following function classes:

   (a) Interval $[a, b]$.
   (b) Disc in $\mathbb{R}^2$.
   (c) Half space in $\mathbb{R}^d$.
   (d) Axis-aligned rectangles.

2. Show that the classifier f_θ that returns 1 if the input number x is larger than $\sin(\theta x)$ and 0 otherwise can shatter any finite subset of the set $\{2^{-m} \mid m \in \mathbb{N}\}$.
3. Prove the following properties of Rademacher complexity:

   (a) (Monotonicity) If $\mathcal{F} \subset \mathcal{G}$, then $Rad_N(\mathcal{F}) \leq Rad_N(\mathcal{G})$.
   (b) (Convex hull) Let $conv(\mathcal{F})$ be the convex hull of $\mathcal{F}$. Then $Rad_N(\mathcal{F}) = Rad_N(conv(\mathcal{F}))$.
   (c) (Scale and shift) For any function class $\mathcal{F}$ and $c, d \in \mathbb{R}$. $Rad_N(c\mathcal{F} + d) = |c| Rad_N(\mathcal{F})$.
   (d) (Lipschitz composition) If ϕ is an L-Lipschitz function, then $Rad_N(\phi \cdot \mathcal{F}) \leq L \cdot Rad_N(\mathcal{F})$.

4. Let $\mathcal{F}$ be the class of linear predictors given by $y = \boldsymbol{w}^\top \boldsymbol{x}$ with the restriction of $\|\boldsymbol{w}\|_1 \leq W_1$ and $\|\boldsymbol{x}\|_\infty \leq X_\infty$ for $\boldsymbol{x} \in \mathbb{R}^d$. Then, show that

$$Rad_N(\mathcal{F}) \leq \frac{W_1 X_\infty \sqrt{2 \ln(d)}}{\sqrt{N}}.$$

5. Let $\mathcal{A}$ be a set of N vectors in $\mathbb{R}^m$, and let $\bar{\boldsymbol{a}}$ be the mean of the vectors in $\mathcal{A}$. Then:

$$Rad_N(\mathcal{A}) \leq \max_{\boldsymbol{a} \in \mathcal{A}} \|\boldsymbol{a} - \bar{\boldsymbol{a}}\|_2 \cdot \frac{\sqrt{2 \log N}}{m}.$$

   In particular, if $\mathcal{A}$ is a set of binary vectors,

$$Rad_N(\mathcal{A}) \leq \sqrt{\frac{2 \log N}{m}}.$$

6. For a metric space $\mathcal{S}, \rho$ and $\mathcal{T} \subset \mathcal{S}$ we say that $\hat{\mathcal{T}} \subset \mathcal{S}$ is an ϵ-cover of $\mathcal{T}$, if $\forall t \in \mathcal{T}$, there exists $t' \in \mathcal{T}$ such that $\rho(t, t') \leq \epsilon$. The ϵ-covering number of $\mathcal{T}$ is defined by

$$N(\epsilon, \mathcal{T}, \rho) = \min\{|\mathcal{T}'| : \mathcal{T}' \text{ is an } \epsilon\text{-cover of } \mathcal{T}\}.$$

If $\mathcal{Z}$ is compact w.r.t. metric ρ, $\ell(\mathcal{A}_{\mathcal{S}}, \cdot)$ is Lipschitz continuous with Lipschitz constant $c(\mathcal{S})$, i.e.,

$$|\ell(\mathcal{A}_{\mathcal{S}}, z_1) - \ell(\mathcal{A}_{\mathcal{S}}, z_2)| \leq c(\mathcal{S})\rho(z_1, z_2), \quad \forall z_1, z_2 \in \mathcal{Z},$$

then show that $\mathcal{A}$ is $(K, \epsilon(\mathcal{S}))$-robust, where

$$K = N(\gamma/2, \mathcal{Z}, \rho), \ \ \epsilon(\mathcal{S}) = c(\mathcal{S})\gamma$$

for $\gamma > 0$.

Chapter 13
Generative Models and Unsupervised Learning

13.1 Introduction

The last part of our voyage toward the understanding of the geometry of deep learning concerns perhaps the most exciting aspect of deep learning—*generative models.* Generative models cover a large spectrum of research activities, which include the variational autoencoder (VAE) [174, 175], generative adversarial network (GAN) [88, 176, 177], normalizing flow [178–181], optimal transport (OT) [182–184], etc. This field has evolved very quickly, and at any machine learning conference like NeurIPS, CVPR, ICML, ICLR, etc., you may have seen exciting new developments that far surpass existing approaches. In fact, this may be one of the excuses why the writing of this chapter has been deferred till the last minute, since there could be new updates during the writing.

For example, Fig. 13.1 shows the examples of fake human faces generated by various generative models starting from the GAN[88] in 2014 to styleGAN[89] in 2018. You may be amazed to see how the images become so realistic with so much detail within such a short time period. In fact, this may be another reason why DeepFake by generative models has become a societal problem in the modern deep learning era.

Besides creating fake faces, another reason that a generative model is so important is that it is a systematic means of designing unsupervised learning algorithms. For example, in Yann LeCun's famous cake analogy at NeurIPS 2016, he emphasized the importance of unsupervised learning by saying "If intelligence is a cake, the bulk of the cake is unsupervised learning, the icing on the cake is supervised learning, and the cherry on the cake is reinforcement learning (RL)." Referring to the GAN, Yann LeCun said that it was "the most interesting idea in the last 10 years in machine learning," and predicted that it may become one of the most important engines for modern unsupervised learning.

Despite their popularities, one of the reasons generative models are difficult to understand is that there are so many variations, such as the VAE [174], β-VAE [175],

J. C. Ye, *Geometry of Deep Learning*, Mathematics in Industry 37,
https://doi.org/10.1007/978-981-16-6046-7_13

Fig. 13.1 Four years of face generation using generative models

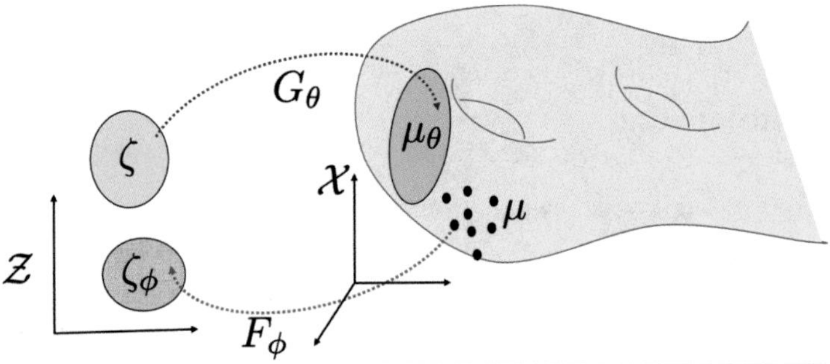

Fig. 13.2 Geometry of generative models

GAN [88], f-GAN [176], W-GAN [177], normalizing flow [178–180], GLOW [181], optimal transport [182–184], cycleGAN [185], W-GAN [177], starGAN [87], CollaGAN [186], to name just a few. Moreover, the modern deep generative models, in particular GANs, have been characterized by the public media as magical black boxes which can generate anything from nothing. Therefore, one of the main goals of this chapter is to demystify the public belief of generative models by providing a coherent geometric picture of generative models.

Specifically, our unified geometric view starts from Fig. 13.2. Here, the ambient image space is $\mathcal{X}$, where we can take samples with the real data distribution μ. If the latent space is $\mathcal{Z}$, the generator G can be treated as a mapping from the latent space to the ambient space, $G : \mathcal{Z} \mapsto \mathcal{X}$, often realized by a deep network with parameter θ, i.e. $G := G_\theta$. Let ζ be a fixed distribution on the latent space, such as uniform or Gaussian distribution. The generator G_θ pushes forward ζ to a distribution $\mu_\theta = G_{\theta\#}\zeta$ in the ambient space $\mathcal{X}$ (don't worry about the term "push-forward" at this point, as it will be explained later). Then, the goal of the generative model training is to make μ_θ as close as possible to the real data distribution μ. Additionally, for the case of autoencoding generative model, the generator works as a decoder, and there exists an additional encoder. More specifically, an encoder F maps from the sample space to the latent space $F : \mathcal{X} \mapsto \mathcal{Z}$, parameterized by ϕ, i.e. $F = F_\phi$ so that the encoder pushes forward μ to a distribution $\zeta_\phi = F_{\phi\#}\mu$ in the latent space. Accordingly, the additional constraint is again to minimize the distance between ζ_ϕ and ζ.

Using this unified geometric model, we can show that various types of generative models such as VAE, β-VAE, GAN, OT, normalizing flow, etc. only differ in their choices of distances between μ_θ and μ or between ζ_ϕ and ζ, and how to train the generator and encoder to minimize the distances.

Therefore, this chapter is structured somewhat differently from the conventional approaches to describing generative models. Rather than directly diving into specific details of each generative model, here we try to first provide a unified theoretical view, and then derive each generative model as a special case. Specifically, we first provide a brief review of probability theory, statistical distances, and optimal transport theory [182, 184]. Using these tools, we discuss in detail how each specific algorithm can be derived by simply changing the choice of statistical distance.

13.2 Mathematical Preliminaries

In this section, we assume that the readers are familiar with basic probability and measure theory [2]. For more background on the formal definition of probability space and related terms from the measure theory, see Chap. 1.

Definition 13.1 (Push-Forward of a Measure) Let $(\mathcal{X}, \mathcal{F}, \mu)$ be a probability space, let $\mathcal{Y}$ be a set, and let $f : \mathcal{X} \mapsto \mathcal{Y}$ be a function. The push-forward of μ by f is the probability measure $\nu : f(\mathcal{F}) \mapsto [0, 1]$ defined by

$$\nu(S) = \mu(f^{-1}(S)), \tag{13.1}$$

which is often denoted by $\nu = f_{\#}\mu$.

As an important example, a *random variable* $X : \Omega \mapsto M$ from a set of possible outcomes Ω to a measurable space M can be regarded as a push-forward of a measure. More specifically, on a probability space $(\Omega, \mathcal{F}, \mu)$, a probability measure ν that a random variable X takes on a set $S \subset M$ is written as

$$\begin{aligned} \nu(S) &:= \nu(\{X \in S\}) \\ &= \mu\left(\{\omega \in \Omega \mid X(\omega) \in S\}\right) \\ &= \mu(X^{-1}(S)). \end{aligned} \tag{13.2}$$

Accordingly, we can regard the random variable X as pushing forward the measure μ on Ω to a measure ν on $\mathbb{R}$.

Example (Push-Forward Measure)
Consider Example 1.4. We now introduce a real-valued random variable:

$$X(\omega) = \begin{cases} 1, & \text{if } \omega = \mathrm{H}, \\ 0, & \text{if } \omega = \mathrm{T}. \end{cases}$$

Then, the push-forward measure $Q = X_{\#}P$ is given by

$$Q(\emptyset) = 0, \quad Q(\{1\}) = 0.5, \quad Q(\{0\}) = 0.5, \quad Q(\{0, 1\}) = 1.$$

We now define the *Radon–Nikodym derivative*, which is a mathematical tool to derive the probability density function (pdf) for the continuous domain, or probability mass function (pmf) for the discrete domain in a rigorous setting. This is also important in deriving the statistical distances, in particular, the divergences. For this, we need to understand the concept of an *absolutely continuous measure*.

Definition 13.2 (Absolutely Continuous Measure) If μ and ν are two measures on any event set $\mathcal{F}$ of Ω, we say that ν is absolutely continuous with respect to μ, or $\nu \ll \mu$, if for every measurable set A, $\mu(A) = 0$ implies $\nu(A) = 0$.

Figure 13.3a shows the case that ν is not absolutely continuous with respect to μ, whereas Fig. 13.3b corresponds to a case where $\nu \ll \mu$. Beside being a prerequisite for the existence of a Radon–Nikodym derivative, the absolute continuity is important since it validates whether the use of a particular divergence is appropriate in designing a specific generative model.

Theorem 13.1 (Radon–Nikodym Theorem) *Let λ and ν be two measures on any event set $\mathcal{F}$ of Ω. If $\lambda \ll \nu$, then there exists a non-negative function g on Ω such that*

$$\lambda(A) = \int_A d\lambda = \int_A g d\nu, \quad A \in \mathcal{F}. \tag{13.3}$$

The function g is called the Radon–Nikodym derivative or density of λ w.r.t. ν and is denoted by $d\lambda/d\nu$. One of the popular Radon–Nikodym derivatives in probability

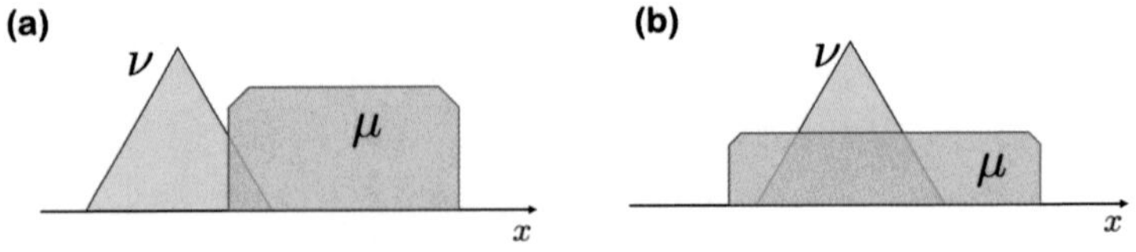

Fig. 13.3 (**a**) ν is not absolute continuous w.r.t. μ. (**b**) $\nu \ll \mu$

theory is the probability density function (pdf) or probability mass function (pmf) as discussed below. The Radon–Nikodym derivative is also a key to defining an f-divergence as a statistical distance measure.

Example (Radon–Nikodym Derivative for Discrete Probability Measure)
Let $a_1 < a_2 < \cdots$ be a sequence of real numbers and let $p_n, n = 1, 2, \cdots,$ be a sequence of positive numbers such that $\sum_{n=1}^{\infty} p_n = 1$. Then,

$$F(x) = \begin{cases} \sum_{i=1}^{n} p_i, & a_n \leq x < a_{n+1} \\ 0, & -\infty < x < a_1 \end{cases}. \tag{13.4}$$

This is often called the discrete *cumulative distribution function (cdf)*, and for this discrete case, it increases stepwise. Then, the corresponding probability measure is

$$P(A) = \sum_{i:a_i \in A} p_i. \tag{13.5}$$

Let ν be the counting measure. Then,

$$P(A) = \int_A f d\nu = \sum_{a_i \in A} f(a_i). \tag{13.6}$$

By inspection of (13.5) and (13.6), we can see that the Radon–Nikodym derivative is given by

$$f(a_i) = p_i, \quad i = 1, 2, \cdots, \tag{13.7}$$

which is often called the probability mass function (pmf).

Example (Radon–Nikodym Derivative for Continuous Probability Measure)
Recall that the continuous domain cumulative distribution function (cdf) F is given by

$$F(x) = \int_{-\infty}^{x} f(y)dy, \quad x \in \mathbb{R}, \tag{13.8}$$

(continued)

where $f(y)$ is the probability density function (pdf). Then, the corresponding probability belonging to an interval A can be computed by

$$P(A) = \int_A f(y)dy \tag{13.9}$$

for any interval A. Therefore, we can easily see that the pdf f is the Radon–Nikodym derivative with respect to the Lebesgue measure.

Although the Radon–Nikodym derivative is used to derive the pdf and pmf, it is a more general concept often used for any integral operation with respect to a measure. The following proposition is quite helpful for evaluating integrals with respect to a push-forward measure.

Proposition 13.1 (Change-of-Variable Formula) *Let $(\mathcal{X}, \mathcal{F}, \mu)$ be a probability space, and let $f : \mathcal{X} \mapsto \mathcal{Y}$ be a function, such that a push-forward measure ν is defined by $\nu = f_\# \mu$. Then, we have*

$$\int_{\mathcal{Y}} g d\nu = \int_{\mathcal{X}} g \circ f d\mu, \tag{13.10}$$

where $\circ$ denotes the function composition.

13.3 Statistical Distances

As discussed before, the distance in the probability space is one of the key concepts for understanding the generative models. In statistics, a statistical distance quantifies the distance between two statistical objects, which can be two random variables, or two probability distributions or samples. The distance can be between an individual sample point and a population or a wider sample of points.

13.3.1 f-Divergence

Defining a metric in the probability space is often complicated, if not impossible. Therefore, relaxed forms of the metric are often used. For example, the statistical distances that satisfy 1) and 2) of Definition 1.1 are referred to as *divergences*, and are quite often used in statistics and machine learning. One of the most widely used forms of divergence in machine learning is f-divergence, which is defined as follows.

Definition 13.3 (**f-Divergence**) Let μ and ν are two probability distributions over a space Ω such that $\mu \ll \nu$. Then, for a convex function f such that $f(1) = 0$, the f-divergence of μ from ν is defined as

$$D_f(\mu||\nu) := \int_\Omega f\left(\frac{d\mu}{d\nu}\right) d\nu, \tag{13.11}$$

where $d\mu/d\nu$ is the Radon–Nikodym derivative w.r.t ν. If $\mu \ll \xi$ and $\nu \ll \xi$ for a common measure ξ on Ω, then their probability densities p and q satisfy $d\mu = pd\xi$ and $d\nu = qd\xi$. In this case the f-divergence can be written as

$$D_f(P||Q) := \int_\Omega f\left(\frac{p(x)}{q(x)}\right) q(x)d\xi(x). \tag{13.12}$$

One thing which is very important and should be treated carefully is the condition $\mu \ll \nu$. For example, if μ is the measure of the original data and ν is the distribution for the generated data, their absolute continuity w.r.t each other should be checked first to choose a right form of divergence.

For the discrete case, when $Q(x)$ and $P(x)$ become the respective probability mass functions, then the f-divergence can be written as

$$D_f(P||Q) := \sum_x Q(x) f\left(\frac{P(x)}{Q(x)}\right). \tag{13.13}$$

Depending on the choice of the convex function f, we can obtain various special cases. Some of the representative special cases are as follows.

13.3.1.1 Kullback–Leibler (KL) Divergence

The corresponding generator f is given by

$$f(t) = t \log t.$$

In the discrete case, KL divergence can be represented by

$$\begin{aligned} D_{KL}(P||Q) &= \sum_x Q(x) \frac{P(x)}{Q(x)} \log \frac{P(x)}{Q(x)} \\ &= \sum_x P(x) \log \frac{P(x)}{Q(x)} \\ &= -\sum_x (P(x) \log Q(x) - P(x) \log P(x)) \\ &= H(P, Q) - H(P), \end{aligned} \tag{13.14}$$

where $H(P, Q)$ is the cross entropy of P and Q, and $H(P)$ is the entropy of P:

$$H(P) = -\sum_{x} P(x) \log P(x), \tag{13.15}$$

$$H(P, Q) = -\sum_{x} P(x) \log Q(x). \tag{13.16}$$

Therefore, KL divergence is often called the *relative entropy*.

13.3.1.2 Jensen–Shannon (JS) Divergence

This corresponds to a special case of f-divergence with the generator

$$f(t) = (t+1)\log\left(\frac{2}{t+1}\right) + t\log t.$$

Using this, we can show that JS divergence is closely related to the KL divergence as:

$$D_{JS}(P||Q) = \frac{1}{2}D_{KL}(P||M) + \frac{1}{2}D_{KL}(Q||M), \tag{13.17}$$

where $M = (P+Q)/2$.

Note that JS divergence has important advantages over KL divergence. Since $M = (P+Q)/2$, we can always guarantee $P \ll M$ and $Q \ll M$. Therefore, the Radon–Nikodym derivative dP/dM and dQ/dM are always well-defined and the f-divergence in (13.11) can be obtained. On the other hand, to use the KL divergence $D_{KL}(P||Q)$ or $D_{KL}(Q||P)$, we should have $P \ll Q$ or $Q \ll P$ respectively, which is difficult to know a prior in practice.

The generators for other forms of f-divergence are defined in Table 13.1. Later, we will show that various types of GAN architecture emerge depending on the choice of the generator.

13.3.2 Wasserstein Metric

Unlike the f-divergence, the Wasserstein metric is a metric that satisfies all four properties of a metric in Definition 1.1. Therefore, this becomes a powerful way of measuring distance in the probability space. For example, to define an f-divergence, we should always check the absolute continuity w.r.t. each other, which is difficult in practice. In the Wasserstein metric, such hassles are no longer necessary.

Let (M, d) be a metric space with a metric d. For $p \geq 1$, let $P_p(M)$ denote the collection of all probability measures μ on M with a finite p-th moment. Then, the

Table 13.1 Various realizations of f-GANs

Divergence name	Generator $f(u)$	Convex conjugate $f^*(t)$	dom f^*	$g_f(v)$	$f^*(g_f(v))$
Kullback–Leiber (KL)	$u \log u$	$\exp(t-1)$	$\mathbb{R}$	v	$\exp(v-1)$
Reverse KL	$-\log u$	$-1-\log(-t)$	$\mathbb{R}_-$	$-\exp(-v)$	$-1+v$
Pearson χ^2	$(u-1)^2$	$\frac{1}{4}t^2+t$	$\mathbb{R}$	v	$\frac{1}{4}v^2+v$
Squared Hellinger	$(\sqrt{u}-1)^2$	$\frac{t}{1-t}$	$t<1$	$1-\exp(-v)$	$\exp(v)-1$
Jensen–Shannon	$(u+1)\log\frac{2}{u+1}+u\log u$	$-\log(2-\exp(t))$	$t<\log(2)$	$\log(2)-\log(1+\exp(-v))$	$-\log(2)-\log\left(1-\frac{1}{1+\exp(-v)}\right)$
GAN	$(u+1)\log\frac{1}{u+1}+u\log u$	$-\log(1-\exp(t))$	$\mathbb{R}_-$	$-\log(1+\exp(-v))$	$-\log\left(1-\frac{1}{1+\exp(-v)}\right)$

p-th Wasserstein distance between two probability measures μ and ν in $P_p(M)$ is defined as

$$W_p(\mu,\nu) := \left(\inf_{\pi \in \Pi(\mu,\nu)} \int_{M\times M} d(x,y)^p d\pi(x,y) \right)^{1/p} \tag{13.18}$$

$$= \left(\inf_{\pi \in \Pi(\mu,\nu)} \mathbb{E}_\pi \left[d(X,Y)^p \right] \right)^{1/p}, \tag{13.19}$$

where $\Pi(\mu,\nu)$ denotes the collection of all measures on $M \times M$ with marginals μ and ν on the first and second factors respectively, and X, Y are the random vectors with the joint distribution π, and $\mathbb{E}_\pi[\cdot,\cdot]$ is the expectation with respect to the joint measure π defined by

$$\mathbb{E}_\pi \left[f(X,Y) \right] = \int_{M\times M} f(x,y) d\pi(x,y). \tag{13.20}$$

When $p = 1$, this is often called the "earth-mover distance" or Wasserstein-1 metric. In the following, we provide some examples where the closed form solution for the Wasserstein distance in (13.18) can be obtained.

Example: 1-D Cases
Let μ and ν denote the 1-D probability measure with the cumulative distribution functions, F and G, respectively. Then, we have

$$W_p(\mu,\nu) = \left(\int_0^1 |F^{-1}(z) - G^{-1}(z)|^p dz \right)^{\frac{1}{p}}. \tag{13.21}$$

Example: Normal Distribution
If $\mu \sim \mathcal{N}(m_1, \Sigma_1)$ and $\nu \sim \mathcal{N}(m_2, \Sigma_2)$ are two normal distributions. Then, we have

$$W_2(\mu,\nu) = \|m_1 - m_2\|^2 + B^2(\Sigma_1, \Sigma_2), \tag{13.22}$$

where

$$B^2(\Sigma_1,\Sigma_2) = \mathrm{Tr}(\Sigma_1) + \mathrm{Tr}(\Sigma_2) - 2\mathrm{Tr}\left[\left(\Sigma_1^{1/2} \Sigma_2 \Sigma_1^{1/2} \right)^{1/2} \right], \tag{13.23}$$

where $\mathrm{Tr}(\cdot)$ denotes the matrix trace.

In general, a direct computation of the distance in (13.18) is often difficult. The following section shows that there exists a more manageable way of computing the Wasserstein metric through the dual formulation. In fact, this leads to the theory of optimal transport [182, 184].

13.4 Optimal Transport

13.4.1 Monge's Original Formulation

Optimal transport provides a mathematical means to operate between two probability measures [182, 184]. Formally, we say that $T : \mathcal{X} \mapsto \mathcal{Y}$ transports a probability measure $\mu \in P(\mathcal{X})$ to another measure $\nu \in P(\mathcal{Y})$, if

$$\nu(B) = \mu\left(T^{-1}(B)\right), \quad \text{for all } \nu\text{-measurable sets } B, \tag{13.24}$$

which is simply the push-forward of the measure, i.e., $\nu = T_{\#}\mu$. See Fig. 13.4 for an example of an optimal transport.

Suppose there is a cost function $c : \mathcal{X}\times\mathcal{Y} \to \mathbb{R}\cup\{\infty\}$ such that $c(x, y)$ represents the cost of moving one unit of mass from $x \in \mathcal{X}$ to $y \in \mathcal{Y}$. Monge's original OT problem [182, 184] is then to find a transport map T that transports μ to ν at the minimum total transportation cost:

$$\begin{aligned} \min_T \ \mathbb{M}(T) &:= \textstyle\int_{\mathcal{X}} c(x, T(x))d\mu(x) \\ \text{subject to} \quad & \nu = T_{\#}\mu. \end{aligned} \tag{13.25}$$

The nonlinear push-forward constraint $\nu = T_{\#}\mu$ is difficult to handle and sometimes leads to a void T due to assignment of indivisible mass [182, 184].

In the following, we provide some examples where the closed form solution for the optimal transport map can be obtained.

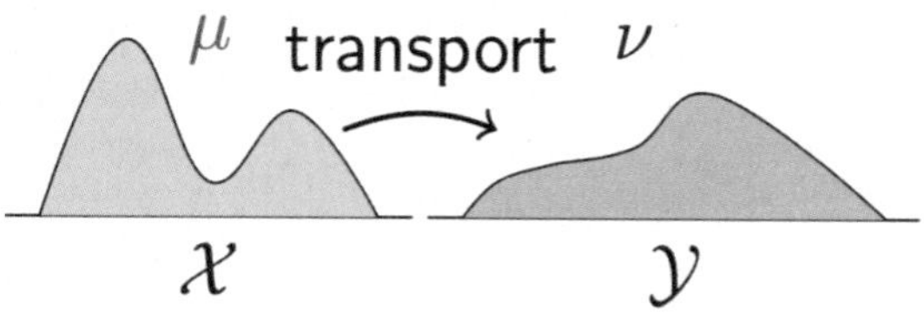

Fig. 13.4 Optimal transport from a distribution (measure) μ to another measure ν

Example: 1-D Cases
Using the change of variable $x = F^{-1}(z)$, the Wasserstein-p metric in (13.21) can be represented by:

$$W_p(\mu, \nu) = \left(\int_0^1 |F^{-1}(z) - G^{-1}(z)|^p dz\right)^{\frac{1}{p}}$$
$$= \left(\int_{\mathbb{R}} |x - G^{-1}(F(x))|^p dF(x)\right)^{\frac{1}{p}}. \tag{13.26}$$

Therefore, for the given transport cost $c(x, y) = |x - y|^p$, we can see that Monge's optimal transport map is given by

$$T(x) = G^{-1}(F(x)).$$

Example: Normal Distribution
If $\mu \sim \mathcal{N}(m_1, \Sigma_1)$ and $\nu \sim \mathcal{N}(m_2, \Sigma_2)$ are two normal distributions. Then, the optimal transport map $T_{\#}\mu = \nu$ is given by

$$T : x \mapsto m_2 + A(x - m_1), \tag{13.27}$$

where

$$A = \Sigma_1^{-1/2}\left(\Sigma_1^{1/2}\Sigma_2\Sigma_1^{1/2}\right)^{1/2}\Sigma_1^{-1/2}. \tag{13.28}$$

In particular, if $\Sigma_1 = \sigma_1 I$ and $\Sigma_2 = \sigma_2 I$, then the optimal transport map is given by

$$T : x \mapsto m_2 + \frac{\sigma_2}{\sigma_1}(x - m_1). \tag{13.29}$$

13.4.2 *Kantorovich Formulation*

Kantorovich relaxed the original OT by considering probabilistic transport that allows mass splitting from a source toward several targets [182, 184]. Specifically, Kantorovich introduced a joint measure $\pi \in P(\mathcal{X} \times \mathcal{Y})$ such that the original problem can be relaxed as

$$\begin{aligned} \min_{\pi} \quad & \int_{\mathcal{X}\times\mathcal{Y}} c(x,y)d\pi(x,y) \\ \text{subject to} \quad & \pi(A\times\mathcal{Y})=\mu(A),\ \ \pi(\mathcal{X}\times B)=\nu(B) \end{aligned} \tag{13.30}$$

for all measurable sets $A \in \mathcal{X}$ and $B \in \mathcal{Y}$. Here, the last two constraints come from the observation that the total amount of mass removed from any measurable set has to be equal to the marginal distributions [182, 184].

Another important advantage of the Kantorovich formulation is the dual formulation, as stated in the following theorem:

Theorem 13.2 (Kantorovich Duality Theorem) *[182, Theorem 5.10, pp.57–59] Let $(\mathcal{X}, \mu)$ and $(\mathcal{Y}, \nu)$ be two probability spaces and let $c : \mathcal{X} \times \mathcal{Y} \to \mathbb{R}$ be a continuous cost function, such that $|c(x, y)| \le c_{\mathcal{X}}(x)+c_{\mathcal{Y}}(y)$ for some $c_{\mathcal{X}} \in L^1(\mu)$ and $c_{\mathcal{Y}} \in L^1(\nu)$, where $L^1(\mu)$ denotes a Lebesgue space with an integral function with the measure μ. Then, there is a duality:*

$$\min_{\pi\in\Pi(\mu,\nu)} \int_{\mathcal{X}\times\mathcal{Y}} c(x,y)d\pi(x,y)$$

$$= \sup_{\varphi\in L^1(\mu)} \left\{ \int_{\mathcal{X}} \varphi(x)d\mu(x) + \int_{\mathcal{Y}} \varphi^c(y)d\nu(y) \right\} \tag{13.31}$$

$$= \sup_{\psi\in L^1(\mu)} \left\{ \int_{\mathcal{X}} \psi^c(x)d\mu(x) + \int_{\mathcal{Y}} \psi(y)d\nu(y) \right\}, \tag{13.32}$$

where

$$\Pi(\mu,\nu) := \{\pi \mid \pi(A\times\mathcal{Y})=\mu(A),\ \ \pi(\mathcal{X}\times B)=\nu(B)\}, \tag{13.33}$$

and the above maximum is taken over the so-called Kantorovich potentials *φ and ψ, whose c-transforms are defined as*

$$\varphi^c(y) := \inf_x \{c(x,y) - \varphi(x)\}, \tag{13.34}$$

$$\psi^c(x) := \inf_y \{c(x,y) - \psi(y)\}. \tag{13.35}$$

In the Kantorovich dual formulation, the computation of the c-transform φ^c is important. In the following, we show several important examples.

Example: The Case $c(x, y) = \|x - y\|$
For any 1-Lipschitz function φ, if $c(x, y) = \|x - y\|$, then we have $\varphi^c = -\varphi$.

Proof From the definition of a c-transform:

$$\varphi^c(y) = \inf_x \{\|x - y\| - \varphi(x)\} \le -\varphi(y),$$

where the last inequality comes by taking $x = y$. In addition,

$$\varphi^c(y) = \inf_x \{\|x - y\| - \varphi(x)\} \ge \inf_x \{\|x - y\| - \|x - y\| - \varphi(y)\} = -\varphi(y)$$

by making use of the 1-Lipschitz behavior of φ. Therefore, $\varphi^c = -\varphi$. □

Example: The Case $c(x, y) = \frac{1}{2}\|x - y\|^2$
For a given transportation cost $c(x, y) = \frac{1}{2}\|x - y\|^2$, we have

$$\varphi^c(x) = \frac{x^2}{2} - \left(\frac{x^2}{2} - \varphi(x)\right)^*,$$

where $(\cdot)^*$ denotes the convex conjugate.

Proof From the definition of c-transform, we have

$$\varphi^c(y) = \inf_x \frac{1}{2}\|x - y\|^2 - \varphi(x) = \inf_x \frac{x^2}{2} + \frac{y^2}{2} - \langle x, y\rangle - \varphi(x),$$

which leads to

$$\frac{y^2}{2} - \varphi^c(y) = \sup_x \langle x, y\rangle - \left(\frac{x^2}{2} - \varphi(x)\right) = \left(\frac{y^2}{2} - \varphi(y)\right)^*.$$

Therefore, we have

$$\varphi^c(x) = \frac{x^2}{2} - \left(\frac{x^2}{2} - \varphi(x)\right)^*.$$

□

In particular, when $c(x, y) = \|x - y\|$, we can reduce possible candidates of φ to 1-Lipschitz functions so that we can simplify φ^c to $-\varphi$ [182]. Using this, the Wasserstein-1 norm can be represented by

$$W_1(\mu, \nu) := \min_{\pi\in\Pi(\mu,\nu)} \int_{\mathcal{X}\times\mathcal{X}} ||x - y||d\pi(x, y) \tag{13.36}$$

$$= \sup_{\varphi\in\text{Lip}_1(\mathcal{X})} \left\{ \int_{\mathcal{X}} \varphi(x)d\mu(x) - \int_{\mathcal{X}} \varphi(y)d\nu(y) \right\}, \tag{13.37}$$

where $\text{Lip}_1(\mathcal{X}) = \{\varphi \in L^1(\mu) : |\varphi(x) - \varphi(y)| \le ||x - y||\}$. Compared to the primal form (13.36) which requires the integration with respect to the joint measures, the dual formulation in (13.37) just requires marginals μ and ν, which make the computation much more tractable. This is why the dual form is more widely used in generative models.

13.4.3 Entropy Regularization

Another way to address optimal transport problems in a computationally feasible way is by using the so-called Sinkhorn distance [183]. Rather than solving the dual problem, the main idea is to use entropy regularization with respect to the joint distribution π so that the optimal transport map can be found by solving a regularized primal problem. As the paper title indicates ("Sinkhorn distances: Lightspeed Computation of Optimal Transport") [183], the introduction of the entropy regularization leads to a computationally efficient optimization problem.

Although the original formulation is for the discrete measure, here we provide a continuous formulation of the Sinkhorn distances to use the similar notation as before. More specifically, the continuous-domain entropy regularized optimal transport is formulated by [187]

$$\inf_{\pi\in\Pi(\mu,\nu),\pi>0} \int_{\mathcal{X}\times\mathcal{Y}} c(x, y)d\pi(x, y) + \gamma \int_{\mathcal{X}\times\mathcal{Y}} \pi(x, y)(\log \pi(x, y) - 1)d(x, y), \tag{13.38}$$

where $\Pi(\mu, \nu)$ denotes the set of joint distributions whose marginal distributions are $\mu(x)$ and $\nu(y)$, respectively. Then, the following proposition shows that the associate dual problem has very interesting formulation.

Proposition 13.2 *The dual of the primal problem in* (13.38) *is given by*

$$\sup_{\phi,\varphi} \int_{\mathcal{X}} \phi(x)d\mu(x) + \int_{\mathcal{Y}} \varphi(y)d\nu(y) - \gamma \int_{\mathcal{X}\times\mathcal{Y}} \exp\left(\frac{-c(x, y) + \phi(x) + \varphi(y)}{\gamma}.\right) d(x, y). \tag{13.39}$$

Proof Using the convex conjugate formulation in Chap. 1, we know that e^x is the convex conjugate of $x \log x - x$ for $x > 0$. Accordingly, we have

$$\begin{aligned}
&\sup_{\phi,\varphi} \int_{\mathcal{X}} \phi d\mu + \int_{\mathcal{Y}} \varphi d\nu - \gamma \int_{\mathcal{X}\times\mathcal{Y}} \exp\left(\frac{-c+\phi+\varphi}{\gamma}\right) d(x, y) \\
&= \sup_{\phi,\varphi} \int_{\mathcal{X}} \phi d\mu + \int_{\mathcal{Y}} \varphi d\nu + \int_{\mathcal{X}\times\mathcal{Y}} \inf_{\pi>0} d\pi(c - \phi - \varphi) + \gamma(\pi \log \pi - \pi) d(x, y) \\
&= \inf_{\pi>0} \int_{\mathcal{X}\times\mathcal{Y}} c\pi + \gamma\pi(\log \pi - 1) d(x, y) \\
&\quad + \inf_{\pi>0} \sup_{\phi,\varphi} \int_{\mathcal{X}} \phi d\mu - \int_{\mathcal{X}\times\mathcal{Y}} \phi d\pi + \int_{\mathcal{Y}} \varphi d\mu - \int_{\mathcal{X}\times\mathcal{Y}} \varphi d\pi.
\end{aligned}$$

Under the constraint that $\pi \in \Pi(\mu, \nu)$, the last four terms vanish. Therefore, we have

$$\begin{aligned}
&\sup_{\phi,\varphi} \int_{\mathcal{X}} \phi(x) d\mu(x) + \int_{\mathcal{Y}} \varphi(y) d\nu(y) - \gamma \int_{\mathcal{X}\times\mathcal{Y}} \exp\left(\frac{-c(x, y) + \phi(x) + \varphi(y)}{\gamma}\right) d(x, y) \\
&= \inf_{\pi\in\Pi(\mu,\nu),\pi>0} \int_{\mathcal{X}\times\mathcal{Y}} c(x, y) d\pi(x, y) + \gamma \int_{\mathcal{X}\times\mathcal{Y}} \pi(x, y)(\log \pi(x, y) - 1) d(x, y).
\end{aligned}$$

This concludes the proof. □

The Sinkhorn distance formulation can then be obtained by the change of variables for the dual problem (13.39). Specifically, for $\phi, \varphi > 0$, consider the following change of variables:

$$\alpha(x) = e^{\frac{\phi(x)}{\gamma}}, \quad \beta(y) = e^{\frac{\varphi(x)}{\gamma}}, \tag{13.40}$$

which leads to

$$\sup_{\alpha,\beta} \gamma \int_{\mathcal{X}} \log \alpha(x) d\mu(x) + \gamma \int_{\mathcal{Y}} \log \beta(y) d\nu(y) - \gamma \int_{\mathcal{X}\times\mathcal{Y}} \alpha(x) \exp\left(-\frac{c(x, y)}{\gamma}\right) \beta(y) d(x, y). \tag{13.41}$$

Using the variational calculus, for a given perturbation $\alpha \to \alpha + \epsilon\delta\alpha$, the first-order variation is given by

$$\int_{\mathcal{X}} \frac{\delta\alpha(x)}{\alpha(x)} \frac{d\mu(x)}{dx} dx - \int_{\mathcal{X}} \delta\alpha(x) \int_{\mathcal{Y}} \exp\left(-\frac{c(x, y)}{\gamma}\right) \beta(y) dy dx \tag{13.42}$$

$$= \int_{\mathcal{X}} \delta\alpha(x) \left(\frac{1}{\alpha(x)} \frac{d\mu}{dx}(x) - \int_{\mathcal{Y}} \exp\left(-\frac{c(x, y)}{\gamma}\right) \beta(y) dy\right) dx = 0. \tag{13.43}$$

Thus, we have

$$\alpha(x) = \frac{\frac{d\mu}{dx}(x)}{\int_{\mathcal{Y}} \exp\left(-\frac{c(x,y)}{\gamma}\right)\beta(y)dy}. \tag{13.44}$$

Similarly, we have

$$\beta(y) = \frac{\frac{d\nu}{dy}(y)}{\int_{\mathcal{X}} \exp\left(-\frac{c(x,y)}{\gamma}\right)\alpha(x)dx}. \tag{13.45}$$

In fact, the update rule (13.44) and (13.45) are the main iterations for Sinkhorn's fixed point iteration [183].

13.5 Generative Adversarial Networks

With the mathematical backgrounds set, we are now ready to discuss specific forms of the generative models, and explain how they can be derived in a unified theoretical framework. In this section, we will mainly describe the decoder-type generative models, which we simply call generative models. Later, we will explain how this analysis can be extended to the autoencoder-type generative models.

13.5.1 Earliest Form of GAN

The original form of generative adversarial network (GAN) [88] was inspired by the success of discriminative models for classification. In particular, Goodfellow et al. [88] formulated generative model training as a minimax game between a generative network (generator) that maps a random latent vector into the data in the ambient space, and a discriminative network trying to distinguish the generated samples from real samples. Surprisingly, this minimax formation of a deep generative model can transfer the success of deep discriminative models to generative models, resulting in significant improvement in generative model performance [88]. In fact, the success of GANs has generated significant interest in the generative model in general, which has been followed by many breakthrough ideas.

Before we explain the geometric structure of the GAN and its variants from a unified framework, we briefly present the original explanation of the GAN, since it is more intuitive to the general public. Let $\mathcal{X}$ and $\mathcal{Z}$ denote the ambient and latent space equipped with the measure μ and ζ, respectively (recall the geometric picture in Fig. 13.2). Then, the original form of the GAN solves the following minimax game:

$$\min_G \max_D \ell_{GAN}(D, G), \tag{13.46}$$

where

$$\ell_{GAN}(D, G) := \mathbb{E}_\mu \left[\log D(x)\right] + \mathbb{E}_\zeta \left[\log(1 - D(G(z))\right],$$

where $D(x)$ is the discriminator that takes as input a sample and outputs a scalar between $[0, 1]$, $G(z)$ is the generator that maps a latent vector z to the ambient space vector, and

$$\mathbb{E}_\mu \left[\log D(x)\right] = \int_X \log D(x) d\mu(x),$$

$$\mathbb{E}_\zeta \left[\log(1 - D(G(z)))\right] = \int_Z \log(1 - D(G(z))) d\zeta(z).$$

The meaning of (13.46) is that the generator tries to fool the discriminator, while the discriminator wants to maximize the differentiation power between the true and generated samples. In GANs, the discriminator and generator are usually implemented as deep networks which are parameterized by network parameters ϕ and θ, i.e. $D(x) := D_\phi(x)$, $G(z) := G_\theta(z)$. Therefore, (13.46) can be formulated as a minmax problem with respect to θ and ϕ.

Figure 13.5 illustrates some of the samples generated by GANs from this minmax optimization that appeared in their original paper [88]. By current standards, the results look very poor, but when these were published in 2014, they shocked the world and were considered state-of-the art. We can again see the light-speed progress of generative model technology.

Since it was first published, one of the puzzling questions about GANs is the mathematical origin of the minmax problems, and why it is important. In fact, the pursuit of understanding such questions has been very rewarding, and has led to the discovery of numerous key results that are essential toward the understanding of the geometric structure of GANs.

Among them, two most notable results are the f-GAN [176] and Wasserstein GAN (W-GAN) [177], which will be reviewed in the following sections. These works reveal that the GAN indeed originates from minimizing statistical distances using *dual* formulation. These two methods differ only in their choices of statistical distances and associated dual formulations.

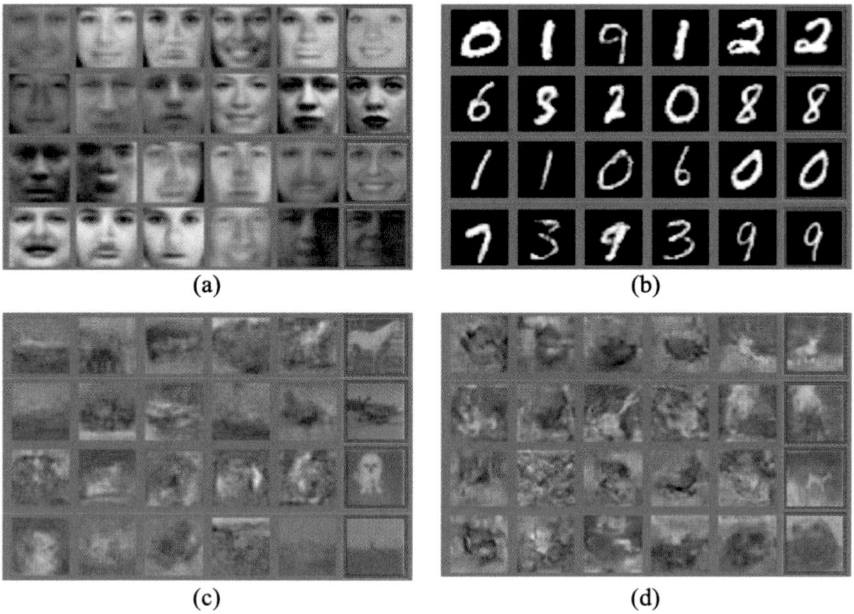

Fig. 13.5 Examples of GAN-generated samples in [88]. The rightmost columns show the nearest training example of the neighboring sample, in order to demonstrate that the model has not memorized the training set. These images show actual samples from the model distributions, not conditional means given samples of hidden units. (**a**) TFD, (**b**) MNIST, (**c**) CIFAR-10 (fully connected model), (**d**) CIFAR-10

13.5.2 *f*-GAN

The f-GAN [176] was perhaps one of the most important theoretical results in the early history of GANs, and clearly demonstrates the importance of the statistical distances and dual formulation. As the name suggests, the f-GAN starts with f-divergence.

Recall that f-divergence is defined by

$$D_f(\mu||\nu) = \int_\Omega f\left(\frac{d\mu}{d\nu}\right) d\nu \tag{13.47}$$

if $\mu \ll \nu$. The main idea of the f-GAN (which includes the original GAN) is to use f-divergence as a statistical distance between the real data distribution $\mathcal{X}$ with the measure μ and the synthesized data distribution in the ambient space $\mathcal{X}$ with the measure $\nu := \mu_\theta$ so that the probability measure ν gets closer to μ (see Fig. 13.2, where μ_θ is now considered as ν for notational simplicity). The key observation is that instead of directly minimizing the f-divergence, something very interesting emerges if we formulate its dual problems. More specifically, the author exploits the following dual formulation of the f-divergence [176], whose proof is repeated here for educational purposes. Recall the following definition of a convex conjugate (for more detail, see Chap. 1):

Definition 13.4 ([6]) For a given function $f : I \mapsto \mathbb{R}$, its convex conjugate is defined by

$$f^*(u) = \sup_{\tau \in I}\{u\tau - f(\tau)\}. \tag{13.48}$$

If f is a convex function, the convex conjugate of its convex conjugate is the function itself, i.e.

$$f(u) = f^{**}(u) = \sup_{\tau \in I^*}\{u\tau - f^*(\tau)\}, \tag{13.49}$$

if $f^* : I^* \mapsto \mathbb{R}$. This is the property we need in the following lemma.

Lemma 13.1 ([176]) *Let $\mu \ll \nu$. Then, for any class of functions τ mapping from $\mathcal{X}$ to $\mathbb{R}$, we have the lower bound*

$$D_f(\mu||\nu) \geq \sup_{\tau \in I^*} \int_{\mathcal{X}} \tau(x)d\mu(x) - \int_{\mathcal{X}} f^*(\tau(x))d\nu(x), \tag{13.50}$$

where $f^* : I^* \mapsto \mathbb{R}$ is the convex conjugate of f.

Proof The proof is a simple consequence of the convex conjugate. More specifically, we have

$$\begin{aligned}
D_f(\mu||\nu) &= \int_{\mathcal{X}} f\left(\frac{d\mu}{d\nu}\right) d\nu \\
&= \int_{\mathcal{X}} \sup_{\tau \in I^*}\left\{\tau \frac{d\mu}{d\nu} - f^*(\tau)\right\} d\nu \\
&\geq \sup_{\tau \in I^*} \int_{\mathcal{X}} \left\{\tau \frac{d\mu}{d\nu} - f^*(\tau)\right\} d\nu \\
&= \sup_{\tau \in I^*} \int_{\mathcal{X}} \tau d\mu - f^*(\tau) d\nu \\
&= \sup_{\tau \in I^*} \int_{\mathcal{X}} \tau(x)d\mu(x) - \int_{\mathcal{X}} f^*(\tau(x))d\nu(x).
\end{aligned}$$

□

The lower bound in (13.50) is tight and can be achieved at

$$\tau = f'\left(\frac{d\mu}{d\nu}\right) = f'\left(\frac{p(x)}{q(x)}\right), \tag{13.51}$$

where the last equality holds when $d\mu = pd\xi$ and $d\nu = qd\xi$ for common measure ξ [176].

While the lower bound in (13.50) is intuitive, one of the complications in the derivation of the f-GAN is that the function τ should be within the domain of f^*, i.e. $\tau \in I^*$. To address this, the authors in [176] proposed the following trick:

$$\tau(x) = g_f(V(x)), \tag{13.52}$$

where $V : \mathcal{X} \mapsto \mathbb{R}$ without any constraint on the output range, and $g_f : \mathbb{R} \mapsto I^*$ is an *output activation function* that maps the output to the domain of f^*. Then, the f-GAN can be formulated as follows:

$$\min_G \max_{g_f} \ell_{fGAN}(G, g_f), \tag{13.53}$$

where

$$\ell_{fGAN}(G, g_f) := \mathbb{E}_\mu\left[g_f(V(x))\right] - \mathbb{E}_\zeta\left[f^*(g_f(V(G(z))))\right].$$

For example, if we choose

$$f(t) = -(t+1)\log(t+1) + t\log t,$$

then its convex conjugate is given by

$$\begin{aligned} f^*(u) &= \sup_{t\in\mathbb{R}_+} \{ut + (t+1)\log(t+1) - t\log t\} \\ &= -\log(1-e^u). \end{aligned}$$

The domain of the conjugate function f^* should be $\mathbb{R}_-$ in order to make the $1-e^u > 0$. One of the functions g_f to allow this is given by

$$g_f(V) = \log\left(\frac{1}{1+e^{-V}}\right) = \log \mathrm{Sig}(V),$$

where $\mathrm{Sig}(\cdot)$ is the sigmoid function. Accordingly, we have

$$f^*(g_f(V)) = -\log\left(1 - e^{\log \mathrm{Sig}(V)}\right) = -\log(1 - \mathrm{Sig}(V)).$$

Therefore, if we use a discriminator with the sigmoid being the last layer we have $D(x) = \mathrm{Sig}(V(x))$ and this leads to the following f-GAN cost function:

$$\begin{aligned}
&\sup_{\tau\in I^*}\int_{\mathcal{X}}\tau(x)d\mu(x)-\int_{\mathcal{X}}f^*(\tau(x))d\nu(x)\\
&=\sup_{g_f,V}\int_{\mathcal{X}}g_f(V(x))d\mu(x)-\int_{\mathcal{X}}f^*(g_f(V(x)))d\nu(x)\\
&=\sup_{D}\int_{\mathcal{X}}\log D(x)d\mu(x)+\int_{\mathcal{X}}\log(1-D(x))d\nu(x).
\end{aligned}$$

Finally, the measure ν is for the samples from latent space $\mathcal{Z}$ with the measure ζ by generator $G(z)$, $z \in \mathcal{Z}$, so ν is the push-forward measure $G_{\#}\zeta$ (see Fig. 13.2). Using the change-of-variable formula in Proposition 13.1, the final loss function is given by

$$\begin{aligned}
\ell(D,G) :=& \sup_{D}\int_{\mathcal{X}}\log D(x)d\mu(x)\\
&+\int_{\mathcal{Z}}\log(1-D(G(z)))d\zeta(x).
\end{aligned}$$

This is equivalent to the original GAN cost function.

By changing the generator f, we can now obtain various types of GAN variants. Table 13.1 summarizes various forms of the f-GAN.

13.5.3 Wasserstein GAN (W-GAN)

Note that the f-GAN interprets the GAN training as a statistical distance minimization in the form of dual formulation. However, its main limitation is that the f-divergence is not a metric, limiting the fundamental performance.

A similar minimization idea is employed for the Wasserstein GAN, but now with a real metric in probability space. More specifically, the W-GAN minimizes the following Wasserstein-1 norm:

$$W_1(P,Q) := \min_{\pi\in\Pi(\mu,\nu)}\int_{\mathcal{X}\times\mathcal{X}}||x-x'||d\pi(x,x'), \tag{13.54}$$

where $\mathcal{X}$ is the ambient space, μ and ν are measures for the real data and generated data, respectively, and $\pi(x,x')$ is the joint distribution with the marginals μ and ν, respectively (recall the definition of $\Pi(\mu,\nu)$ in (13.33)).

Similar to the f-GAN, rather than solving the complicated primal problem, a dual problem is solved. Recall that the Kantorivich dual formulation leads to the following dual formulation of the Wasserstein 1-norm:

$$W_1(\mu, \nu) = \sup_{\varphi \in \mathrm{Lip}_1(\mathcal{X})} \left\{ \int_{\mathcal{X}} \varphi(x) d\mu(x) - \int_{\mathcal{X}} \varphi(x') d\nu(x') \right\}, \tag{13.55}$$

where $\mathrm{Lip}_1(\mathcal{X})$ denotes the 1-Lipschitz function space with domain $\mathcal{X}$. Again, the measure ν is for the generated samples from latent space $\mathcal{Z}$ with the measure ζ by generator $G(z)$, $z \in \mathcal{Z}$, so ν can be considered as the push-forward measure $\nu = G_{\#}\mu$. Using the change-of-variable formula in Proposition 13.1, the final loss function is given by

$$W_1(\mu, \nu) = \sup_{\varphi \in \mathrm{Lip}_1(\mathcal{X})} \left\{ \int_{\mathcal{X}} \varphi(x) d\mu(x) - \int_{\mathcal{Z}} \varphi(G(z)) d\zeta(z) \right\}. \tag{13.56}$$

Therefore, the Wasserstein 1-norm minimization problem can be equivalently represented by the following minmax formulation:

$$\begin{aligned} &\min_{\nu} W_1(\mu, \nu) \\ &= \min_{G} \max_{\varphi \in \mathrm{Lip}_1(\mathcal{X})} \left\{ \int_{\mathcal{X}} \varphi(x) d\mu(x) - \int_{\mathcal{Z}} \varphi(G(z)) d\zeta(z) \right\}, \end{aligned}$$

where $G(z)$ is called the generator, and the Kantorovich potential φ is called the discriminator.

Therefore, imposing a1-Lipschitz condition on the discriminator is necessary in the W-GAN [177]. There are many approaches to address this. For example, in the original W-GAN paper [177], weight clipping was used to impose a 1-Lipschitz condition. Another method is to use spectral normalization [188], which utilizes the power iteration method to impose a constraint on the largest singular value of the weight matrix in each layer. Yet another popular method is the W-GAN with the gradient penalty (WGAN-GP), where the gradient of the Kantorovich potential is constrained to be 1 [189]. Specifically, the following modified loss function is used for the minmax problem:

$$\begin{aligned} &\ell_{W-GAN}(G; \varphi) \\ &= \left(\int_{\mathcal{X}} \varphi(x) d\mu(x) - \int_{\mathcal{Z}} \varphi(G(z)) d\zeta(z) \right) \\ &- \eta \int_{\mathcal{X}} (\|\nabla_{\tilde{x}} \varphi(x)\|_2 - 1)^2 d\mu(x), \end{aligned} \tag{13.57}$$

where $\eta > 0$ is the regularization parameter to impose a 1-Lipschitz property on the discriminators, and $\tilde{x} = \alpha x + (1 - \alpha)G(z)$ with α being random variables from the uniform distribution between [0, 1] [189].

13.5.4 StyleGAN

As mentioned before, one of the most exciting developments in CVPR 2019 was the introduction of novel generative adversarial network (GAN) called StyleGAN by Nvidia [89], which can produce very realistic high-resolution images.

Aside from various sophisticated tricks, StyleGAN also introduced important innovations from a theoretical perspective. For example, one of the main breakthroughs of styleGAN comes from AdaIN. The neural network in Fig. 13.6 generates the latent codes that are used as style image feature vectors. Then,

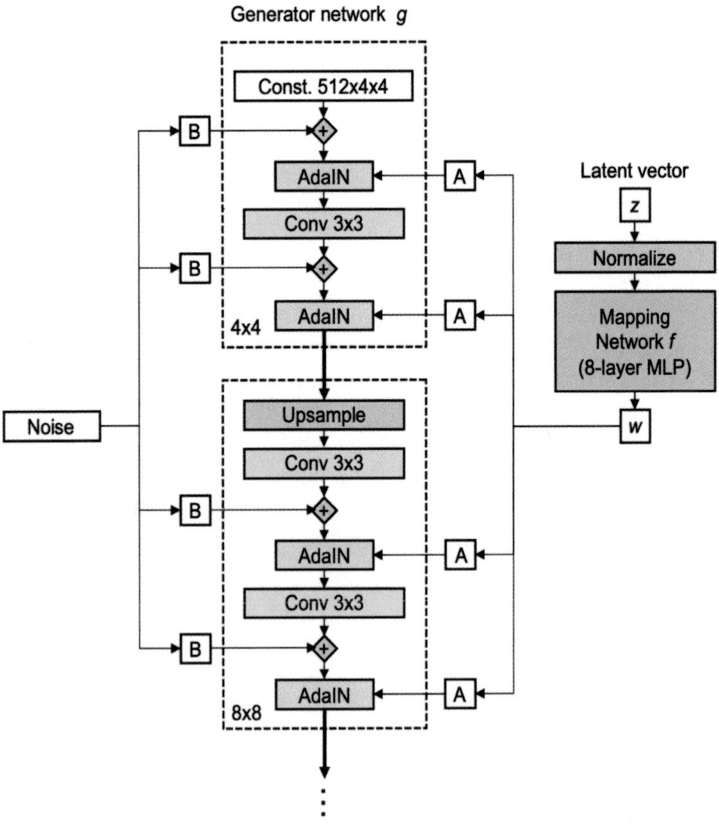

Fig. 13.6 Architecture of StyleGAN

the AdaIN layer combines the style features and the content features together to generate more realistic features at each resolution.

Yet another breakthrough idea is that SytleGAN introduces noise into each layer to create stochastic variation, as shown in Fig. 13.6. Recall that most of the GANs starts with the simple latent vector z in the latent space as an input to the generator. On the other hand, the noise at each layer of StyleGAN can be considered as a more complicated latent space, so that a mapping from a more complicated input latent space to the data domain produces more realistic images. In fact, by introducing a more complicated latent space, styleGAN enables local changes in the pixel level and targets stochastic variation in generating local variants of features.

13.6 Autoencoder-Type Generative Models

Although we have already discussed the generative model such as the GAN, historically the autoencoder-type generative model precedes the GAN-type models. In fact, the autoencoder-type generative model goes back to the denoising autoencoder [190], which is a deterministic form of encoder–decoder networks.

The real *generative* autoencoder model in fact originates from the variational autoencoder (VAE) [174], which enables the generation of the target samples by changing latent variables using random samples. Another breakthrough in the VAE comes from the normalizing flow [178–181], which significantly improves the quality of generated samples by allowing invertible mapping. In this section, we review the two ideas in a unified geometric framework. To do this, we first explain the important concept in variational inference—the evidence lower bound (ELBO) or the variational lower bound [191].

13.6.1 ELBO

In variational inference such as VAE, our model distribution $p_\theta(x)$ is obtained by combining a simple distribution $p(z)$ with a family of conditional distributions $p_\theta(x|z)$, so that our objective is written as

$$\begin{aligned} \log p_\theta(x) &= \log\left(\int p_\theta(x,z)dz\right) \\ &= \log\left(\int p_\theta(x|z)p(z)dz\right). \end{aligned} \tag{13.58}$$

Here, the goal is to find the parameter θ to maximize the loglikelihood using the given data set $x \in \mathcal{X}$.

Although $p(z)$ and $p_\theta(x|z)$ will generally be simple by choice, it may be impossible to compute $\log p_\theta(x)$ analytically due to the need to solve the integral inside the logarithm. A trick to address this problem is to introduce a distribution $q_\phi(z|x)$ parameterized by ϕ and conditioned on x such that

$$\begin{aligned}\log p_\theta(x) &= \log\left(\int p_\theta(x|z)\frac{p(z)}{q_\phi(z|x)}q_\phi(z|x)dz\right)\\&\geq \int \log\left(p_\theta(x|z)\frac{p(z)}{q_\phi(z|x)}\right)q_\phi(z|x)dz,\end{aligned}$$

where we use Jensen's inequality [192]. Accordingly, we have

$$\begin{aligned}\log p_\theta(x) &\geq \int \log p_\theta(x|z)q_\phi(z|x)dz - \int \log\left(\frac{q_\phi(z|x)}{p(z)}\right)q_\phi(z|x)dz\\&= \int \log p_\theta(x|z)q_\phi(z|x)dz - D_{KL}(q_\phi(z|x)||p(z)),\end{aligned}$$

which is often called the evidence lower bound (ELBO) or the variational lower bound [191].

Since the choice of posterior $q_\phi(z|x)$ could be arbitrary, the goal of the variational inference is to find q_ϕ to maximize the ELBO, or, equivalently, minimize the following loss function:

$$\ell_{ELBO}(x;\theta,\phi) := -\int \log p_\theta(x|z)q_\phi(z|x)dz + D_{KL}(q_\phi(z|x)||p(z)), \tag{13.59}$$

where the first term is the likelihood term and the second KL term can be interpreted as the penalty term. Then, variational inference tries to find θ and ϕ to minimize the loss for a given x, or average loss for all x.

13.6.2 *Variational Autoencoder (VAE)*

Using the ELBO, we are now ready to derive the VAE. However, our derivation is somewhat different from the original derivation of the VAE [174], since the original derivation makes it difficult to show the link to normalizing flow [178–181]. The following derivation originates from the f-VAE [193].

Specifically, among the various choices of $q_\phi(z|x)$ for the ELBO, we choose the following form:

$$q_\phi(z|x) = \int \delta(z - F_\phi^x(u))r(u)du, \tag{13.60}$$

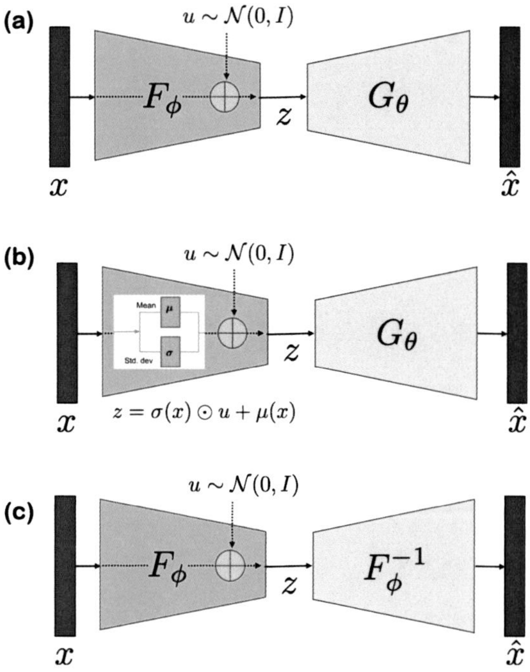

Fig. 13.7 Variational autoencoder architecture: (**a**) general form, (**b**) original VAE, and (**c**) invertible flow

where $r(u)$ is the standard Gaussian, and $F_\phi^x(u)$ is the encoder function for a given x which has another noisy input u. See Fig. 13.7a for the concept of the encoder $F_\phi^x(u)$. For the given encoder function, we have the following key result for the ELBO loss.

Proposition 13.3 *For the given encoder in* (13.60)*, the ELBO loss in* (13.59) *can be represented by*

$$\begin{aligned}\ell_{ELBO}(x;\theta,\phi) := &-\int \log p_\theta(x|F_\phi^x(u))r(u)du\\ &+\int \log\left(\frac{r(u)}{p(F_\phi^x(u))}\right)r(u)du\\ &-\int \log\left|\det\left(\frac{\partial F_\phi^x(u)}{\partial u}\right)\right|r(u)du.\end{aligned} \tag{13.61}$$

Proof Let us start with the ELBO:

$$\ell_{ELBO}(x;\theta,\phi) := \int \log\left(p_\theta(x|z)\frac{p(z)}{q_\phi(z|x)}\right)q_\phi(z|x)dz,$$

which can be represented by

$$\ell_{ELBO}(x,\phi) := \int \left(\log\left(p_\theta(x|z)p(z)\right) - \log q_\phi(z|x)\right) q_\phi(z|x)dz. \tag{13.62}$$

Using the encoder representation in (13.60), the first term of (13.62) becomes

$$\begin{aligned}
&\int\int \log\left(p_\theta(x|z)p(z)\right)\delta(z - F_\phi^x(u))r(u)dudz \\
&= \int \log\left(p_\theta(x|F_\phi^x(u))p(F_\phi^x(u))\right) r(u)du \\
&= \int \log p_\theta(x|F_\phi^x(u))r(u)du + \int \log p(F_\phi^x(u))r(u)du.
\end{aligned}$$

Similarly, the second term of (13.62) becomes

$$\begin{aligned}
&\int\int \log\left(\int \delta(z - F_\phi^x(u'))r(u')du'\right)\delta(z - F_\phi^x(u))r(u)dudz \\
&= \int \log\left(\int \delta(F_\phi^x(u) - F_\phi^x(u'))r(u')du'\right) r(u)du.
\end{aligned}$$

Now, using the following change of variables:

$$v = F_\phi^x(u'), \quad u' = H_x(v),$$

the corresponding Jacobian determinant is given by

$$\det\left(\frac{du'}{dv}\right) = \frac{1}{\det\left(\frac{dv}{du'}\right)} = \frac{1}{\det\left(\frac{\partial F_\phi^x(u')}{\partial u'}\right)}.$$

Then, we have

$$\begin{aligned}
&\int \log\left(\int \delta(F_\phi^x(u) - F_\phi^x(u'))r(u')du'\right) r(u)du \\
&= \int \log\left(\int \delta(F_\phi^x(u) - v)\frac{r(H_x(v))}{\left|\det\left(\frac{\partial F_\phi^x(u')}{\partial u'}\right)\right|}dv\right) r(u)du
\end{aligned}$$

$$= \int \log \left(\frac{r(H_x(F_\phi^x(u)))}{\left| \det \left(\frac{\partial F_\phi^x(u')}{\partial u'} \right) \right|_{v=F_\phi^x(u)}} \right) r(u)du$$

$$= \int \log r(u) r(u)du - \int \log \left| \det \left(\frac{\partial F_\phi^x(u)}{\partial u} \right) \right| r(u)du.$$

By collecting terms together, we have

$$\begin{aligned} \ell_{ELBO}(x, \phi) := &- \int \log p_\theta(x|F_\phi^x(u)) r(u)du \\ &+ \int \log \left(\frac{r(u)}{p(F_\phi^x(u))} \right) r(u)du \\ &- \int \log \left| \det \left(\frac{\partial F_\phi^x(u)}{\partial u} \right) \right| r(u)du. \end{aligned}$$

This concludes the proof. □

Proposition 13.3 is a universal result that can be applied to the VAE, normalizing flow, etc. Major differences between them come from the choice of the encoder $F_\phi^x(u)$. In particular, for the case of the VAE [174], the following form of the encoder function $F_\phi^x(u)$ is used:

$$z = F_\phi^x(u) = \mu_\phi(x) + \sigma_\phi(x) \odot u, \quad u \sim \mathcal{N}(0, I_d), \tag{13.63}$$

where I_d is the $d \times d$ identity matrix and d is the dimension of the latent space. This was referred to as the reparameterization trick in the original VAE paper [174]. Under this choice, the second term in (13.61) becomes

$$\begin{aligned} &\int \log \left(\frac{r(u)}{p(F_\phi^x(u))} \right) r(u)du \\ &= - \int \frac{1}{2} \|u\|^2 r(u)du + \int \frac{1}{2} \|\mu(x) + \sigma(x) \odot u\|^2 r(u)du \\ &= \frac{1}{2} \sum_{i=1}^{d} (\sigma_i^2(x) + \mu_i^2(x) - 1), \end{aligned} \tag{13.64}$$

whereas the third term becomes

$$-\int \log \left|\det\left(\frac{\partial F_\phi^x(u)}{\partial u}\right)\right| r(u)du = -\frac{1}{2}\sum_{i=1}^{d} \log \sigma_i^2(x). \tag{13.65}$$

Finally, the first term in (13.61) is the likelihood term, which can be represented as follows by assuming the Gaussian distribution:

$$\begin{aligned} &-\int \log p_\theta(x|F_\phi^x(u))r(u)du \\ &= \int \frac{1}{2}\|x - G_\theta(F_\phi^x(u))\|^2 r(u)du \\ &= \frac{1}{2}\int \|x - G_\theta(\mu_\phi(x) + \sigma_\phi(x) \odot u)\|^2 r(u)du. \end{aligned} \tag{13.66}$$

Therefore, the encoder and decoder parameter optimization problem for the VAE can be obtained as follows:

$$\min_{\theta,\phi} \ell_{VAE}(\theta, \phi),$$

where

$$\begin{aligned} \ell_{VAE}(\theta, \phi) = &\frac{1}{2}\int_X \int \|x - G_\theta(\mu_\phi(x) + \sigma_\phi(x) \odot u)\|^2 r(u)du d\mu(x) \\ &+ \frac{1}{2}\sum_{i=1}^{d} \int_X (\sigma_i^2(x) + \mu_i^2(x) - \log \sigma_i^2(x) - 1)d\mu(x). \end{aligned} \tag{13.67}$$

Once the neural network is trained, one of the very important advantages of the VAE is that we can simply control the decoder output by changing the random samples. More specifically, the decoder output is now given by

$$\hat{x}(u) = G_\theta(\mu_\phi(x) + \sigma_\phi(x) \odot u), \tag{13.68}$$

which has an explicit dependency on the random variable u. Therefore, for a given x, we can change the output by drawing sample u.

13.6.3 β-VAE

By inspection of VAE loss in (13.67), we can easily see that the first term represents the distance between the generated samples and the real ones, whereas the second term is the KL distance between the real latent space measure and posterior distribution. Therefore, VAE loss is a measure of the distances that considers both latent space and the ambient space between real and generated samples.

In fact, this observation nicely fits into our geometric view of the autoencoder illustrated in Fig. 13.2. Here, the ambient image space is $\mathcal{X}$, the real data distribution is μ, whereas the autoencoder output data distribution is μ_θ. The latent space is $\mathcal{Z}$. In the autoencoder, the generator G_θ corresponds to the decoder, which is a mapping from the latent space to the sample space, $G_\theta : \mathcal{Z} \mapsto \mathcal{X}$, realized by a deep network. Then, the goal of the decoder training is to make the push-forward measure $\mu_\theta = G_{\theta\#}\zeta$ as close as possible to the real data distribution μ. Additionally, an encoder F_ϕ maps from the real data in $\mathcal{X}$ to the latent space $F_\phi : \mathcal{X} \mapsto \mathcal{Z}$ so that the encoder pushes forward the measure μ to a distribution $\zeta_\phi = F_{\#}\mu$ in the latent space. Therefore, the VAE design problem can be formulated by minimizing the sum of the both distances, which are measured by average sample distance and KL distance, respectively.

Rather than giving uniform weights for both distances, β-VAE [175] relaxes this constraint of the VAE. Following the same incentive in the VAE, we want to maximize the probability of generating real data, while keeping the distance between the real and estimated posterior distributions small (say, under a small constant). This leads to the following β-VAE cost function:

$$\begin{aligned} &\ell_{\beta-VAE}(\theta, \phi) \\ =&\frac{1}{2}\int_{\mathcal{X}}\int \|x - G_\theta(\mu_\phi(x) + \sigma_\phi(x) \odot u)\|^2 r(u) du d\mu(x) \\ &+ \frac{\beta}{2}\sum_{i=1}^{d}\int_{\mathcal{X}} (\sigma_i^2(x) + \mu_i^2(x) - \log\sigma_i^2(x) - 1) d\mu(x), \end{aligned} \tag{13.69}$$

where β now controls the importance of the distance measure in the latent space. When $\beta = 1$, it is the same as the VAE. When $\beta > 1$, it applies a stronger constraint on the latent space.

As a higher β imposes more constraint on the latent space, it turns out that the latent space is more interpretable and controllable, which is known as the *disentanglement*. More specifically, if each variable in the inferred latent representation z is only sensitive to one single generative factor and relatively invariant to other factors, we will say this representation is disentangled or factorized. One benefit that often comes with disentangled representation is good interpretability and easy generalization to a variety of tasks. For some conditionally independent generative factors, keeping them disentangled is the most efficient representation, and β-VAE provides more disentangled representation. For example, the generated faces from the original VAE have various directions, whereas they are toward specific directions in the β-VAE, implying that factors for the face directions is successfully disentangled [175].

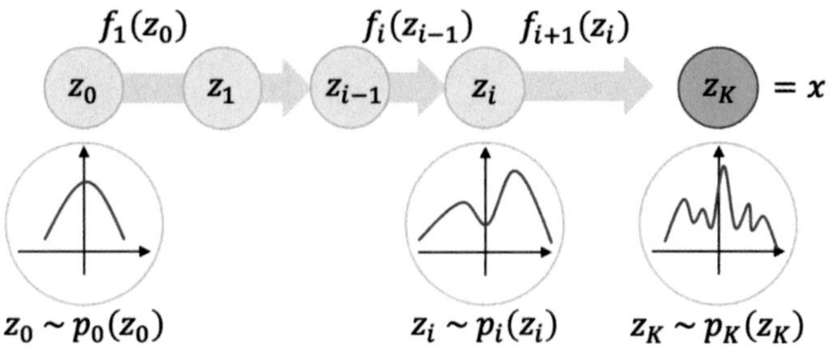

Fig. 13.8 Concept of normalizing flow

13.6.4 Normalizing Flow, Invertible Flow

The normalizing flow (NF) [178–181] is a modern way of overcoming the limitation of VAE. As shown in Fig. 13.8, normalizing flow transforms a simple distribution into a complex one by applying a sequence of invertible transformation functions. Flowing through a chain of transformations, we repeatedly substitute the variable for the new one according to the change-of-variables theorem and eventually obtain a probability distribution of the final target variable. Such a sequence of invertible transformations is the origin of the name "normalizing flow" [179].

The derivation of the cost function for a normalizing flow also starts with the same ELBO and encoder model in (13.60). However, the normalizing flow chooses a different encoder function:

$$z = F_\phi^x(u) = F_\phi(\sigma u + x), \tag{13.70}$$

where F_ϕ is an invertible function. Here, the invertibility is the key component, so the algorithm is often called the invertible flow. Specifically, if we choose the decoder as the inverse of the encoder function, i.e. $G_\theta = F_\phi^{-1}$, a very interesting phenomenon happens. More specifically, the first term in (13.61) can be simplified as follows:

$$\begin{aligned} & -\int \log p_\theta(x|F_\phi^x(u))r(u)du \\ &= \frac{1}{2}\int \|x - G_\theta(F_\phi^x(u))\|^2 r(u)du \\ &= \frac{1}{2}\int \|x - G_\theta(F_\phi(\sigma u + x))\|^2 r(u)du \\ &= \frac{1}{2}\int \|\sigma u\|^2 r(u)du = \frac{\sigma^2}{2}, \end{aligned}$$

which becomes a constant. Therefore, it is no longer necessary to consider the decoder part in the parameter estimation. Accordingly, aside from the constant term, the ELBO loss in (13.61) can be simplified as

$$\begin{aligned}\ell_{flow}(x,\phi) = &-\int \log\left(p(F_\phi^x(u))\right) r(u)du \\ &-\int \log\left|\det\left(\frac{\partial F_\phi^x(u)}{\partial u}\right)\right| r(u)du,\end{aligned} \tag{13.71}$$

where we have also removed the $\int \log r(u)r(u)du$ term since this is also a constant. For the Gaussian assumption for $p(z)$, (13.71) can be further simplified as

$$\begin{aligned}\ell_{flow}(x,\phi) =&\frac{1}{2}\int \|F_\phi(\sigma u + x)\|^2 r(u)du \\ &-\int \log\left|\det\left(\frac{\partial F_\phi(\sigma u + x)}{\partial u}\right)\right| r(u)du.\end{aligned} \tag{13.72}$$

Now the main technical difficulty of NF arises from the last term, which involves a complicated determinant calculation for a huge matrix. As discussed before, NF mainly focuses on the encoder function F_ϕ (and, likewise, the decoder G), which is composed of a sequence of transformations:

$$F_\phi(u) = (h_K \circ h_{K-1} \circ \cdots \circ h_1)(u), \tag{13.73}$$

Using the change-of-variable formula,

$$\frac{\partial F_\phi(u)}{\partial u} = \frac{h_K}{\partial h_{K-1}} \cdots \frac{h_2}{\partial h_1}\frac{h_1}{\partial u}, \tag{13.74}$$

we have

$$\log\left|\det\left(\frac{\partial F_\phi(u)}{\partial u}\right)\right| = \sum_{i=1}^{K} \log\left|\det\left(\frac{\partial h_i}{\partial h_{i-1}}\right)\right|, \tag{13.75}$$

where $h_0 = u$. Therefore, most of the current research efforts for NF have focused on how to design an invertible block such that the determinant calculation is simple. Now, we review a few representative techniques.

NICE (nonlinear independent component estimation) [178] is based on learning a non-linear bijective transformation between the data space and a latent space. The architecture is composed of a series of blocks defined as follows, where x_1 and x_2 are a partition of the input in each layer, and y_1 and y_2 are partitions of the output.

Then, the NICE update is given by

$$\begin{aligned} y_1 &= x_1, \\ y_2 &= x_2 + \mathcal{F}(x_1), \end{aligned} \tag{13.76}$$

where $\mathcal{F}(\cdot)$ is a neural network. Then, the block inversion can be readily done by

$$\begin{aligned} x_1 &= y_1, \\ x_2 &= y_2 - \mathcal{F}(y_1). \end{aligned} \tag{13.77}$$

Furthermore, it is easy to see that its Jacobian has a unit determinant and the cost function in (13.72) and its gradient can be tractably computed.

However, this architecture imposes some constraints on the functions the network can represent; for instance, it can only represent volume-preserving mappings. Follow-up work [180] addressed this limitation by introducing a new reversible transformation. More specifically, they extend the space of such models using real-valued non-volume-preserving (real NVP) transformations using the following operation [180]:

$$\begin{aligned} y_1 &= x_1, \\ y_2 &= x_2 \odot \exp(s(x_1)) + t(x_1), \end{aligned} \tag{13.78}$$

where s denotes point-wise scaling, t is referred to as a translation network, and $\odot$ is the element-wise multiplication. Then, the corresponding Jacobian matrix is given by

$$\frac{\partial y}{\partial x} = \begin{bmatrix} I_d & 0 \\ \frac{\partial y_2}{\partial x_1} & \operatorname{diag}(\exp(s(x_1))) \end{bmatrix}. \tag{13.79}$$

Given the observation that this Jacobian is triangular, we can efficiently compute its determinant as

$$\det\left(\frac{\partial y}{\partial x}\right) = \exp\left(\sum_j s(x_1[j])\right), \tag{13.80}$$

where $x_1[j]$ denotes the j-th element of x_1. The inverse of the transform can also be easily implemented by

$$\begin{aligned} x_1 &= y_1, \\ x_2 &= (y_2 - t(y_1)) \odot \exp(-s(y_1)). \end{aligned} \tag{13.81}$$

The corresponding block architecture is illustrated in Fig. 13.9.

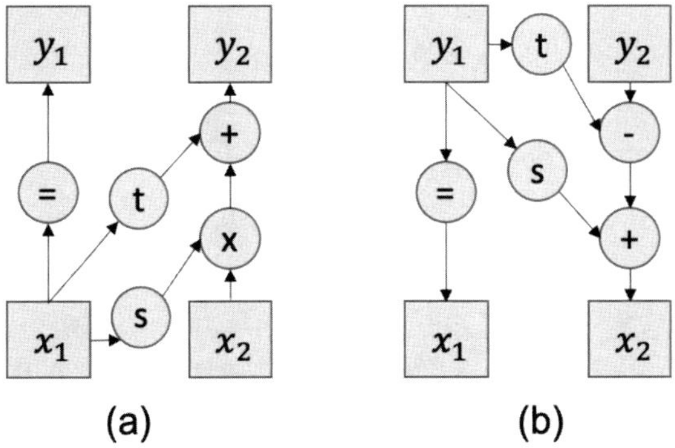

Fig. 13.9 Forward and inverse architecture of a building block in real NVP transform [180]. (**a**) Forward propagation. (**b**) Inverse propagation

Fig. 13.10 Example of normalizing flow using GLOW [181]. Figure courtesy of https://openai.com/blog/glow/

Due to the successive applications of transforms, one of the important advantages of NF is the gradual changes of the distribution. Figure 13.10 shows examples using GLOW—the generative flow using 1×1 invertible convolution [181]. As the name indicates, GLOW has additional 1×1 invertible convolution blocks to increase the expressiveness of the network.

13.7 Unsupervised Learning via Image Translation

So far, we have discussed generative models that generate samples from noise. Generative models are also useful to convert one distribution to another. This is why generative models become the main workhorse for unsupervised learning tasks.

Among the various unsupervised learning tasks, in this section we are mainly focusing on image translation, which is a very active area of research.

13.7.1 *Pix2pix*

Pix2pix [194] was presented in 2016 by researchers from Berkeley in their work "Image-to-Image Translation with Conditional Adversarial Networks." This is not unsupervised learning per se, as it requires matched data sets, but it opens a new era of image translation, so we review this here.

Most of the problems in image processing and computer vision can be posed as "translating" an input image into a corresponding output image. For example, a scene may be rendered as an RGB image, a gradient field, an edge map, a semantic label map, etc. In analogy to automatic language translation, we define automatic image-to-image translation as the task of translating one possible representation of a scene into another, given a large amount of training data.

Pix2pix uses a generative adversarial network (GAN) [88] to learn a function to map from an input image to an output image. The network is made up of two main pieces, the generator, and the discriminator. The generator transforms the input image to get the output image. The discriminator measures the similarity of the generated image to the target image from the data set, and tries to guess if this was produced by the generator.

For example, in Fig. 13.11, the generator produces a photo-realistic shoe image from a sketch, and the discriminator tries to differentiate whether the generated images are the real photo from the sketch or the fake one.

The nice thing about pix2pix is that it is generic and does not require the user to define any relationship between the two types of images. It makes no assumptions about the relationship and instead learns the objective during training, by comparing the defined inputs and outputs during training and inferring the objective. This makes pix2pix highly adaptable to a wide variety of situations, including ones where it is not easy to verbally or explicitly define the task we want to model.

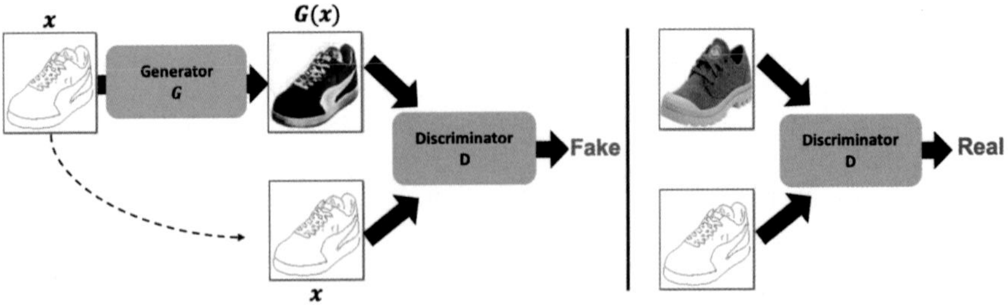

Fig. 13.11 Discriminator concept in pix2pix

That said, one downside of pix2pix is that it requires paired data sets to learn their relationship, and these are often difficult to obtain in practice. This issue is largely addressed by cycleGAN [185], which is the topic of the following section.

13.7.2 CycleGAN

Image-to-image translation is an important task in computer vision and graphics problems. Examples include:

- Translating summer landscapes to winter landscapes (or the reverse).
- Translating paintings to photographs (or the reverse).
- Translating horses to zebras (or the reverse).

As discussed before, pix2pix [194] is designed for such tasks, but it requires paired examples, specifically, a large data set of many examples of input images in the domain $\mathcal{X}$ (e.g. sketches of shoes) and the same images with the desired modification that can be used as the expected output images in $\mathcal{Y}$ (e.g. photos of shoes) (see the left column of Fig. 13.12). The requirement for a paired training data set is a limitation. These data sets are challenging and are even impossible to collect, e.g. photos of zebras and horses with exactly the same poses, size, etc.

Rather the unpaired situation in Fig. 13.12 is more realistic, where the collection of the images in $\mathcal{X}$ (for example, photos) and the unpaired collection of images in $\mathcal{Y}$ (for example, Monet's paintings) are available. Then, the goal of the image translation is to convert the distribution in $\mathcal{X}$ and $\mathcal{Y}$ and vice versa. In fact, the cycleGAN by Zhu et al. [185] demonstrated that such unpaired image translation is indeed possible.

The cycleGAN problem nicely fits into our geometric view of the autoencoder in Fig. 13.2, which is redrawn in Fig. 13.13 using a domain $\mathcal{Y}$. Accordingly, optimal transport (OT) [182, 184] provides a rigorous mathematical tool to understand the geometry of unsupervised learning by cycleGAN.

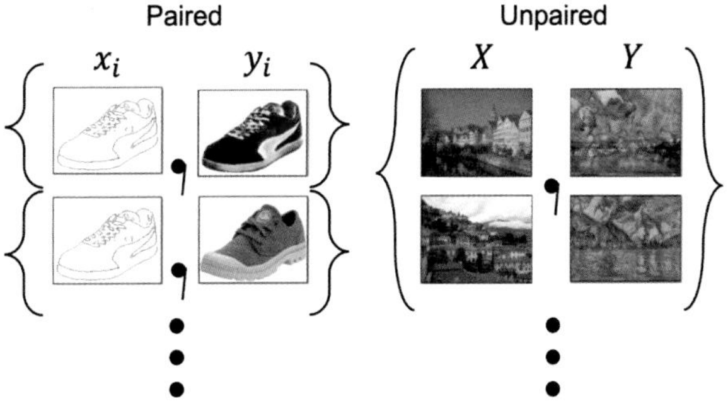

Fig. 13.12 Paired vs. unpaired image translation

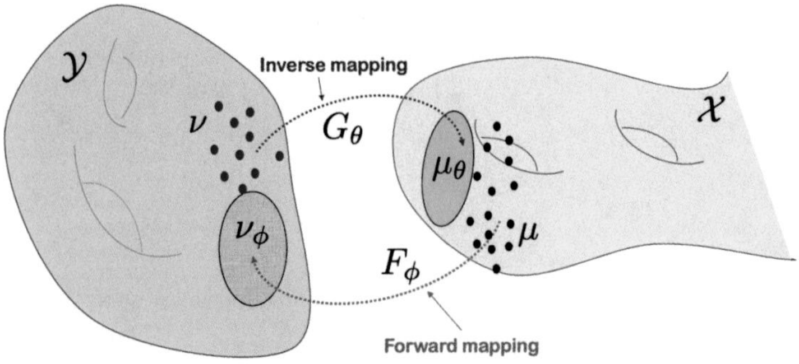

Fig. 13.13 Geometric view of CycleGAN-based unsupervised learning

Here, the target image space $\mathcal{X}$ is equipped with a probability measure μ, whereas the original image space $\mathcal{Y}$ has a probability measure ν. Since there are no paired data, the goal of unsupervised learning is to match the probability distributions rather than each individual sample. This can be done by finding transportation maps that transport the measure μ to ν, and vice versa. More specifically, the transportation from a measure space $(\mathcal{Y}, \nu)$ to another measure space $(\mathcal{X}, \mu)$ is done by a generator $G_\theta : \mathcal{Y} \mapsto \mathcal{X}$, realized by a deep network parameterized with θ. Then, the generator G_θ "pushes forward" the measure ν in $\mathcal{Y}$ to a measure μ_θ in the target space $\mathcal{X}$ [182, 184]. Similarly, the transport from $(\mathcal{X}, \mu)$ to $(\mathcal{Y}, \nu)$ is performed by another neural network generator F_ϕ, so that the generator F_ϕ pushes forward the measure μ in $\mathcal{X}$ to ν_ϕ in the original space $\mathcal{Y}$. Then, the optimal transport map for unsupervised learning can be achieved by minimizing the statistical distances $\text{dist}(\mu_\theta, \mu)$ between μ and μ_θ, and $\text{dist}(\nu_\phi, \nu)$ between ν and ν_ϕ, and our proposal is to use the Wasserstein-1 metric as a means to measure the statistical distance.

More specifically, for the choice of a metric $d(x, x') = \|x - x'\|$ in $\mathcal{X}$, the Wasserstein-1 metric between μ and μ_θ can be computed by Villani [182], Peyré et al. [184]

$$W_1(\mu, \mu_\theta) = \inf_{\pi \in \Pi(\mu,\nu)} \int_{\mathcal{X}\times\mathcal{Y}} \|x - G_\theta(y)\| d\pi(x, y). \tag{13.82}$$

Similarly, the Wasserstein-1 distance between ν and ν_ϕ is given by

$$W_1(\nu, \nu_\phi) = \inf_{\pi \in \Pi(\mu,\nu)} \int_{\mathcal{X}\times\mathcal{Y}} \|F_\phi(x) - y\| d\pi(x, y). \tag{13.83}$$

Rather than minimizing (13.82) and (13.83) separately with distinct joint distributions, a better way of finding the transportation map is to minimize them together with the same joint distribution π:

$$\inf_{\pi\in\Pi(\mu,\nu)} \int_{\mathcal{X}\times\mathcal{Y}} \|x - G_\theta(y)\| + \|F_\phi(x) - y\| d\pi(x, y). \tag{13.84}$$

One of the most important contributions of [195] is to show that the primal formulation of the unsupervised learning in (13.84) can be represented by a dual formulation:

$$\min_{\phi,\theta} \max_{\psi,\varphi} \ell_{cycleGAN}(\theta, \phi; \psi, \varphi), \tag{13.85}$$

where

$$\ell_{cycleGAN}(\theta, \phi; \psi, \varphi) := \lambda\ell_{cycle}(\theta, \phi) + \ell_{Disc}(\theta, \phi; \psi, \varphi), \tag{13.86}$$

where $\lambda > 0$ is the hyper-parameter, and the cycle-consistency term is given by

$$\begin{aligned}\ell_{cycle}(\theta, \phi) &= \int_{\mathcal{X}} \|x - G_\theta(F_\phi(x))\| d\mu(x) \\ &\quad + \int_{\mathcal{Y}} \|y - F_\phi(G_\theta(y))\| d\nu(y),\end{aligned}$$

whereas the second term is

$$\begin{aligned}\ell_{Disc}(\theta, \phi; \psi, \varphi) &= \max_{\varphi} \int_{\mathcal{X}} \varphi(x) d\mu(x) - \int_{\mathcal{Y}} \varphi(G_\theta(y)) d\nu(y) \\ &\quad + \max_{\psi} \int_{\mathcal{Y}} \psi(y) d\nu(y) - \int_{\mathcal{X}} \psi(F_\phi(x)) d\mu(x).\end{aligned} \tag{13.87}$$

Here, φ, ψ are often called Kantorovich potentials and satisfy the 1-Lipschitz condition (i.e.

$$\begin{aligned}|\varphi(x) - \varphi(x')| &\le \|x - x'\|, \ \forall x, x' \in \mathcal{X}, \\ |\psi(y) - \psi(y')| &\le \|y - y'\|, \ \forall y, y' \in \mathcal{Y}.\end{aligned}$$

In the machine learning context, the 1-Lipschitz potentials φ and ψ correspond to the Wasserstein-GAN (W-GAN) discriminators [177]. Specifically, φ corresponds to a discriminator to differentiate fake samples in real and generated images in $\mathcal{X}$, whereas ψ is a discriminator to tell the fake and real samples in the domain $\mathcal{Y}$. Moreover, the cycle-consistency term ℓ_{cycle} works to impose the one-to-one

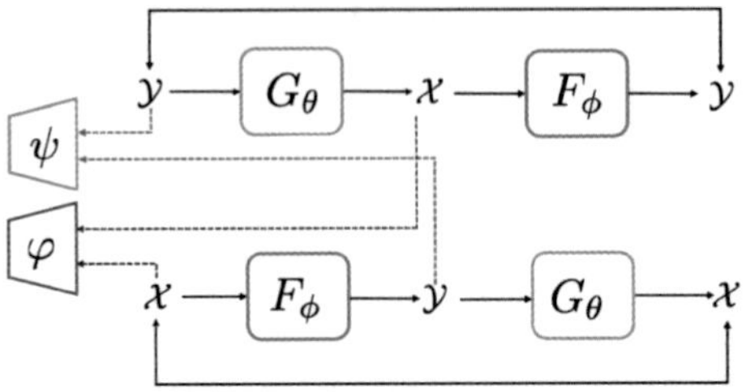

Fig. 13.14 CycleGAN network architecture

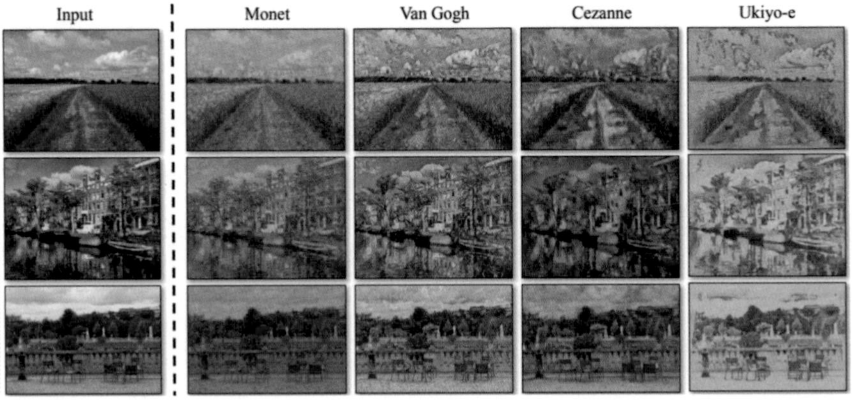

Fig. 13.15 Unsupervised style transfer in paintings

correspondence between the original and target domain, removing the mode-collapsing behaviors of GANs. The corresponding network architecture can be represented in Fig. 13.14. Specifically, φ tries to find the difference between the true image x and the generated image $G_\theta(y)$, whereas ψ attempts to find the fake measurement data that are generated by the synthetic measurement procedure $F_\phi(x)$. In fact, this formulation is equivalent to the cycleGAN formulation [185] except for the use of 1-Lipschitz discriminators.

CycleGAN has been very successful for various unsupervised learning tasks. Figure 13.15 shows the examples of unsupervised style transfers between two different styles of paintings.

13.7.3 StarGAN

In Fig. 13.15, one downside of cycleGAN is that we need to train separate generators for each pair of domains. For example, if there are N different styles in the paintings, there should be $N(N-1)$ distinct generators to translate the images (see Fig. 13.16a).

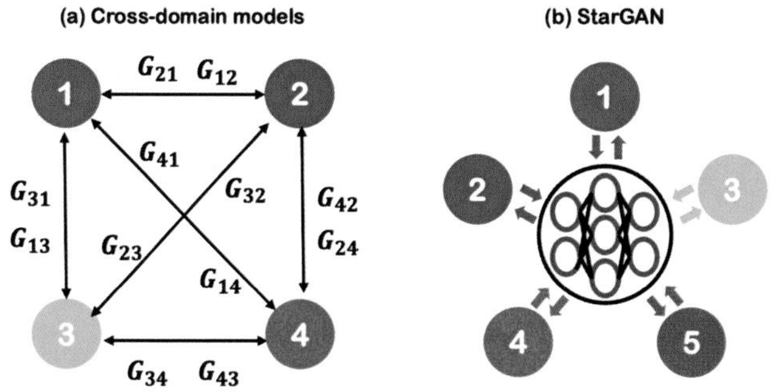

Fig. 13.16 Multi-domain translation: (**a**) cycleGAN, and (**b**) starGAN

To overcome the limitations of the scalability of cycleGAN, starGAN was proposed [87]. Specifically, as shown in Fig. 13.16b, one generator is trained such that it can translate into multiple domains by adding a mask vector that represents a target domain. This mask vector is augmented along the channel direction using one-hot vector encoding.

Given training data from two different domains, these models learn to translate images from one domain to the other. For example, changing the hair color (attribute) of a person from black (attribute value) to blond (attribute value). We denote a domain as a set of images sharing the same attribute value. People with black hair compose one domain and people with blond hair compose another domain. Here, the discriminator has two things to do. It should be able to identify whether an image is fake or not. With the help of an auxiliary classifier, the discriminator can also predict the domain of the image given as input to the discriminator (see Fig. 13.17).

With the auxiliary classifier, the discriminator learns the mapping of the original image and its corresponding domain from the data set. When the generator generates a new image conditioned on a target domain c (say blond hair), the discriminator can predict the generated image's domain so G will generate new images till the discriminator can predict it as target domain c (blond hair). Figure 13.18 shows such an example of multi-domain translation using a single starGAN generator.

13.7.4 Collaborative GAN

In many applications requiring multiple inputs to obtain the desired output, if any of the input data is missing, it often introduces large amounts of bias. Although many techniques have been developed for imputing missing data, image imputation is still difficult due to the complicated nature of natural images. To address this problem, a novel framework collaborative GAN (CollaGAN) [186] was proposed.

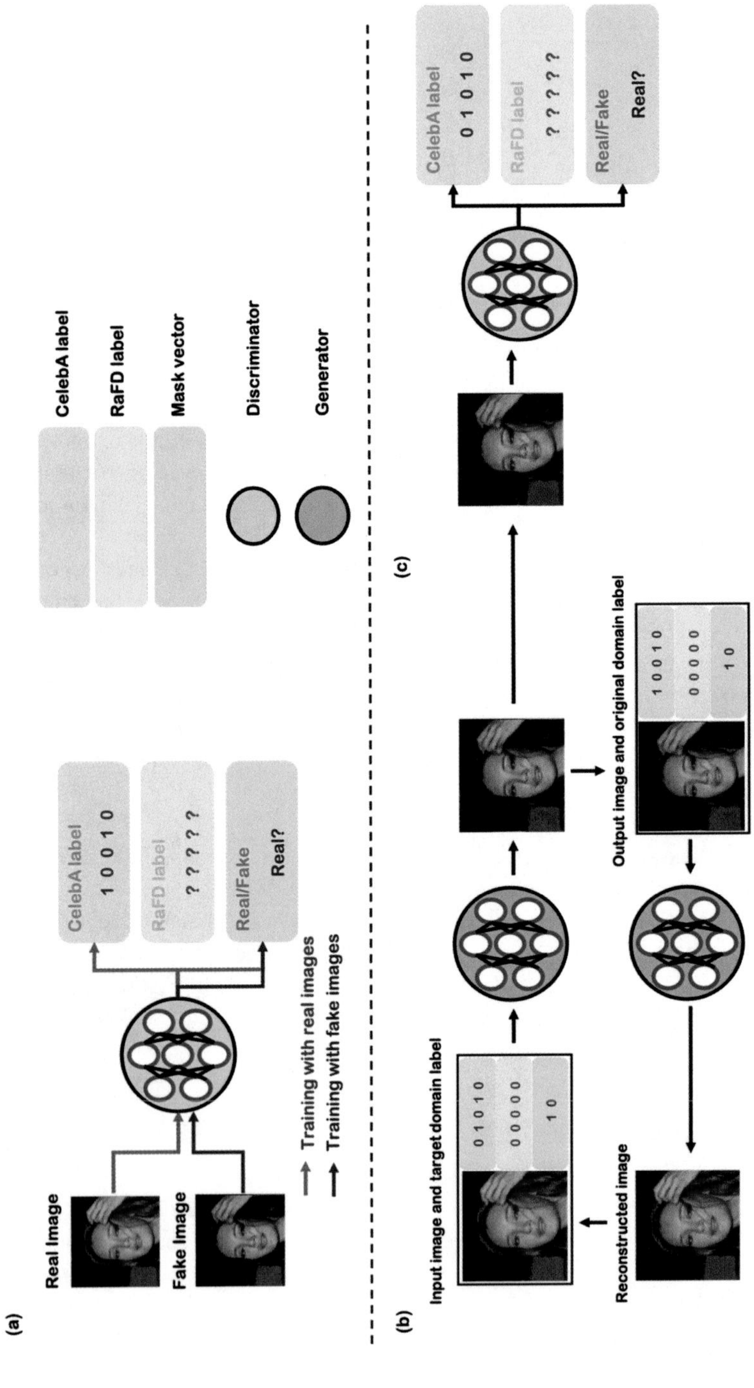

Fig. 13.17 Generator and discriminator architecture of starGAN [87]. **(a)** Training discriminator. **(b)** Training generator **(c)** Fooling the discriminator

Fig. 13.18 Examples of multi-domain translation using a single StarGAN generator

Specifically, CollaGAN converts an image imputation problem to a multi-domain images-to-image translation task so that a single generator and discriminator network can successfully estimate the missing data using the remaining clean data set. More specifically, CycleGAN and StarGAN are interested in transferring one image to another, as shown in Fig. 13.19a,b without considering the remaining domain data set. However, in image imputation problems, the missing data occurs infrequently, and the goal is to estimate the missing data by utilizing the other clean data set. Therefore, an image imputation problem can be correctly described as in Fig. 13.19c, where one generator can estimate the missing data using the remaining

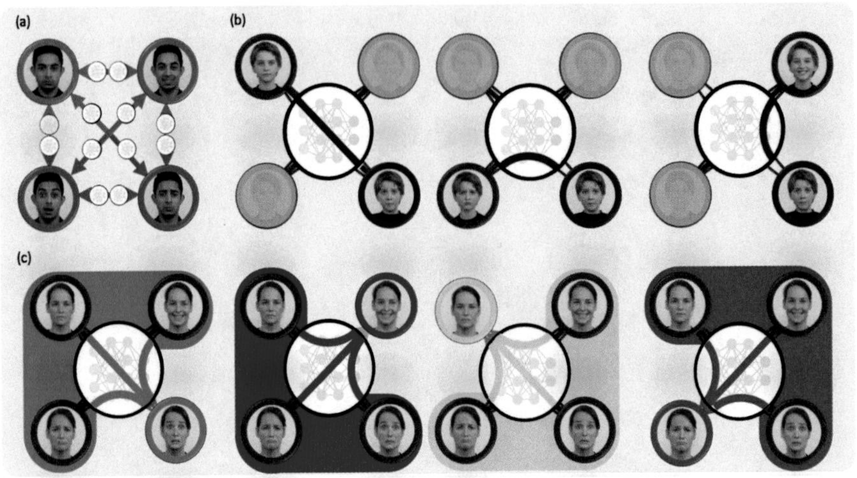

Fig. 13.19 Comparison with various multi-domain translation architecture. (**a**) Cross-domain models. (**b**) StarGAN. (**c**) Collaborative GAN

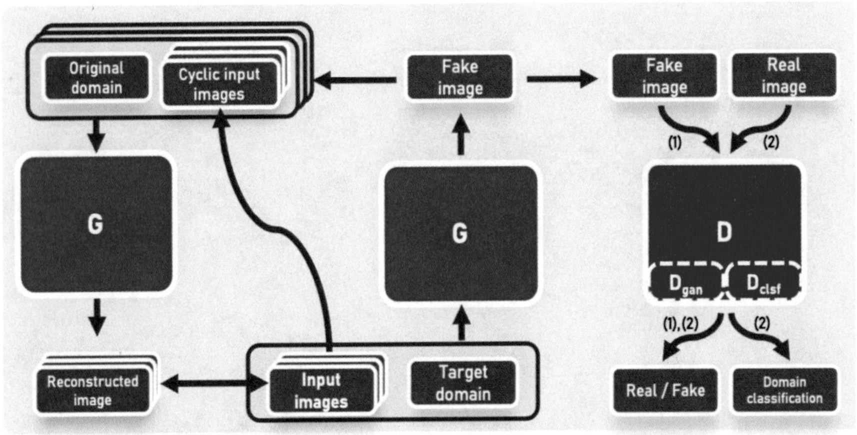

Fig. 13.20 CollaGAN generator and discriminator architecture

clean data set. Since the missing data domain is not difficult to estimate a priori, the imputation algorithm should be designed such that one algorithm can estimate the missing data in *any* domain by exploiting the data for the rest of the domains.

Due to the specific applications, CollaGAN is not an unsupervised learning method. However, one of the key concepts in CollaGAN is the cycle consistency for multiple inputs, which is useful for other applications. Specifically, since the inputs are multiple images, the cycle loss should be redefined. In particular, for the N-domain data, from a generated output, we should be able to generate $N - 1$ new combinations as the other inputs for the backward flow of the generator (Fig. 13.20

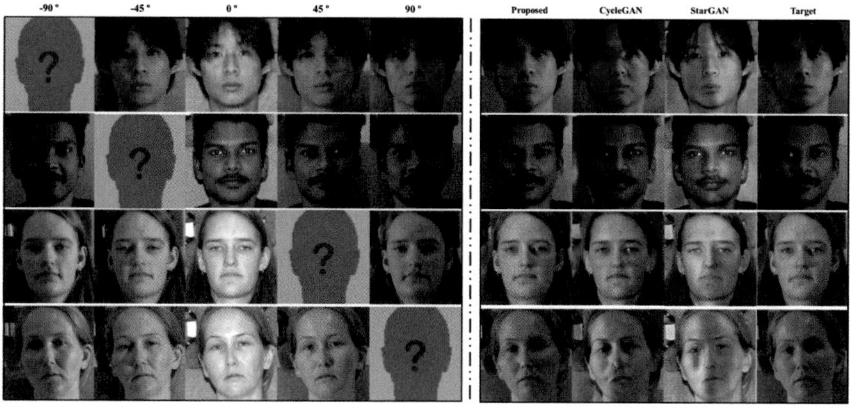

Fig. 13.21 Missing image imputation results from CollaGAN

middle). For example, when $N = 4$, there are three combinations of multi-input and single-output so that we can reconstruct the three images of original domains using backward flow of the generator. In regard to the discriminator, the discriminator should have a classifier header as well as the discriminator part similar to that of StarGAN.

Figure 13.21 shows an example of missing domain imputation, where CollaGAN produces very realistic images.

13.8 Summary and Outlook

So far we have discussed exciting field of deep learning—generative models. This is nonetheless an inclusive review, as there are so many exciting other algorithms. Here, the main emphasis is to provide a unified mathematical view to understand the various algorithms. As emphasized in the chapter, this field is important not only due to the fancy applications, but also from firm mathematical backgrounds that are grounded. As Yann LeCun said, unsupervised learning is the core of deep learning, so there will be many exciting new applications and opportunities for developing new theory, so young researchers are invited to participate in this exciting field.

13.9 Exercises

1. Show the following equality:

$$D_{JS}(P||Q) = \frac{1}{2}D_{KL}(P||M) + \frac{1}{2}D_{KL}(Q||M), \tag{13.88}$$

where $M = (P + Q)/2$.

2. Show that for JS divergence, absolute continuity is not necessary.
3. For the following generator function $f(u)$, derive (1) the f-divergence form, and (2) the f-GAN formulation from the definition of f-divergence using convex dual.

   (a) $f(u) = (u+1)\log\frac{2}{u+1} + u\log u$.
   (b) $f(u) = u\log u$.
   (c) $f(u) = (u-1)^2$.

4. Let μ and ν denote 1-D probability measures with the cumulative distribution functions F and G, respectively. Show that the Wasserstein-p distance between μ and ν is given by (13.21).
5. Prove Eq. (13.22).
6. Prove Eq. (13.26).
7. Derive the optimal transport map T in (13.27) between two Gaussian distributions.
8. Show that the AdaIN can be interpreted as the optimal transport between two i.i.d. Gaussian distributions.
9. Let the transport cost $c(x, y) : \mathcal{X}\times\mathcal{Y} \to \mathbb{R}\cup\{\infty\}$ be given by $c(x, y) = h(x-y)$ with h strictly convex.

   (a) Show that there exists a Kantorovich potential φ such that the optimal transport plan T that transports the measure μ in $\mathcal{X}$ to ν in $\mathcal{Y}$ can be represented as

   $$T(x) = x - (\nabla h)^{-1}\nabla\varphi(x). \tag{13.89}$$

   where $(\nabla h)^{-1}$ denotes the inverse function of ∇h.
   (b) As a special case, if $h(x-y) = \frac{1}{2}\|x-y\|^2$, show that the optimal transport map can be represented by

   $$y = T(x) = \nabla u(x),$$

   where $u(x) := x^2/2 - \varphi(x)$ is convex for some function $\varphi(x)$.

10. Prove (13.64).
11. Prove (13.66).
12. For the given reparametrization trick in the VAE

$$z = F_\phi^x(u) = \mu_\phi(x) + \sigma_\phi(x) \odot u, \quad u \sim \mathcal{N}(0, I_d), \tag{13.90}$$

where $x \in \mathbb{R}^n$, $z, u \in \mathbb{R}^d$ and $\mu_\phi(\cdot), \sigma_\phi(\cdot) : \mathbb{R}^n \mapsto \mathbb{R}^d$ and $\odot$ is the element-wise multiplication, show the following equality:

$$-\int \log \left|\det\left(\frac{\partial F_\phi^x(u)}{\partial u}\right)\right| r(u)du = -\frac{1}{2}\sum_{i=1}^{d} \log \sigma_i^2(x),$$

where $r(u)$ is the probability density function.

13. What are the advantages and disadvantages of the β-VAE over the VAE?
14. Consider the NICE update for the normalizing flow given by

$$y_1 = x_1, \quad y_2 = x_2 + \mathcal{F}(y_1). \tag{13.91}$$

(a) Why does the Jacobian term become the identity? Please derive explicitly.
(b) Suppose we are interested in a more expressive network given by

$$y_1 = x_1 + \mathcal{G}(x_2), \quad y_2 = x_2 + \mathcal{F}(y_1) \tag{13.92}$$

for some function $\mathcal{G}$. What is the inverse operation? How can you make the corresponding normalizing flow cost function simple in terms of Jacobian calculation? You may want to split the update into two steps to simplify the derivation.

Chapter 14
Summary and Outlook

With the tremendous success of deep learning in recent years, the field of data science has undergone unprecedented changes that can be considered a “revolution”. Despite the great successes of deep learning in various areas, there is a tremendous lack of rigorous mathematical foundations which enable us to understand why deep learning methods perform well. In fact, the recent development of deep learning is largely empirical, and the theory that explains the success remains seriously behind. For this reason, until recently, deep learning was viewed as pseudoscience by rigorous scientists, including mathematicians.

In fact, the success of deep learning appears very mysterious. Although sophisticated network architectures have been proposed by many researchers in recent years, the basic building blocks of deep neural networks are the convolution, pooling and nonlinearity, which from a mathematical point of view are regarded as very primitive tools from the “Stone Age”. However, one of the most mysterious aspects of deep learning is that the cascaded connection of these “Stone Age” tools results in superior performance that far exceeds the sophisticated mathematical tools. Nowadays, in order to develop high-performance data processing algorithms, we do not have to hire highly educated doctoral students or postdocs, but only give TensorFlow and many training data to undergraduate students. Does it mean a dark age of mathematics? Then, what is the role of the mathematicians in this data-driven world?

A popular explanation for the success of deep neural networks is that the neural network was developed by mimicking the human brain and is therefore destined for success. In fact, as discussed in Chap. 5, one of the most famous numerical experiments is the emergence of the hierarchical features from a deep neural network when it is trained to classify human faces. Interestingly, this phenomenon is similarly observed in human brains, where hierarchical features of the objects emerge during visual information processing. Based on these numerical observations, some of the artificial neural network “hardliners” even claim that instead of mathematics we need to investigate the biology of the brain to design more sophisticated artificial neural networks and to understand the working principle of

J. C. Ye, *Geometry of Deep Learning*, Mathematics in Industry 37,
https://doi.org/10.1007/978-981-16-6046-7_14

artificial neural networks. However, when neuroscientists (especially computational neuroscientists) were asked why the brain extracts such hierarchical features, it was surprising to find out that they usually rely on numerical simulations with artificial neural networks to explain how hierarchical properties arise in the brain. From a mathematical point of view, this is a typical example of "circular proof", an apparent logical fallacy.

Then, how can we fill in the gap between empirical success and the lack of the theory? In fact, one of the lessons we learn from the history of science is that the gap between the empirical observation and the lack of theory is not the limiting factor, but suggests the birth of a new science. For example, during the "golden age of physics" in the early twentieth century, some of the most exciting empirical discoveries in physics were quantum phenomena. Experimental physicists discovered many exotic quantum phenomena that could not be explained by either Newtonian or relativistic physics. In fact, there was a serious lag in the theoretical physics that could explain newly discovered quantum phenomena. Mathematical models were further developed, questioned, and refuted by the empirical observations. Even the greatest Albert Einstein said that he could not believe quantum physics since "God does not play dice with the universe." During these intense intellectual efforts to explain the seemingly unexplainable empirical observations, the new theory of quantum mechanics was rigorously formed, which led to numerous Nobel laureates; and new mathematics such as functional analysis, harmonic analysis, etc., has become mainstream in the modern mathematics. In fact, these efforts by scientists completely changed the landscape of physics and mathematics.

Similarly, now there is a great need to develop mathematical theories to explain the enormous empirical success of deep neural networks. In fact, computer scientists and engineers who work on the implementation are like the experimental physicists who give endless inspiration, and the mathematicians and signal processors are like theoretical physicists who try to find the unified mathematical theory to explain the empirical discoveries. Therefore, contrary to the false belief that we are in the dark age of mathematics, we are now actually living in the "golden age", ready to discover the beautiful mathematical theory of deep learning that can completely change the field of mathematics. Therefore, this book has aimed to explore the mathematical theory of deep learning to crack open the black box of deep learning and open a new age of mathematics.

The field of deep learning is interdisciplinary in nature, and includes mathematics, data science, physics, biology, medicine, etc. Therefore, collaborative research efforts between mathematics and other fields are crucial. This is because empirical results not only give the inspiration for the mathematical theory, but provide a means to verify whether a mathematical theory is correct. Therefore, although this book primarily focuses on discovering the fundamental mathematical principles of deep learning, it is hoped that it will play an instrumental role promoting the basic sciences in physics, biology, chemistry, geophysics, etc. using deep learning, and enable readers to be inspired by new empirical problems to obtain better mathematical models.

Chapter 15
Bibliography

1. R. J. Duffin and A. C. Schaeffer, "A class of nonharmonic Fourier series," *Transactions of the American Mathematical Society*, vol. 72, no. 2, pp. 341–366, 1952.
2. P. R. Halmos, *Measure theory*. Springer, 2013, vol. 18.
3. W. H. Press, S. A. Teukolsky, W. T. Vetterling, and B. P. Flannery, *Numerical recipes 3rd edition: The art of scientific computing*. Cambridge University Press, 2007.
4. R. A. Horn, R. A. Horn, and C. R. Johnson, *Topics in matrix analysis*. Cambridge University Press, 1994.
5. K. Petersen, M. Pedersen *et al.*, "The matrix cookbook, vol. 7," *Technical University of Denmark*, vol. 15, 2008.
6. S. Boyd, S. P. Boyd, and L. Vandenberghe, *Convex optimization*. Cambridge University Press, 2004.
7. O. Russakovsky, J. Deng, H. Su, J. Krause, S. Satheesh, S. Ma, Z. Huang, A. Karpathy, A. Khosla, M. Bernstein *et al.*, "ImageNet large scale visual recognition challenge," *International Journal of Computer Vision*, vol. 115, no. 3, pp. 211–252, 2015.
8. J. Deng, W. Dong, R. Socher, L.-J. Li, K. Li, and L. Fei-Fei, "ImageNet: A large-scale hierarchical image database," in *2009 IEEE Conference on Computer Vision and Pattern Recognition*. IEEE, 2009, pp. 248–255.
9. A. Krizhevsky, I. Sutskever, and G. E. Hinton, "ImageNet classification with deep convolutional neural networks," in *Advances in Neural Information Processing Systems*, 2012, pp. 1097–1105.
10. V. Vapnik, *The nature of statistical learning theory*. Springer Science & Business Media, 2013.
11. B. Schölkopf, A. J. Smola, F. Bach *et al.*, *Learning with kernels: support vector machines, regularization, optimization, and beyond*. MIT Press, 2002.
12. D. G. Lowe, "Distinctive image features from scale-invariant keypoints," *International Journal of Computer Vision*, vol. 60, no. 2, pp. 91–110, 2004.
13. H. Bay, T. Tuytelaars, and L. Van Gool, "SURF: Speeded up robust features," in *European Conference on Computer Vision (ECCV)*. Springer, 2006, pp. 404–417.
14. W. D. Penny, K. J. Friston, J. T. Ashburner, S. J. Kiebel, and T. E. Nichols, *Statistical parametric mapping: the analysis of functional brain images*. Elsevier, 2011.
15. B. Schölkopf, R. Herbrich, and A. J. Smola, "A generalized representer theorem," in *International conference on computational learning theory*. Springer, 2001, pp. 416–426.
16. G. Salton and M. McGill, *Introduction to Modern Information Retrieval*. McGraw Hill Book Company, 1983.

J. C. Ye, *Geometry of Deep Learning*, Mathematics in Industry 37,
https://doi.org/10.1007/978-981-16-6046-7_15

17. E. R. Kandel, J. H. Schwartz, T. M. Jessell, S. Siegelbaum, and A. Hudspeth, *Principles of neural science*. McGraw-Hill New York, 2000, vol. 4.
18. G. M. Shepherd, *Neurobiology*. Oxford University Press, 1988.
19. J. G. Nicholls, A. R. Martin, B. G. Wallace, and P. A. Fuchs, *From neuron to brain*. Sinauer Associates Sunderland, MA, 2001, vol. 271.
20. D. H. Hubel and T. N. Wiesel, "Receptive fields of single neurones in the cat's striate cortex," *The Journal of Physiology*, vol. 148, no. 3, pp. 574–591, 1959.
21. Y. LeCun, B. Boser, J. S. Denker, D. Henderson, R. E. Howard, W. Hubbard, and L. D. Jackel, "Backpropagation applied to handwritten zip code recognition," *Neural Computation*, vol. 1, no. 4, pp. 541–551, 1989.
22. M. Riesenhuber and T. Poggio, "Hierarchical models of object recognition in cortex," *Nature Neuroscience*, vol. 2, no. 11, pp. 1019–1025, 1999.
23. R. Q. Quiroga, L. Reddy, G. Kreiman, C. Koch, and I. Fried, "Invariant visual representation by single neurons in the human brain," *Nature*, vol. 435, no. 7045, pp. 1102–1107, 2005.
24. V. Nair and G. E. Hinton, "Rectified linear units improve restricted Boltzmann machines," in *Proceedings of the 27th International Conference on Machine Learning (ICML-10)*, 2010, pp. 807–814.
25. J. Duchi, E. Hazan, and Y. Singer, "Adaptive subgradient methods for online learning and stochastic optimization," *Journal of Machine Learning Research*, vol. 12, no. 7, pp. 2121–2159, 2011.
26. T. Tieleman and G. Hinton, "Lecture 6.5-RMSprop: Divide the gradient by a running average of its recent magnitude," *COURSERA: Neural Networks for Machine Learning*, vol. 4, no. 2, pp. 26–31, 2012.
27. D. P. Kingma and J. Ba, "Adam: A method for stochastic optimization," *arXiv preprint arXiv:1412.6980*, 2014.
28. D. E. Rumelhart, G. E. Hinton, and R. J. Williams, "Learning representations by back-propagating errors," *Nature*, vol. 323, no. 6088, pp. 533–536, 1986.
29. I. M. Gelfand, R. A. Silverman *et al.*, *Calculus of variations*. Courier Corporation, 2000.
30. C. Szegedy, W. Liu, Y. Jia, P. Sermanet, S. Reed, D. Anguelov, D. Erhan, V. Vanhoucke, and A. Rabinovich, "Going deeper with convolutions," in *Proceedings of the IEEE Conference on Computer Vision and Pattern Recognition*, 2015, pp. 1–9.
31. K. Simonyan and A. Zisserman, "Very deep convolutional networks for large-scale image recognition," *arXiv preprint arXiv:1409.1556*, 2014.
32. J. Johnson, A. Alahi, and L. Fei-Fei, "Perceptual losses for real-time style transfer and super-resolution," in *European Conference on Computer Vision (ECCV)*, 2016, pp. 694–711.
33. K. He, X. Zhang, S. Ren, and J. Sun, "Deep residual learning for image recognition," in *Proceedings of the IEEE Conference on Computer Vision and Pattern Recognition*, 2016, pp. 770–778.
34. H. Li, Z. Xu, G. Taylor, C. Studer, and T. Goldstein, "Visualizing the loss landscape of neural nets," in *Advances in Neural Information Processing Systems*, 2018, pp. 6389–6399.
35. J. C. Ye and W. K. Sung, "Understanding geometry of encoder-decoder CNNs," in *International Conference on Machine Learning*, 2019, pp. 7064–7073.
36. Q. Nguyen and M. Hein, "Optimization landscape and expressivity of deep CNNs," *arXiv preprint arXiv:1710.10928*, 2017.
37. G. Huang, Z. Liu, L. Van Der Maaten, and K. Q. Weinberger, "Densely connected convolutional networks," in *Proceedings of the IEEE Conference on Computer Vision and Pattern Recognition*, 2017, pp. 4700–4708.
38. O. Ronneberger, P. Fischer, and T. Brox, "U-Net: Convolutional networks for biomedical image segmentation," in *International Conference on Medical Image Computing and Computer-Assisted Intervention*. Springer, 2015, pp. 234–241.
39. K. H. Jin, M. T. McCann, E. Froustey, and M. Unser, "Deep convolutional neural network for inverse problems in imaging," *IEEE Transactions on Image Processing*, vol. 26, no. 9, pp. 4509–4522, 2017.

40. Y. Han and J. C. Ye, "Framing U-Net via deep convolutional framelets: Application to sparse-view CT," *IEEE Transactions on Medical Imaging*, vol. 37, no. 6, pp. 1418–1429, 2018.
41. S. Ioffe and C. Szegedy, "Batch normalization: Accelerating deep network training by reducing internal covariate shift," *arXiv preprint arXiv:1502.03167*, 2015.
42. J. C. Ye, Y. Han, and E. Cha, "Deep convolutional framelets: A general deep learning framework for inverse problems," *SIAM Journal on Imaging Sciences*, vol. 11, no. 2, pp. 991–1048, 2018.
43. J. Bruna and S. Mallat, "Invariant scattering convolution networks," *IEEE Transactions on Pattern Analysis and Machine Intelligence*, vol. 35, no. 8, pp. 1872–1886, 2013.
44. I. Goodfellow, Y. Bengio, and A. Courville, *Deep learning*. MIT Press, 2016.
45. N. Srivastava, G. Hinton, A. Krizhevsky, I. Sutskever, and R. Salakhutdinov, "Dropout: a simple way to prevent neural networks from overfitting," *The Journal of Machine Learning Research*, vol. 15, no. 1, pp. 1929–1958, 2014.
46. D. L. Donoho, "Compressed sensing," *IEEE Trans. Information Theory*, vol. 52, no. 4, pp. 1289–1306, 2006.
47. E. J. Candès and B. Recht, "Exact matrix completion via convex optimization," *Found. Comput. Math.*, vol. 9, no. 6, pp. 717–772, 2009.
48. G. Cybenko, "Approximation by superpositions of a sigmoidal function," *Mathematics of Control, Signals and Systems*, vol. 2, no. 4, pp. 303–314, 1989.
49. S. Ryu, J. Lim, S. H. Hong, and W. Y. Kim, "Deeply learning molecular structure-property relationships using attention-and gate-augmented graph convolutional network," *arXiv preprint arXiv:1805.10988*, 2018.
50. T. Mikolov, I. Sutskever, K. Chen, G. S. Corrado, and J. Dean, "Distributed representations of words and phrases and their compositionality," in *Advances in Neural Information Processing Systems*, 2013, pp. 3111–3119.
51. T. Mikolov, K. Chen, G. Corrado, and J. Dean, "Efficient estimation of word representations in vector space," *arXiv preprint arXiv:1301.3781*, 2013.
52. W. L. Hamilton, R. Ying, and J. Leskovec, "Representation learning on graphs: Methods and applications," *arXiv preprint arXiv:1709.05584*, 2017.
53. B. Perozzi, R. Al-Rfou, and S. Skiena, "DeepWalk: Online learning of social representations," in *Proceedings of the 20th ACM SIGKDD International Conference on Knowledge Discovery and Data Mining*, 2014, pp. 701–710.
54. A. Grover and J. Leskovec, "Node2vec: Scalable feature learning for networks," in *Proceedings of the 22nd ACM SIGKDD International Conference on Knowledge Discovery and Data Mining*, 2016, pp. 855–864.
55. M. M. Bronstein, J. Bruna, Y. LeCun, A. Szlam, and P. Vandergheynst, "Geometric deep learning: going beyond Euclidean data," *IEEE Signal Processing Magazine*, vol. 34, no. 4, pp. 18–42, 2017.
56. T. N. Kipf and M. Welling, "Semi-supervised classification with graph convolutional networks," *arXiv preprint arXiv:1609.02907*, 2016.
57. K. Xu, W. Hu, J. Leskovec, and S. Jegelka, "How powerful are graph neural networks?" *arXiv preprint arXiv:1810.00826*, 2018.
58. W. Hamilton, Z. Ying, and J. Leskovec, "Inductive representation learning on large graphs," in *Advances in Neural Information Processing Systems*, 2017, pp. 1024–1034.
59. C. Morris, M. Ritzert, M. Fey, W. L. Hamilton, J. E. Lenssen, G. Rattan, and M. Grohe, "Weisfeiler and Leman go neural: Higher-order graph neural networks," in *Proceedings of the AAAI Conference on Artificial Intelligence*, vol. 33, 2019, pp. 4602–4609.
60. Z. Chen, S. Villar, L. Chen, and J. Bruna, "On the equivalence between graph isomorphism testing and function approximation with GNNs," in *Advances in Neural Information Processing Systems*, 2019, pp. 15 868–15 876.
61. P. Barceló, E. V. Kostylev, M. Monet, J. Pérez, J. Reutter, and J. P. Silva, "The logical expressiveness of graph neural networks," in *International Conference on Learning Representations*, 2019.

62. M. Grohe, "word2vec, node2vec, graph2vec, x2vec: Towards a theory of vector embeddings of structured data," *arXiv preprint arXiv:2003.12590*, 2020.
63. N. Shervashidze, P. Schweitzer, E. J. Van Leeuwen, K. Mehlhorn, and K. M. Borgwardt, "Weisfeiler–Lehman graph kernels," *Journal of Machine Learning Research*, vol. 12, no. 77, pp. 2539–2561, 2011.
64. J. L. Ba, J. R. Kiros, and G. E. Hinton, "Layer normalization," *arXiv preprint arXiv:1607.06450*, 2016.
65. D. Ulyanov, A. Vedaldi, and V. Lempitsky, "Instance normalization: The missing ingredient for fast stylization," *arXiv preprint arXiv:1607.08022*, 2016.
66. Y. Wu and K. He, "Group normalization," in *Proceedings of the European Conference on Computer Vision (ECCV)*, 2018, pp. 3–19.
67. X. Huang and S. Belongie, "Arbitrary style transfer in real-time with adaptive instance normalization," in *Proceedings of the IEEE International Conference on Computer Vision*, 2017, pp. 1501–1510.
68. J. Hu, L. Shen, and G. Sun, "Squeeze-and-excitation networks," in *Proceedings of the IEEE Conference on Computer Vision and Pattern Recognition*, 2018, pp. 7132–7141.
69. P. Veličković, G. Cucurull, A. Casanova, A. Romero, P. Lio, and Y. Bengio, "Graph attention networks," *arXiv preprint arXiv:1710.10903*, 2017.
70. X. Wang, R. Girshick, A. Gupta, and K. He, "Non-local neural networks," in *Proceedings of the IEEE Conference on Computer Vision and Pattern Recognition*, 2018, pp. 7794–7803.
71. H. Zhang, I. Goodfellow, D. Metaxas, and A. Odena, "Self-attention generative adversarial networks," in *International conference on machine learning*. PMLR, 2019, pp. 7354–7363.
72. T. Xu, P. Zhang, Q. Huang, H. Zhang, Z. Gan, X. Huang, and X. He, "AttnGAN: Fine-grained text to image generation with attentional generative adversarial networks," in *Proceedings of the IEEE Conference on Computer Vision and Pattern Recognition*, 2018, pp. 1316–1324.
73. A. Vaswani, N. Shazeer, N. Parmar, J. Uszkoreit, L. Jones, A. N. Gomez, Ł. Kaiser, and I. Polosukhin, "Attention is all you need," in *Advances in Neural Information Processing Systems*, 2017, pp. 5998–6008.
74. J. Devlin, M.-W. Chang, K. Lee, and K. Toutanova, "BERT: Pre-training of deep bidirectional transformers for language understanding," *arXiv preprint arXiv:1810.04805*, 2018.
75. A. Radford, J. Wu, R. Child, D. Luan, D. Amodei, and I. Sutskever, "Language models are unsupervised multitask learners," *OpenAI Blog*, vol. 1, no. 8, p. 9, 2019.
76. T. B. Brown, B. Mann, N. Ryder, M. Subbiah, J. Kaplan, P. Dhariwal, A. Neelakantan, P. Shyam, G. Sastry, A. Askell *et al.*, "Language models are few-shot learners," *arXiv preprint arXiv:2005.14165*, 2020.
77. L. A. Gatys, A. S. Ecker, and M. Bethge, "Image style transfer using convolutional neural networks," in *Proceedings of the IEEE Conference on Computer Vision and Pattern Recognition*, 2016, pp. 2414–2423.
78. Y. Taigman, A. Polyak, and L. Wolf, "Unsupervised cross-domain image generation," *arXiv preprint arXiv:1611.02200*, 2016.
79. Y. Li, C. Fang, J. Yang, Z. Wang, X. Lu, and M.-H. Yang, "Universal style transfer via feature transforms," in *Advances in Neural Information Processing Systems*, 2017, pp. 386–396.
80. Y. Li, M.-Y. Liu, X. Li, M.-H. Yang, and J. Kautz, "A closed-form solution to photorealistic image stylization," in *Proceedings of the European Conference on Computer Vision (ECCV)*, 2018, pp. 453–468.
81. D. Y. Park and K. H. Lee, "Arbitrary style transfer with style-attentional networks," in *Proceedings of the IEEE Conference on Computer Vision and Pattern Recognition*, 2019, pp. 5880–5888.
82. J. Yoo, Y. Uh, S. Chun, B. Kang, and J.-W. Ha, "Photorealistic style transfer via wavelet transforms," in *Proceedings of the IEEE International Conference on Computer Vision*, 2019, pp. 9036–9045.
83. T. Park, M.-Y. Liu, T.-C. Wang, and J.-Y. Zhu, "Semantic image synthesis with spatially-adaptive normalization," in *Proceedings of the IEEE Conference on Computer Vision and Pattern Recognition*, 2019, pp. 2337–2346.

84. X. Huang, M.-Y. Liu, S. Belongie, and J. Kautz, "Multimodal unsupervised image-to-image translation," in *Proceedings of the European Conference on Computer Vision (ECCV)*, 2018, pp. 172–189.
85. J.-Y. Zhu, R. Zhang, D. Pathak, T. Darrell, A. A. Efros, O. Wang, and E. Shechtman, "Toward multimodal image-to-image translation," in *Advances in Neural Information Processing Systems*, 2017, pp. 465–476.
86. H.-Y. Lee, H.-Y. Tseng, J.-B. Huang, M. Singh, and M.-H. Yang, "Diverse image-to-image translation via disentangled representations," in *Proceedings of the European Conference on Computer Vision (ECCV)*, 2018, pp. 35–51.
87. Y. Choi, M. Choi, M. Kim, J.-W. Ha, S. Kim, and J. Choo, "StarGAN: Unified generative adversarial networks for multi-domain image-to-image translation," in *Proceedings of the IEEE Conference on Computer Vision and Pattern Recognition*, 2018, pp. 8789–8797.
88. I. Goodfellow, J. Pouget-Abadie, M. Mirza, B. Xu, D. Warde-Farley, S. Ozair, A. Courville, and Y. Bengio, "Generative adversarial nets," in *Advances in Neural Information Processing Systems*, 2014, pp. 2672–2680.
89. T. Karras, S. Laine, and T. Aila, "A style-based generator architecture for generative adversarial networks," in *Proceedings of the IEEE Conference on Computer Vision and Pattern Recognition*, 2019, pp. 4401–4410.
90. K. Zhang, W. Zuo, Y. Chen, D. Meng, and L. Zhang, "Beyond a Gaussian denoiser: Residual learning of deep CNN for image denoising," *IEEE Transactions on Image Processing*, vol. 26, no. 7, pp. 3142–3155, 2017.
91. M. Bear, B. Connors, and M. A. Paradiso, *Neuroscience: Exploring the brain*. Jones & Bartlett Learning, LLC, 2020.
92. K. Greff, R. K. Srivastava, J. Koutník, B. R. Steunebrink, and J. Schmidhuber, "LSTM: a search space odyssey," *IEEE Transactions on Neural Networks and Learning Systems*, vol. 28, no. 10, pp. 2222–2232, 2016.
93. J. Pérez, J. Marinković, and P. Barceló, "On the Turing completeness of modern neural network architectures," in *International Conference on Learning Representations*, 2018.
94. J.-B. Cordonnier, A. Loukas, and M. Jaggi, "On the relationship between self-attention and convolutional layers," *arXiv preprint arXiv:1911.03584*, 2019.
95. G. Marcus and E. Davis, "GPT-3, bloviator: OpenAI's language generator has no idea what it's talking about," *Technology Review*, 2020.
96. A. Dosovitskiy, L. Beyer, A. Kolesnikov, D. Weissenborn, X. Zhai, T. Unterthiner, M. Dehghani, M. Minderer, G. Heigold, S. Gelly *et al.*, "An image is worth 16×16 words: Transformers for image recognition at scale," *arXiv preprint arXiv:2010.11929*, 2020.
97. G. Kwon and J. C. Ye, "Diagonal attention and style-based GAN for content-style disentanglement in image generation and translation," *arXiv preprint arXiv:2103.16146*, 2021.
98. J. Xie, L. Xu, and E. Chen, "Image denoising and inpainting with deep neural networks," in *Advances in Neural Information Processing Systems*, 2012, pp. 341–349.
99. C. Dong, C. C. Loy, K. He, and X. Tang, "Image super-resolution using deep convolutional networks," *IEEE Transactions on Pattern Analysis and Machine Intelligence*, vol. 38, no. 2, pp. 295–307, 2015.
100. J. Kim, J. K. Lee, and K. Lee, "Accurate image super-resolution using very deep convolutional networks," in *Proceedings of the IEEE Conference on Computer Vision and Pattern Recognition*, 2016, pp. 1646–1654.
101. M. Telgarsky, "Representation benefits of deep feedforward networks," *arXiv preprint arXiv:1509.08101*, 2015.
102. R. Eldan and O. Shamir, "The power of depth for feedforward neural networks," in *29th Annual Conference on Learning Theory*, 2016, pp. 907–940.
103. M. Raghu, B. Poole, J. Kleinberg, S. Ganguli, and J. S. Dickstein, "On the expressive power of deep neural networks," in *Proceedings of the 34th International Conference on Machine Learning*. JMLR, 2017, pp. 2847–2854.
104. D. Yarotsky, "Error bounds for approximations with deep ReLU networks," *Neural Networks*, vol. 94, pp. 103–114, 2017.

105. R. Arora, A. Basu, P. Mianjy, and A. Mukherjee, "Understanding deep neural networks with rectified linear units," *arXiv preprint arXiv:1611.01491*, 2016.
106. S. Mallat, *A wavelet tour of signal processing*. Academic Press, 1999.
107. D. L. Donoho, "De-noising by soft-thresholding," *IEEE Transactions on Information Theory*, vol. 41, no. 3, pp. 613–627, 1995.
108. Y. C. Eldar and M. Mishali, "Robust recovery of signals from a structured union of subspaces," *IEEE Transactions on Information Theory*, vol. 55, no. 11, pp. 5302–5316, 2009.
109. R. Yin, T. Gao, Y. M. Lu, and I. Daubechies, "A tale of two bases: Local-nonlocal regularization on image patches with convolution framelets," *SIAM Journal on Imaging Sciences*, vol. 10, no. 2, pp. 711–750, 2017.
110. J. C. Ye, J. M. Kim, K. H. Jin, and K. Lee, "Compressive sampling using annihilating filter-based low-rank interpolation," *IEEE Transactions on Information Theory*, vol. 63, no. 2, pp. 777–801, 2016.
111. K. H. Jin and J. C. Ye, "Annihilating filter-based low-rank Hankel matrix approach for image inpainting," *IEEE Transactions on Image Processing*, vol. 24, no. 11, pp. 3498–3511, 2015.
112. K. H. Jin, D. Lee, and J. C. Ye, "A general framework for compressed sensing and parallel MRI using annihilating filter based low-rank Hankel matrix," *IEEE Transactions on Computational Imaging*, vol. 2, no. 4, pp. 480–495, 2016.
113. J.-F. Cai, B. Dong, S. Osher, and Z. Shen, "Image restoration: total variation, wavelet frames, and beyond," *Journal of the American Mathematical Society*, vol. 25, no. 4, pp. 1033–1089, 2012.
114. N. Lei, D. An, Y. Guo, K. Su, S. Liu, Z. Luo, S.-T. Yau, and X. Gu, "A geometric understanding of deep learning," *Engineering*, 2020.
115. B. Hanin and D. Rolnick, "Complexity of linear regions in deep networks," in *International Conference on Machine Learning*. PMLR, 2019, pp. 2596–2604.
116. B. Hanin and D. Rolnick. "Deep ReLU networks have surprisingly few activation patterns," *Advances in Neural Information Processing Systems*, vol. 32, pp. 361–370, 2019.
117. X. Zhang and D. Wu, "Empirical studies on the properties of linear regions in deep neural networks," *arXiv preprint arXiv:2001.01072*, 2020.
118. G. F. Montufar, R. Pascanu, K. Cho, and Y. Bengio, "On the number of linear regions of deep neural networks," in *Advances in Neural Information Processing Systems*, 2014, pp. 2924–2932.
119. Z. Allen-Zhu, Y. Li, and Z. Song, "A convergence theory for deep learning via over-parameterization," in *International Conference on Machine Learning*. PMLR, 2019, pp. 242–252.
120. S. Du, J. Lee, H. Li, L. Wang, and X. Zhai, "Gradient descent finds global minima of deep neural networks," in *International Conference on Machine Learning*. PMLR, 2019, pp. 1675–1685.
121. D. Zou, Y. Cao, D. Zhou, and Q. Gu, "Stochastic gradient descent optimizes over-parameterized deep ReLU networks," *arXiv preprint arXiv:1811.08888*, 2018.
122. H. Karimi, J. Nutini, and M. Schmidt, "Linear convergence of gradient and proximal-gradient methods under the Polyak-łojasiewicz condition," in *Joint European Conference on Machine Learning and Knowledge Discovery in Databases*. Springer, 2016, pp. 795–811.
123. Q. Nguyen, "On connected sublevel sets in deep learning," in *International Conference on Machine Learning*. PMLR, 2019, pp. 4790–4799.
124. C. Liu, L. Zhu, and M. Belkin, "Toward a theory of optimization for over-parameterized systems of non-linear equations: the lessons of deep learning," *arXiv preprint arXiv:2003.00307*, 2020.
125. Z. Allen-Zhu, Y. Li, and Y. Liang, "Learning and generalization in overparameterized neural networks, going beyond two layers," *arXiv preprint arXiv:1811.04918*, 2018.
126. M. Soltanolkotabi, A. Javanmard, and J. D. Lee, "Theoretical insights into the optimization landscape of over-parameterized shallow neural networks," *IEEE Transactions on Information Theory*, vol. 65, no. 2, pp. 742–769, 2018.

127. S. Oymak and M. Soltanolkotabi, "Overparameterized nonlinear learning: Gradient descent takes the shortest path?" in *International Conference on Machine Learning*. PMLR, 2019, pp. 4951–4960.
128. S. S. Du, X. Zhai, B. Poczos, and A. Singh, "Gradient descent provably optimizes over-parameterized neural networks," *arXiv preprint arXiv:1810.02054*, 2018.
129. I. Safran, G. Yehudai, and O. Shamir, "The effects of mild over-parameterization on the optimization landscape of shallow ReLU neural networks," *arXiv preprint arXiv:2006.01005*, 2020.
130. A. Jacot, F. Gabriel, and C. Hongler, "Neural tangent kernel: convergence and generalization in neural networks," in *Proceedings of the 32nd International Conference on Neural Information Processing Systems*, 2018, pp. 8580–8589.
131. S. Arora, S. S. Du, W. Hu, Z. Li, R. Salakhutdinov, and R. Wang, "On exact computation with an infinitely wide neural net," *arXiv preprint arXiv:1904.11955*, 2019.
132. Y. Li, T. Luo, and N. K. Yip, "Towards an understanding of residual networks using neural tangent hierarchy (NTH)," *arXiv preprint arXiv:2007.03714*, 2020.
133. Y. Nesterov, *Introductory lectures on convex optimization: A basic course*. Springer Science & Business Media, 2003, vol. 87.
134. Z.-Q. Luo and P. Tseng, "Error bounds and convergence analysis of feasible descent methods: a general approach," *Annals of Operations Research*, vol. 46, no. 1, pp. 157–178, 1993.
135. J. Liu, S. Wright, C. Ré, V. Bittorf, and S. Sridhar, "An asynchronous parallel stochastic coordinate descent algorithm," in *International Conference on Machine Learning*. PMLR, 2014, pp. 469–477.
136. I. Necoara, Y. Nesterov, and F. Glineur, "Linear convergence of first order methods for non-strongly convex optimization," *Mathematical Programming*, vol. 175, no. 1, pp. 69–107, 2019.
137. H. Zhang and W. Yin, "Gradient methods for convex minimization: better rates under weaker conditions," *arXiv preprint arXiv:1303.4645*, 2013.
138. B. T. Polyak, "Gradient methods for minimizing functionals," *Zhurnal Vychislitel'noi Matematiki i Matematicheskoi Fiziki*, vol. 3, no. 4, pp. 643–653, 1963.
139. S. Lojasiewicz, "A topological property of real analytic subsets," *Coll. du CNRS, Les équations aux dérivées partielles*, vol. 117, pp. 87–89, 1963.
140. B. D. Craven and B. M. Glover, "Invex functions and duality," *Journal of the Australian Mathematical Society*, vol. 39, no. 1, pp. 1–20, 1985.
141. K. Kawaguchi, "Deep learning without poor local minima," *arXiv preprint arXiv:1605.07110*, 2016.
142. H. Lu and K. Kawaguchi, "Depth creates no bad local minima," *arXiv preprint arXiv:1702.08580*, 2017.
143. Y. Zhou and Y. Liang, "Critical points of neural networks: Analytical forms and landscape properties," *arXiv preprint arXiv:1710.11205*, 2017.
144. C. Yun, S. Sra, and A. Jadbabaie, "Small nonlinearities in activation functions create bad local minima in neural networks," *arXiv preprint arXiv:1802.03487*, 2018.
145. D. Li, T. Ding, and R. Sun, "Over-parameterized deep neural networks have no strict local minima for any continuous activations," *arXiv preprint arXiv:1812.11039*, 2018.
146. N. P. Bhatia and G. P. Szegö, *Stability Theory of Dynamical Systems*. Springer Science & Business Media, 2002.
147. B. Neyshabur, R. Tomioka, and N. Srebro, "Norm-based capacity control in neural networks," in *Conference on Learning Theory*. PMLR, 2015, pp. 1376–1401.
148. P. Bartlett, D. J. Foster, and M. Telgarsky, "Spectrally-normalized margin bounds for neural networks," *arXiv preprint arXiv:1706.08498*, 2017.
149. V. Nagarajan and J. Z. Kolter, "Deterministic PAC-Bayesian generalization bounds for deep networks via generalizing noise-resilience," *arXiv preprint arXiv:1905.13344*, 2019.
150. C. Wei and T. Ma, "Data-dependent sample complexity of deep neural networks via Lipschitz augmentation," *arXiv preprint arXiv:1905.03684*, 2019.

151. S. Arora, R. Ge, B. Neyshabur, and Y. Zhang, "Stronger generalization bounds for deep nets via a compression approach," in *International Conference on Machine Learning*. PMLR, 2018, pp. 254–263.
152. N. Golowich, A. Rakhlin, and O. Shamir, "Size-independent sample complexity of neural networks," in *Conference On Learning Theory*. PMLR, 2018, pp. 297–299.
153. B. Neyshabur, S. Bhojanapalli, and N. Srebro, "A pac-Bayesian approach to spectrally-normalized margin bounds for neural networks," *arXiv preprint arXiv:1707.09564*, 2017.
154. M. Belkin, D. Hsu, S. Ma, and S. Mandal, "Reconciling modern machine-learning practice and the classical bias–variance trade-off," *Proceedings of the National Academy of Sciences*, vol. 116, no. 32, pp. 15 849–15 854, 2019.
155. M. Belkin, D. Hsu, and J. Xu, "Two models of double descent for weak features," *SIAM Journal on Mathematics of Data Science*, vol. 2, no. 4, pp. 1167–1180, 2020.
156. L. G. Valiant, "A theory of the learnable," *Communications of the ACM*, vol. 27, no. 11, pp. 1134–1142, 1984.
157. W. Hoeffding, "Probability inequalities for sums of bounded random variables," in *The Collected Works of Wassily Hoeffding*. Springer, 1994, pp. 409–426.
158. N. Sauer, "On the density of families of sets," *Journal of Combinatorial Theory, Series A*, vol. 13, no. 1, pp. 145–147, 1972.
159. Y. Jiang, B. Neyshabur, H. Mobahi, D. Krishnan, and S. Bengio, "Fantastic generalization measures and where to find them," *arXiv preprint arXiv:1912.02178*, 2019.
160. P. L. Bartlett, N. Harvey, C. Liaw, and A. Mehrabian, "Nearly-tight VC-dimension and pseudodimension bounds for piecewise linear neural networks." *Journal of Machine Learning Research*, vol. 20, no. 63, pp. 1–17, 2019.
161. P. L. Bartlett and S. Mendelson, "Rademacher and Gaussian complexities: Risk bounds and structural results," *Journal of Machine Learning Research*, vol. 3, pp. 463–482, 2002.
162. A. Blumer, A. Ehrenfeucht, D. Haussler, and M. K. Warmuth, "Learnability and the Vapnik–Chervonenkis dimension," *Journal of the ACM (JACM)*, vol. 36, no. 4, pp. 929–965, 1989.
163. D. A. McAllester, "Some PAC-Bayesian theorems," *Machine Learning*, vol. 37, no. 3, pp. 355–363, 1999.
164. P. Germain, A. Lacasse, F. Laviolette, and M. Marchand, "PAC-Bayesian learning of linear classifiers," in *Proceedings of the 26th Annual International Conference on Machine Learning*, 2009, pp. 353–360.
165. C. Zhang, S. Bengio, M. Hardt, B. Recht, and O. Vinyals, "Understanding deep learning requires rethinking generalization," *arXiv preprint arXiv:1611.03530*, 2016.
166. A. Bietti and J. Mairal, "On the inductive bias of neural tangent kernels," *arXiv preprint arXiv:1905.12173*, 2019.
167. B. Neyshabur, R. Tomioka, and N. Srebro, "In search of the real inductive bias: On the role of implicit regularization in deep learning," *arXiv preprint arXiv:1412.6614*, 2014.
168. D. Soudry, E. Hoffer, M. S. Nacson, S. Gunasekar, and N. Srebro, "The implicit bias of gradient descent on separable data," *The Journal of Machine Learning Research*, vol. 19, no. 1, pp. 2822–2878, 2018.
169. S. Gunasekar, J. Lee, D. Soudry, and N. Srebro, "Implicit bias of gradient descent on linear convolutional networks," *arXiv preprint arXiv:1806.00468*, 2018.
170. L. Chizat and F. Bach, "Implicit bias of gradient descent for wide two-layer neural networks trained with the logistic loss," in *Conference on Learning Theory*. PMLR, 2020, pp. 1305–1338.
171. S. Gunasekar, J. Lee, D. Soudry, and N. Srebro, "Characterizing implicit bias in terms of optimization geometry," in *International Conference on Machine Learning*. PMLR, 2018, pp. 1832–1841.
172. H. Xu and S. Mannor, "Robustness and generalization," *Machine Learning*, vol. 86, no. 3, pp. 391–423, 2012.
173. A. W. Van Der Vaart and J. A. Wellner, "Weak convergence," in *Weak Convergence and Empirical Processes*. Springer, 1996, pp. 16–28.

174. D. P. Kingma and M. Welling, "Auto-encoding variational Bayes," *arXiv preprint arXiv:1312.6114*, 2013.
175. I. Higgins, L. Matthey, A. Pal, C. Burgess, X. Glorot, M. Botvinick, S. Mohamed, and A. Lerchner, "β-VAE: Learning basic visual concepts with a constrained variational framework," *The International Conference on Learning Representations*, vol. 2, no. 5, p. 6, 2017.
176. S. Nowozin, B. Cseke, and R. Tomioka, "f-GAN: Training generative neural samplers using variational divergence minimization," in *Advances in Neural Information Processing Systems*, 2016, pp. 271–279.
177. M. Arjovsky, S. Chintala, and L. Bottou, "Wasserstein GAN," *arXiv preprint arXiv:1701.07875*, 2017.
178. L. Dinh, D. Krueger, and Y. Bengio, "NICE: Non-linear independent components estimation," *arXiv preprint arXiv:1410.8516*, 2014.
179. D. J. Rezende and S. Mohamed, "Variational inference with normalizing flows," *arXiv preprint arXiv:1505.05770*, 2015.
180. L. Dinh, J. Sohl-Dickstein, and S. Bengio, "Density estimation using real NVP," *arXiv preprint arXiv:1605.08803*, 2016.
181. D. P. Kingma and P. Dhariwal, "GLOW: Generative flow with invertible 1×1 convolutions," in *Advances in Neural Information Processing Systems*, 2018, pp. 10 215–10 224.
182. C. Villani, *Optimal transport: old and new*. Springer Science & Business Media, 2008, vol. 338.
183. M. Cuturi, "Sinkhorn distances: Lightspeed computation of optimal transport," in *Advances in Neural Information Processing Systems*, 2013, pp. 2292–2300.
184. G. Peyré, M. Cuturi *et al.*, "Computational optimal transport," *Foundations and Trends in Machine Learning*, vol. 11, no. 5–6, pp. 355–607, 2019.
185. J.-Y. Zhu, T. Park, P. Isola, and A. A. Efros, "Unpaired image-to-image translation using cycle-consistent adversarial networks," in *Proceedings of the IEEE international conference on computer vision*, 2017, pp. 2223–2232.
186. D. Lee, J. Kim, W.-J. Moon, and J. C. Ye, "CollaGAN: Collaborative GAN for missing image data imputation," in *Proceedings of the IEEE Conference on Computer Vision and Pattern Recognition*, 2019, pp. 2487–2496.
187. C. Clason, D. A. Lorenz, H. Mahler, and B. Wirth, "Entropic regularization of continuous optimal transport problems," *Journal of Mathematical Analysis and Applications*, vol. 494, no. 1, p. 124432, 2021.
188. T. Miyato, T. Kataoka, M. Koyama, and Y. Yoshida, "Spectral normalization for generative adversarial networks," *arXiv preprint arXiv:1802.05957*, 2018.
189. I. Gulrajani, F. Ahmed, M. Arjovsky, V. Dumoulin, and A. C. Courville, "Improved training of Wasserstein GANs," in *Advances in Neural Information Processing Systems*, 2017, pp. 5767–5777.
190. P. Vincent, H. Larochelle, I. Lajoie, Y. Bengio, and P.-A. Manzagol, "Stacked denoising autoencoders: Learning useful representations in a deep network with a local denoising criterion," *Journal of Machine Learning Research*, vol. 11, no. 12, pp. 3371–3408, 2010.
191. M. J. Wainwright, M. I. Jordan *et al.*, "Graphical models, exponential families, and variational inference," *Foundations and Trends® in Machine Learning*, vol. 1, no. 1–2, pp. 1–305, 2008.
192. T. M. Cover and J. A. Thomas, *Elements of information theory*. John Wiley & Sons, 2012.
193. J. Su and G. Wu, "f-VAEs: Improve VAEs with conditional flows," *arXiv preprint arXiv:1809.05861*, 2018.
194. P. Isola, J.-Y. Zhu, T. Zhou, and A. A. Efros, "Image-to-image translation with conditional adversarial networks," in *Proceedings of the IEEE Conference on Computer Vision and Pattern Recognition*, 2017, pp. 1125–1134.
195. B. Sim, G. Oh, J. Kim, C. Jung, and J. C. Ye, "Optimal transport driven CycleGAN for unsupervised learning in inverse problems," *SIAM Journal on Imaging Sciences*, vol. 13, no. 4, pp. 2281–2306, 2020.

Index

J. C. Ye, *Geometry of Deep Learning*, Mathematics in Industry 37,
https://doi.org/10.1007/978-981-16-6046-7